Lecture Notes in Computer Scier

Edited by G. Goos and J. Hartmanis

Advisory Board: W. Brauer D. Gries J. Stc

G. Cohen   P. Charpin (Eds.)

EUROCODE '90

International Symposium on Coding Theory
and Applications
Udine, Italy, November 5-9, 1990
Proceedings

Springer-Verlag
Berlin Heidelberg NewYork
London Paris Tokyo
Hong Kong Barcelona
Budapest

Series Editors

Gerhard Goos
GMD Forschungsstelle
Universität Karlsruhe
Vincenz-Priessnitz-Straße 1
W-7500 Karlsruhe, FRG

Juris Hartmanis
Department of Computer Science
Cornell University
Upson Hall
Ithaca, NY 14853, USA

Volume Editors

Gérard Cohen
Ecole Nationale Supérieures des Télécommunications
46, rue Barrault, 75634 Paris Cedex 13, France

Pascale Charpin
INRIA, B. P. 105, 78153 Le Chesnay, France

CR Subject Classification (1991): E.4, E.3

ISBN 3-540-54303-1 Springer-Verlag Berlin Heidelberg New York
ISBN 0-387-54303-1 Springer-Verlag New York Berlin Heidelberg

Printing and binding: Druckhaus Beltz, Hemsbach/Bergstr.
2145/3140-543210 - Printed on acid-free paper

Preface

Eurocode 90, held in Udine, Italy, 5-9 November 1990, is a continuation as well as an extension of the previous colloquia "Trois Journées sur le codage". The previous ones took place in Cachan, France in 1986 and Toulon, France in 1988 ; their proceedings appeared as Lecture Notes in Computer Science, Volume 311 (G. Cohen, P. Godlewski, eds.) and Volume 388 (G. Cohen, J. Wolfmann, eds.) respectively. The Udine meeting gathered approximately one hundred scientists and engineers.

These colloquia are characterized by a very broad spectrum, ranging from algebraic geometry to implementation of coding algorithms. We would like to thank the referees (see enclosed list) with a special mention for P. Camion, J. Conan, A. Thiong Ly and J. Wolfmann, as well as C. Dubois, secretary of "projet code INRIA" and N. Le Ruyet for help in editing.

1 Algebraic Codes

The construction of spherical codes is now a classical problem related to important special cases, as showed by Delsarte et al. (1977). In their invited paper, Ericson and Zinoviev propose a new construction, improving the method of generalized concatenation. Following the work of A. Dür, two papers deal with the structure of Reed-Solomon codes. Elia and Taricco present new results on code automorphism groups which imply some properties on covering radius and coset weight distribution of RS-codes. Berger gives a new basis describing primitive cyclic codes of length q - 1 over F_q ; as an application he obtains directly the group of some automorphisms of the RS-codes. Beth, Lazić and Senk present a very simple construction of an infinite sequence of self-dual codes ; properties of the first four codes imply a conjecture on the distance distribution.

The following two papers are devoted to open problems on Reed-Muller codes. Carlet shows that the weight of an RM-code of any order is related to the weight distribution of an RM-code of order 3 and greater length. Langevin studies the covering radius of the RM-code of order 1 and length 2^m, for small odd m ; he obtains a bound for m = 9.

In his paper, Rodier constructs codewords in the dual of binary BCH-codes of length $2^m - 1$, for an infinite number of m ; he can disprove a conjectured improvement of the Carlitz-Uchiyama bound. Augot, Charpin and Sendrier present an algebraic point of view in order to prove or disprove the existence of words of given weight in binary primitive cyclic codes of short length.

2 Combinatorial Codes

The next three papers are devoted to less classical coding problems. Burger, Chabanne and Girault deal with the construction of Gray codes with an additional constraint that 0-1 transitions should be evenly distributed, to provide, e.g., uniform wearing of memories. Mabogunje and Farell construct unequal error protection codes based on array codes and give simulation results for their bit error rates. Cohen, Gargano and Vaccaro propose t-unidirectional error detecting codes with high rates, for both systematic and nonsystematic cases, together with linear time encoding and decoding algorithms.

Two papers are devoted to graphs and finite fields. Montpetit presents some results in graphs which extend combinatorial results in coding theory. Astié-Vidal and Dugat propose a construction of homogeneous tournaments based on Galois fields.

3 Geometric Codes

In 1989, Pellikaan gave an algorithm which decodes geometric codes up to $\lfloor (d-1)/2 \rfloor$-errors, where d is the designed distance of the code. Le Brigand shows how this algorithm can be performed, using some results about the Jacobian of a hyperelliptic curve. Rotillon and Thiong Ly describe an effective decoding procedure for some geometric codes on the Klein quartic. Gallager proved, in 1963, that almost all binary linear codes meet the Gilbert-Varshamov bound. In her paper, Voss presents a generalisation of this result. Moreover she proves that almost all linear codes in the class of subfield subcodes of certain geometric Goppa codes meet the Gilbert-Varshamov bound. Perret constructs nonlinear geometric codes for which the distance is lowerbounded by use of multiplicative character sums.

4 Protection of Information

The invited paper by Girault is devoted to a survey of a large variety of recent identification schemes. Harari introduces a secret key coding scheme which relies on a subset of a particular set of random codes. Patarin presents some improvements to the work of Luby and Rackoff on pseudorandom permutations.

The paper by Fell deals with the effects of bit change errors on the linear complexity of finite sequences. Chassé considers the situation where the cells of a LFSR over F_q are disturbed by sequences of elements of F_q. Creutzburg obtains valuable results for the determination of convenient parameters for complex number-theoretic transforms.

Domingo-Ferrer and Huguet-Rotger describe a cryptographic scheme for program protection by means of a coding procedure, using a one-way function and a public-key signature.

5 Convolutional codes

The basic principles and limitations of decoding techniques for convolutional codes are presented by Haccoun. Visualizing the decoding process as being a search procedure through the tree or trellis representation of the code, he gives methods to circumvent the inherent shortcomings of Viterbi and sequential decoding. Sfez and Battail describe a weighted-output Viterbi algorithm used in a concatenated scheme where the inner code is convolutional, and simulate it over Gaussian and Rayleigh channels. Baldini Filho and Farrel present a multilevel convolutional coding method over rings, suitable for coded modulation, and show curves of performance for 4-PSK and 8-PSK.

6 Information Theory

This section starts with an invited paper by Sgarro offering a Shannon-theoretic coding theorem for authentication codes. Next paper, by the same author together with Fioretto, continues work on fractional entropy, a kind of measure of uncertainty for list coding. The section ends with two papers on source coding: Battail and Guazzo compare the respective merits of three algorithms (Huffman-Gallager, Lempel-Ziv and Guazzo) ; Capocelli and De Santis derive a tight upper bound on the redundancy of Huffman codes, in terms of the minimum codeword length, and use it to improve a bound due to Gallager.

7 Modulation

The invited paper by Calderbank describes how binary covering codes can be used to design non-equiprobable signaling schemes for use in high-speed modems, gaining in low-dimensional space as would shaping the boundaries of the signal constellation in higher dimensions. The next paper, by Battail, De Oliveira and Weidong, considers a combination of an MDS code over a large alphabet and a one-to-one mapping of the alphabet into a symmetric constellation, for combined coding and multilevel modulation. Two modulation-coding schemes are compared by Sfez, Belfiore, Leeuwin and Fihel for low-rate digital land mobile radio communication. Finally, Leeuwin, Belfiore and Kawas Kaleh derive a Chernoff upper bound for the pairwise error probability over a correlated Rayleigh channel.

8 Application of coding

Darmon and Sadot propose a hybrid ARQ+FEC system using a convolutional code with large constraint length and sequential decoding, combined with a modified Go-Back-N ARQ protocol, adapted to a two-way troposcatter (Rayleigh) channel. The last paper, by Politano and Deprey, describes a VLSI implementation of a Reed-Solomon coder-decoder.

List of Referees

D. Augot, G. Battail, F. Bayen, J.C. Belfiore, P. Camion, C. Carlet, G. Castagnoli, P. Charpin, G. Chassé, G. Cohen, J. Conan, B. Courteau, M. Darmon, J.L. Dornstetter, H. Fell, L. Gargano, M. Girault, C. Goutelard, S. Harari, D. Haccoun, S. Lebel, D. Le Brigand, A. Lobstein, G. Longo, H.F. Mattson, P. Langevin, A. Montpetit, F. Morain, G. Norton, J. Patarin, J.J. Quisquater, P. Sadot, N. Sendrier, P. Solé, H. Stichtenoth, A. Thiong Ly, U. Vaccaro, J. Wolfmann, G. Zémor.

June 1991

Gérard Cohen
Pascale Charpin

Contents

1. Algebraic Codes

2. Combinatorial Codes

3. Geometric Codes

4. Protection of Information

5. Convolutional Codes

6. Information Theory

7. Modulation

8. Applications of Coding

SECTION 1

ALGEBRAIC CODES

CONCATENATED SPHERICAL CODES
Codes sphériques concaténés

Th. Ericson (Sweden)

V.A. Zinoviev (USSR)

A new construction for spherical codes in R^n is given. The construction is compared with known constructions for finite dimensions n and for the asymptotical case.

Denote by S^n the unit sphere in Euclidean n-space R^n. A finite set $\mathcal{X} \subset S^n$ of size $M = |\mathcal{X}|$ is a spherical code with parameters (n,ρ,M), if

$$\rho = \rho(\mathcal{X}) = \min_{x \neq y} \rho(x,y); \ x, y \in \mathcal{X} \ ,$$

where $\rho(x,y)$ denotes squared Euclidean norm:

$$\rho(x,y) = \sum_{i=1}^{n} (x_i - y_i)^2 \ .$$

Denote by $M(n,\rho)$ the maximal size of a spherical n-code with minimum distance ρ. The construction of spherical codes with maximal size is a classical problem which includes several important special cases (see [1] and references therein).

In several papers the idea of using generalized concatenation for the construction of spherical codes has been developed (see [2,3] and references therein). Here we improve these methods for both finite lengths codes and for the asymptotic case.

The main idea of generalized concatenation is to use partitioning of the inner code. We add here two new ideas:

- the initial code is a set of points uniformly located in R^1;

- equal-weight codes are used to control the norm of the resulting code vector in R^n (here we develop the ideas of papers [4,5]).

We need some facts about block codes. A binary block (n,d,N) - code is an arbitrary subset $C \subset E^n$, where $E = \{0,1\}$, n is the length, N is the size, and d is the minimal (Hamming) distance. Denote by (n,w,d,N) the parameters of a block (n,d,N)-code with the fixed (Hamming) weight w.

We give two simple constructions (called "4-points and 8-points modulations") rather than the general construction, that needs more space.

Theorem 1 ("4-points modulation"): Let us have 4 points in R^1; $\times = \{-3,-1,1,3\}$, which we number in the following way:

$$x(1,1) = -3, \ x(0,0) = -1, \ x(1,0) = 1, \ x(0,1) = 3 \ .$$

Take two (binary) codes A_1: (n,d_1,N_1) and A_2: (n,w,d_2,N_2) and let $a = (a_1,\ldots, a_n) \in A_1$ and $b = (b_1,\ldots,b_n) \in A_2$ be any two code words. Then the set of all vectors

$$(\frac{1}{\sqrt{\eta}}x(a_1,b_1),\ldots, \frac{1}{\sqrt{\eta}}x(a_n,b_n)) \ ,$$

where $a \in A_1$ and $b \in A_2$ run over all values and $\eta = n + 8w$, form a spherical code $\times$: (n,ρ,M) with parameters

$$\rho \geq \min \ (\frac{4d_1}{\eta}, \frac{16d_2}{\eta}), \quad M = N_1 N_2 \ .$$

Example 1. Let us compare our construction with generalized concatenated codes (GCC) as described in [3]. Our "4-points modulation" corresponds to QAM for $\mu=4$ and $l=2$ (16 points, uniformly placed as rectangular lattice in R^2, that gives 4 points for one dimension). For n=16 and $\rho=2/5$ GCC gives $M=2^{20}$ (see table 1 in [3]), but the code is not spherical. Using in the last step the equal-weight code A_4: (8,4,2,70) (instead of A'_4: $(8,1,2^8)$) gives a spherical code with $M > 2^{18}$. In our construction take codes A_1: $(16,4,2^{11})$ and A_2: $(16,3,2,35 \cdot 16)$. Then we have $\eta=40$, which gives the spherical code $\times$: (16,2/5,M) with $M>2^{20}$.

For n=32 and $\rho=1/5$ GCC gives $M=2^{47}$ (table 1 in [3]); but the code is not spherical. Using in the last step the equal weight code (16,8,2,N), $N=\binom{16}{8} = 12870 \sim 2^{13.5}$ (instead of $(16,1,2^{16})$) gives a spherical code with $M \sim 2^{44.5}$. Take in our construction codes A_1: $(32,4,2^{26})$ and A_2: $(32,6,2,N_2)$, $N_2 = \binom{32}{6} \sim 2^{20}$. Then we have $\eta=80$, which gives the spherical code X: $(32,1/5,M \sim 2^{46})$.

For n=32 and $\rho=2/5$ GCC gives $M=2^{32}$ (table 1 in [3]) and the code is not spherical; the same calculation gives $M \sim 2^{30.5}$ for the corresponding spherical code. Take for our construction codes A_1: $(32,6,2^{21})$ and A_2: $(32,3,2,N_2)$, $N_2= \binom{32}{3} = 155 \cdot 2^5 > 2^{12}$. This gives (for $\eta=56$) the spherical code X: $(32,3/7,M)$ with $M > 2^{33}$.

Theorem 2 ("8 points modulation") Let us have 8 points in R^1; $\mathsf{X} = \{-7,-5,-3,-1,1,3,5,7\}$, which we number in the following way:

$x(1,1,1) = -7$, $x(0,0,1) = -5$, $x(1,0,0) = -3$, $x(0,1,0) = -1$,
$x(1,1,0) = 1$, $x(0,0,0) = 3$, $x(1,0,1) = 5$, $x(0,1,1) = 7$.

Let us have three (binary) codes A_1: (n,d_1,N_1), A_2: (n,w_2,d_2,N_2) and A_3: (n,w_3,d_3,N_3) and let $a = (a_1,...,a_n) \in A_1$, $b = (b_1,...,b_n) \in A_2$, $c = (c_1,...,c_n) \in A_3$ be any code words. The code A_3 is the direct sum of two codes A_{30} and A_{31} of lengths $n-w_2$ and w_2 respectively. The code word $c \in A_3$ is represented so that the part corresponding to A_{30} is represented in those positions where the code word b has zeroes, while the part corresponding to A_{31} is represented in the positions where b has ones. Then the set of all vectors

$$\left(\frac{1}{\sqrt{\eta}}x(a_1,b_1,c_1),...,\frac{1}{\sqrt{\eta}}x(a_n,b_n,c_n)\right),$$

where $a \in A_1$, $b \in A_2$, $c \in A_3$ run over all values and $\eta = \frac{n}{4} + 2w_2 + 4w_{3,1} + 12w_{3,0}$, form a spherical code X: (n,ρ,M) with parameters

$$\rho \geq \min\left(\frac{4d_1}{\eta}, \frac{16d_2}{\eta}, \frac{64d_3}{\eta}\right), \quad M = N_1 N_2 N_3 .$$

Example 2 Our "8-point modulation" corresponds to QAM for $\mu=6$ and $l=2$ (64 points, uniformly placed as rectangular lattice in R^2, that gives 8 points for one dimension). For n=64 and $\rho=2/21$ GCC gives $M=2^{116}$ (table 2 in [3]), but this code is not spherical. Take in our construction codes A_1: $(64,16,2^{28})$, A_2: $(64,24,4,N_2)$ and A_3: $(64,6,2,N_3)$, where $N_2 > 2^{51.80}$ (if we take the words of weight 32 of the extended Hamming code $(64,4,2^{57})$) and $N_3 = \binom{24}{8}\binom{40}{6} > 2^{41.36}$, this gives for $\eta = 2^9$ the spherical code $\mathcal{X}$: $(64,2/21,M)$ with $M > 2^{121.16}$.

REFERENCES

[1] P. Delsarte, J.M.Goethals, J.J. Seidel, "Spherical codes and designs", Geometriae Dedicata, Vol. 6, pp. 363-388, 1977.

[2] V.V. Ginsburg, "Multi-dimensional signals for continuous channels", Problems of Information Transmission, Vol.20, No. 1, pp. 20-34, 1984.

[3] V.A. Zinoviev, V.V. Zyablov, S.L. Portnoy, "Concatenated methods for construction and decoding of codes in Euclidean space", Institute for Problems of Information Transmission of the USSR Academy of Sciences, Moscow, Preprint, 1987.

[4] S.M. Dodunekov, Th. Ericson, V.A. Zinoviev, "Concatenation methods for construction of spherical codes in n-dimensional Euclidean space", Linköping University, Dept. of Electrical Engineering, Report LiTH-ISY-I-1075, March 1990.

[5] A.R. Calderbank, L.H. Ozarow, "Non-equiprobable signalling on the Gaussian channel", Mathematical Sciences Research Center, AT & T Bell Laboratories, 1989.

Th. Ericson
Department of Electrical Engineering
Linköping Institute of Technology
S-581 83 Linköping, Sweden

V.A. Zinoviev
Institute for Problems of Information Transmission
Academy of Sciences of USSR
Ermolova str. 19, GSP-4, Moscow
101447, USSR

A Note on Automorphism Groups of Codes and Symbol Error Probability Computation

M. Elia *
Dipartimento di Elettronica
Politecnico di Torino
Corso Duca degli Abruzzi 24
10129 Torino - Italy

G. Taricco
Centro Ricerche FIAT
Strada Torino 50
10143 Orbassano - Italy

Abstract

In this paper we consider code automorphism groups and their effects on the computation of the symbol error probability after complete decoding. In particular both Maximum Likelihood and Unique Coset Leader decoding strategies are investigated when applied to Reed-Solomon codes. Therefore referring to these codes several explicit formulas for the evaluation of the symbol error probability are obtained.

1 - Introduction

An important problem concerning error-correcting codes is the search of closed-form expressions for the Symbol Error Rate (hereafter denoted by SER) after complete decoding in terms of the error probability of memoryless symmetric channels. The prominent combinatorial nature of the problem, [1,8], requires prohibitive calculations for many codes, nevertheless several of these cases can be approached by exploiting the existence of a code automorphism group. Also the computation of word error probabilities and error detection probabilities can be of relevant interest, however it is known, see [1,8], that both these parameters can be directly obtained, respectively, from the code weight distribution and the coset leader weight distribution. Although much is known about automorphism groups of codes, [1,3,6], it is not always evident how to apply this powerful instrument to error probability computations.

In this paper we briefly dwell on general relations between codes and their groups and we will apply the results to analyze the performance of Reed-Solomon (RS) codes. In particular, Section 2 collects results concerning automorphism groups of codes, standard arrays, complete decoding and SER evaluation. Section 3 considers the relation between group automorphisms and generating polynomials of cyclic codes. Finally, Section 4 presents results concerning SER computations for special classes of RS codes.

*His work was financially supported by the Italian CNR (Consiglio Nazionale delle Ricerche) under grant n. 87.02717.07.

2 - Decoding and Automorphism Groups

Through the paper we will always refer to a systematic linear block code (n, k, d), which is a k-dimensional subspace of an n-dimensional vector space over a finite field $\mathcal{F} = GF(p^m)$ (p is a prime number), with d denoting the minimum Hamming distance among codewords. For sake of reference, let us recall some definitions and concepts concerning automorphism group and complete decoding of codes.

2.1 - Automorphism Groups

Any permutation of the coordinate entries changes a code C into a code C', both codes sharing a large number of properties; in particular transformations that leave C unchanged lead to the notion of code automorphism.

Definition 2.1 - *An automorphism σ of a code C is an invertible mapping:*

$$\sigma : C \mapsto C \quad ,$$

of C onto itself, satisfying the following properties:

- $\sigma(\alpha \mathbf{c}) = \alpha \sigma(\mathbf{c})$;
- $\sigma(\mathbf{c}_1 + \mathbf{c}_2) = \sigma(\mathbf{c}_1) + \sigma(\mathbf{c}_2)$;

where α is an element of the finite field $\mathcal{F}$.

Code automorphisms form a group under composition [1]. Two notions of code automorphism groups are of interest to us:

- **The automorphism group** of C, denoted $Aut(C)$, which is the set of automorphisms σ that preserve the Hamming metric, i.e., $wt(\sigma(\mathbf{c})) = wt(\mathbf{c})$, where $wt(\mathbf{c})$ denotes the Hamming weight of the vector $\mathbf{c}$.

- **The permutation automorphism group** of C, denoted $PAut(C)$, which is the subgroup of $Aut(C)$ formed by transformations that are permutations of coordinate places.

2.2 - Complete Decoding

Complete decoding means that every received word is uniquely assigned to a codeword, even if its distance from every codeword is greater than the packing radius t [1]. In the following, we will assume that all codes will be completely decoded over memoryless symmetric channels with symbol error probability p_{se}. Let us recall that a variety of complete decoding strategies can be defined by using the partition of the vector space $\mathcal{F}^n$ into *cosets* of C, [1]. The *coset leader* of any coset $\ell_i + C$ $i = 0, \ldots, q^{n-k} - 1$ is the *correctable error pattern*.
This partition is called *standard array*, [1], and the set of coset leaders defines the *decoding strategy*. In particular, we will be interested in two strategies:

i) **Maximum Likelihood (ML)** decoding where the coset leaders are chosen among the coset words of smallest weight;

ii) **Unique Coset Leader (UCL)** decoding where the coset leaders are chosen either to be the coset word of smallest weight if the weight is not greater than the *packing radius* [1], or to be the unique word with all zero in information positions [4].

It is known [1] that for any complete decoding which is amenable to a standard array decoding, the estimated error patterns do not depend on the transmitted codeword. Therefore we will assume, without loss of generality, that the all-zero codeword is transmitted.

2.3 - Automorphism groups and Standard arrays

Let us consider the effect of the automorphism group over the standard array. Let $\ell + C$ be any coset: then we have:

$$\sigma(\ell + C) = \sigma(\ell) + C \quad \forall \ \sigma \in Aut(C) \quad ;$$

this relation shows that every automorphism maps cosets into cosets, possibly with leaders of the same weight. This property allows us to define the action of the automorphism group on the set of coset leaders, therefore partitioning this set into equivalent classes (or orbits). Each class will be identified by an element called *coset leader prototype*. Note that the group $Aut(C)$ preserves the coset weight distribution, which turns out to be the same for all cosets in the same class.

2.4 - Symbol Error Probability after Complete Decoding

The symbol error rate p_s is defined as the probability that an information symbol is incorrect after complete decoding. Assuming that the all-zero codeword is transmitted, and letting $\hat{\mathbf{x}} = (\hat{x}_1, \hat{x}_2, \ldots, \hat{x}_n)$ to denote a decoded codeword, therefore p_s is given by [1]:

$$p_s = \frac{1}{k} \sum_{j=1}^{k} Prob\{\hat{x}_j \neq 0\} = \frac{1}{k} \sum_{e} f(e) Prob\{e\} \quad , \tag{1}$$

where $f(e)$ is the number of incorrect information symbols after the decoding when the error vector is e. In order to give an explicit expression for p_s, let us recall that the probability $Prob(e)$ of an error vector e of weight h on a memoryless channel is $Prob(h) = p_{sc}^h (1 - p_{sc})^{n-h}$. Therefore, by splitting in equation (1) the sum over e into two sums, we have:

$$p_s = \frac{1}{k} \sum_{e} f(e) Prob\{e\} = \frac{1}{k} \sum_{h=1}^{n} \left[\sum_{wt(e)=h} f(e) \right] Prob(h) \quad .$$

Setting $B_h = \sum_{wt(e)=h} f(e)$, we can write

$$p_s = \frac{1}{k} \sum_{h=1}^{n} B_h Prob(h) \quad . \tag{2}$$

where B_h counts the total number of residual errors after decoding in information positions due to all possible error vectors of weight h. By expanding the expression (2) with respect to p_{sc}, we get an alternative polynomial expression:

$$p_s = \sum_{i=t+1}^{n} E_i p_{sc}^i \quad ,$$

where $t = \lfloor \frac{d-1}{2} \rfloor$. The algebraic computations necessary for evaluating the E_i's from the B_h's are cumbersome and require the use of computer programs for symbolic manipulations.

Finally let us observe that for codes admitting a transitive automorphism group, [8], the evaluation of B_h can be considerably simplified. In fact, due to the transitive property, each symbol in every codeword plays the same role irrespective of being an information or check symbol, (this does not mean that whenever a transitive automorphism group exists we can exchange the role of information and parity check symbols). Therefore, it may be simpler to count the total number of residual errors as

$$\beta_h = \sum_{wt(e)=h} F(e)$$

where $F(e)$ is the total number of incorrect symbols after decoding, when the error vector is e, thus B_h is obtained as $\frac{k}{n}\beta_h$.

3 - Automorphism groups and cyclic codes

Code automorphisms are fully characterized, [3], by the action over the code generating matrix, therefore they are also completely characterized by the action over the generating polynomial of a cyclic code. Let $\mathcal{A} = \{\alpha_1, \ldots, \alpha_r\}$ be the set of the roots of the generating polynomial $g(x)$. These roots that belong to the exponent n can be represented by powers of β which possibly belongs to an extended field $GF(q^l)$.

For sake of reference we summarize a number of polynomial transformations that may be code automorphisms:

- $\sigma_\alpha : c(x) \mapsto \alpha c(x)$ $\alpha \in GF(q)$, scalar magnification (always an automorphism);

- $\gamma : c(x) \mapsto xc(x) \bmod x^n - 1$, cyclic shift (it is indeed an automorphism of any cyclic code);

- $\pi_\alpha : c(x) \mapsto c(\alpha x)$ $\alpha \in GF(q)$;

- $\sigma_i : c(x) \mapsto c(x^i) \bmod x^n - 1$ $i \in Z_n$, may alter the word weight;

- $\kappa = \sigma_p$,   the Frobenius mapping, p is the characteristic of $GF(q)$;

- $\iota = \sigma_{-1}$,   the reciprocal transformation.

It need not to be proved that the automorphism group, [3], of a cyclic code of length n always contains the cyclic group C_n.

The next elementary proposition will be useful in the following.

Lemma 3.1 *In any finite field, let α be an element that belongs to the exponent n and let $u = (n, s)$ the greatest common divisor of n and s. Then, the greatest common divisor of $x^s - \alpha^s$ and $x^n - 1$ is $x^u - \alpha^u$.*

Proof - First, observe that if $u \mid s$ then $(x^u - \alpha^u) \mid (x^s - \alpha^s)$. If we write $n = \lambda s + r$, with $0 \le r < s$, the following chain of relations is immediately verified:

$$\begin{aligned}(x^s - \alpha^s,\ x^n - 1) &= (x^s - \alpha^s,\ \alpha^s x^{n-s} - 1) = \ldots = (x^s - \alpha^s,\ \alpha^{\lambda s} x^{n-\lambda s} - 1) \\ &= (x^s - \alpha^s,\ \alpha^{n-r} x^r - 1) = (x^s - \alpha^s,\ x^r - \alpha^r)\ .\end{aligned} \quad (3)$$

Iterating the process, as in the Euclidean algorithm, we finally get:

$$(x^s - \alpha^s,\ x^n - 1) = (x^s - \alpha^s,\ x^u - \alpha^u) = x^u - \alpha^u\ .$$

Using lemma 3.1, it is not hard to prove the following proposition.

Theorem 3.1 - *Let us consider a (n, k, d) cyclic code C generated by $g(x)$, therefore*

    i) If there exists an integer s, relatively prime with n, such that

$$A_s = \{\alpha_1^s, \dots, \alpha_r^s\} = A \quad ,$$

    then σ_s is a code automorphism.

    ii) The order of σ_s is the minimum integer m such that

$$s^{m+1} = s \quad \text{in} \quad \mathbb{Z}_n \quad .$$

Proof - It is easy to show that the image of C under σ_s satisfies the inclusion relation

$$\sigma_s(C) \subseteq C \quad .$$

In fact, we have:

$$\sigma_s(g(x)) = g(x^s) = \prod_{\alpha \in A} (x^s - \alpha) = \prod_{\alpha \in A} (x^s - \alpha^s) = \mu(x) \cdot g(x) \quad .$$

To complete the proof we have to show the reverse inclusion

$$C \subseteq \sigma_s(C) \quad .$$

Given that $(s, n) = 1$, by lemma 3.1 we know that $\mu(x)$ and $x^n - 1$ are relatively prime; therefore a polynomial $a(x)$ exists such that

$$a(x)\mu(x) = 1 \bmod x^n - 1 \quad ,$$

which in turn yields

$$a(x)g(x^s) = a(x)\mu(x)g(x) = g(x) \bmod x^n - 1 \quad .$$

That is the assertion *i)*. To prove the assertion *ii)* let us consider iterated applications of σ_s. We have

$$\sigma_{s^i}(c(x)) = c(x^{s^i}) = \sigma_s(c(x^{s^{i-1}})) \quad ,$$

so that:

$$\sigma_s(c(x)) = \sigma_{s^{m+1}}(c(x)) \quad .$$

When σ_s is applicable to cyclic codes, it gives information on structure and order of the automorphism group. In fact, let us consider the following commuting relation between γ and σ:p

$$\sigma_s[\gamma(c(x))] = \sigma_s[xc(x)] = x^s c(x^s) = \gamma^s \sigma_s(c(x)) \quad ,$$

which can be formally written in terms of operators as:

$$\gamma^s \sigma_s = \sigma_s \gamma \quad .$$

As a consequence of this relation, γ and σ_s generate a non-commutative group of order $m \cdot n$.

4 - Reed-Solomon codes

In this Section we consider the application of automorphism groups to exploit several characteristics of RS codes, which are traditionally defined as $(n,\ k,\ d)$ cyclic codes over $GF(q)$ with $n = q - 1$, $k = n - d + 1$ and generator polynomial of the type:

$$g(x) = \prod_{i=1}^{d-1}(x - \alpha^i) \quad ; \tag{4}$$

where α is a primitive element in $GF(q)$. RS codes are Maximum Distance Separable codes since the Singleton bound is satisfied with equality, i.e. $d = n - k + 1$. They are able to correct up to $t = \lfloor \frac{d-1}{2} \rfloor$ errors. Moreover a closed expression for the weight distribution of the codewords is known, [1]. Let A_i denote the number of weight-i codewords, then $A_0 = 1$ and we have

$$A_i = \binom{n}{i}(q-1)\sum_{j=0}^{i-d}(-1)^j\binom{i-1}{j}q^{i-d-j}, \quad \text{for } n \geq i \geq d \ . \tag{5}$$

In certain cases it may be useful to consider a slightly wider definition of RS codes by taking generating polynomials of the form

$$g'(x) = \prod_{i=1}^{d-1}(x - \alpha^{b+(i-1)h}) \quad . \tag{6}$$

It can be shown that, under proper conditions, the coset weight distribution is the same for RS codes of the same size but with different generating polynomials.

Proposition 4.1 - *The RS codes C and C', with generating polynomials $g(x)$ and $g'(x)$ respectively shown in (4) and (6), have the same code weight distribution and coset weight distribution provided that n and h are relatively prime.*

Proof - Applying successively the mappings $\pi_{\alpha^{b-h}}$ and σ_h to $g'(x)$, we have

$$\pi_{\alpha^{b-h}}(\sigma_h(g'(x))) = g'(\alpha^{b-h}x^h) = \alpha^{(d-1)(b-h)}\prod_{i=1}^{d-1}(x^h - \alpha^{ih}) = \mu(x)\,g(x) \quad ,$$

showing that C' is mapped into C. Assuming that $(n, h) = 1$, this mapping is also onto, moreover any $c'(x)$ and its image $c'(\alpha^b x^h)$ have the same weight because the operation corresponds to scrambling the non-zero positions of $c'(x)$.

Notice that C and C', in proposition 4.1, have generally *different* automorphism groups. Consider, for example, two different RS codes $(7, 5, 3)$ over $GF(8)$ with generator polynomials $g(x) = (x - \alpha)(x - \alpha^2)$ and $g'(x) = (x - \alpha^3)(x - \alpha^4)$, respectively: we have that $\sigma_{-1}(C') = C'$ while $\sigma_{-1}(C) \neq C$.

4.1 - Some special results

The automorphism groups of RS codes have been obtained in [6], however in this Section we report some special properties of these groups that are useful to SER computation. Since RS codes are cyclic, they always have a non trivial automorphism group, with C_n as a subgroup. Moreover there are RS codes with a special symmetry for any parameter combination as stated in the following Proposition.

Proposition 4.2 *For every $1 \leq t \leq \frac{n-1}{2}$ there exists a RS code $(n, n - 2t, 2t + 1)$ with automorphism group that contains the automorphism σ_{-1} defined as*

$$\sigma_{-1}(c(x)) = c(x^{n-1}) \quad .$$

12

Proof - Let us consider the generating polynomials

$$g(x) = \prod_{i=1}^{t} (x - \alpha^{\frac{n-2i+1}{2}})(x - \alpha^{\frac{n+2i-1}{2}}) \ ,$$

which is clearly invariant under the automorphism σ_{-1}. It is immediately verified that $g(x)$ generates a RS code according to equation (6) with $b = \frac{n-2t+1}{2}$, $h = 1$ and $d = 2t+1$.

The covering radius of the RS codes was found in [5]. Let us give here another proof based on our approach. Recall that the covering radius of a code gives the maximum weight of coset leaders in ML complete decoding.

Theorem 4.1 - *The covering radius of a RS code (n,k,d) is $d-1$.*

Proof - Let $g(x) = \prod_{i=1}^{d-1}(x - \alpha^i)$ be the generating polynomial of the RS code C. Let us consider the word $\ell(x) = \prod_{i=1}^{d-2}(x - \alpha^i)$, it must have weight $d-1$ because it generates a $(n, k+1, d-1)$ RS code. At the same code belong all words in the coset $\ell(x) + C$, so that the covering radius of C, i.e., the maximum minimum weight in a coset, is exactly $d-1$.

Now, we consider the problem of evaluating the number of cosets with leader (under ML decoding) of weight $d-1$. The existence of such cosets is guaranteed by Theorem 4.1. First, we will show that the set obtained by the union of some cosets of a RS code with minimum distance d having leaders of weight $d-1$ form a RS code with minimum distance $d-1$. Then, we will prove that no other cosets with coset leader of minimum weight $(d-1)$ exist.

Theorem 4.2 - *An $(n, n-d+2, d-1)$ RS code C' with generating polynomial $\ell(x)$ can be partitioned into cosets of a subcode C which is an $(n, n-d+1, d)$ RS code having generating polynomial $g(x) = \prod_{i=1}^{d-1}(x - \alpha^i)$, i.e.,*

$$C' = \biguplus_{\beta \in GF(q)} [\beta\ell(x) + C]$$

where either $\ell(x) = g(x)/(x - \alpha)$ or $\ell(x) = g(x)/(x - \alpha^{d-1})$.

Proof - Every coset $\beta\ell(x) + C$ is a subset of a RS code with generating polynomial $\ell(x)$ because $\ell(x)$ divides $g(x)$; therefore, their union is a subset of C'. Furthermore the sets are disjoint and their number is q, so that the cardinality of C' is equal to $q \cdot q^{n-d+1} = q^{n-d+2}$, i.e., C' is a subset of a RS code $(n, n-d+2, d-1)$ having the same cardinality of the code: therefore, it is the code itself.

The next theorem can be considered a converse to the last proposition. It states that there are exactly $2n$ cosets with leader of minimum weight $d-1$.

Theorem 4.3 *Given a $(n, n-d+1, d)$ RS code C and a coset leader $\ell(x)$ of weight $d-1$, the set*

$$C' = \biguplus_{\beta \in GF(q)} [\beta\ell(x) + C] \ ,$$

is a RS code $(n, n-d+2, d-1)$.

Proof - Let $g(x) = \prod_{i=1}^{d-1}(x - \alpha^i)$ be the generating polynomial of C. It is straightforward to verify that

$$C' = \biguplus_{\beta \in GF(q)} [\beta \ell(x) + C] \quad ,$$

is a linear code by observing that C' is a closed set, i.e., linear combinations of its codewords are still in the set.

The minimum distance of C' is $d - 1$, i.e., the minimum weight word in the coset, since all the cosets have the same weight distribution, being σ_β an automorphism of C. Therefore, the code C' has parameters $(n, n - d + 2, d - 1)$.

Furthermore, every coset with minimum weight $d - 1$ has exactly one word of weight $d - 1$ that can be represented as a $(d - 2)$-degree monic polynomial. In fact, if there were two ones, their difference should be a codeword of weight less than or equal to $d - 1$, that is impossible, being d the minimum distance of the code. There is at least one because if you consider a $d - 1$-weight word with degree greater than $d - 2$, you can remove the non-zero information positions (coefficients of the monomials with degree greater than $d - 1$) by subtracting the codeword with the same information part.

Now, we will show that C' is a RS code with generating polynomial one of that defined in Proposition 4.2, for which we have either $\ell(\alpha^1) = \ell(\alpha^2) = \ldots = \ell(\alpha^{d-2}) = 0$ or $\ell(\alpha^2) = \ell(\alpha^3) = \ldots = \ell(\alpha^{d-1}) = 0$. If we assume, on the contrary, that

$$\ell(\alpha^i) = \lambda_i \qquad \text{for } i = 1, 2, \ldots, d - 1$$

holds for generic λ's, then we can show that a word $\ell(x) + c(x)$ exists of weight $d - 2$, which should be of weight smaller than the minimum weight word assumed. In fact, consider the equation

$$\sum_{j=1}^{d-2} \xi_j x^{i_j} = \beta \ell(x) + a(x) g(x) \quad , \tag{7}$$

where $\xi_1, \xi_2, \ldots, \xi_{d-2}$ and β are unknown.

Substituting respectively α, $\alpha^2, \ldots$, α^{d-2} for x in equation (7), we get the system of equations:

$$\sum_{j=1}^{d-1} \xi_j \alpha^{r_{ij}} = \beta \lambda_r \quad r = 1, 2, \ldots, d - 2 \quad .$$

Setting $x_j = \alpha^{i_j}$, it can be rewritten in the form

$$\sum_{j=1}^{d-1} \xi_j x_j^r - \beta \lambda_r = 0 \quad r = 1, 2, \ldots, d - 2 \quad .$$

This homogeneous system has a non zero solution if and only if the determinant Δ of the matrix of the coefficients is zero. Δ is the determinant of a matrix that resembles the Vandermonde matrix, and it can be evaluated in closed form. In fact, consider

$$\Delta' = \begin{vmatrix} x_1 & x_2 & \cdots & x_{d-2} & \lambda \\ x_1^2 & x_2^2 & \cdots & x_{d-2}^2 & \lambda^2 \\ \vdots & \vdots & \ddots & \vdots & \vdots \\ x_1^{d-1} & x_2^{d-1} & \cdots & x_{d-2}^{d-1} & \lambda^{d-1} \end{vmatrix} \quad .$$

We can expand Δ' as a Vandermonde determinant:

$$\Delta' = x_1 x_2 \cdots x_{d-2} \lambda (x_1 - \lambda)(x_2 - \lambda) \cdots (x_{d-2} - \lambda) \prod_{1 \le i < j \le d-2} (x_i - x_j)$$

$$= x_1 x_2 \cdots x_{d-2} [\prod_{1 \le i < j \le d-2} (x_i - x_j)] \sum_{i=1}^{d-1} (-1)^{i-1} \sigma_{d-1-i} \lambda^i \quad ,$$

having defined the symmetric functions of $x_1, \ldots, x_{d-2}$ as:

$$\begin{cases} \sigma_0 = 1 \\ \sigma_i = \sum_{1 \le j_1 < \cdots < j_i \le d-2} x_{j_1} \cdots x_{j_i} \quad \text{for } i = 1, \ldots, d-2 \\ \sigma_i = 0 \quad \text{for } i \ge d-1 \end{cases} \quad (8)$$

Resorting to umbral calculus [10], we can formally substitute $\lambda^i \mapsto -\lambda_i$ and obtain:

$$\Delta = x_1 x_2 \cdots x_{d-2} [\prod_{1 \le i < j \le d-2} (x_i - x_j)] \sum_{i=1}^{d-1} (-1)^i \sigma_{d-1-i} \lambda_i \quad .$$

The x_i's are different form each other and from zero by definition. Thus, if $\Delta = 0$, then we have:

$$\sum_{i=1}^{d-1} (-1)^i \sigma_{d-1-i} \lambda_i = 0 \quad .$$

Denoting by σ_i' the elementary symmetric functions of $x_2 x_3 \cdots x_{d-2}$, we have the relations

$$\sigma_i = x_1 \cdot \sigma_{i-1}' + \sigma_i' \quad \text{for } i = 1, 2, \ldots, d-2 \quad ,$$

and in turn we obtain

$$x_1 = - \frac{\sum_{i=2}^{d-1} (-1)^i \sigma_{d-1-i}' \lambda_i}{\sum_{i=1}^{d-2} (-1)^i \sigma_{d-2-i}' \lambda_i} \quad .$$

From this equation, we see that a non-zero solution leading to a $(d-2)$-weight word always exists except in the case that we have:

$$\lambda_i = 0 \quad \text{for } i = 1, \ldots, d-2 \quad \text{and} \quad \lambda_{d-1} \ne 0$$

or

$$\lambda_1 \ne 0 \quad \text{and} \quad \lambda_i = 0 \quad \text{for } i = 2, \ldots, d-1 \quad ,$$

and this proves the assertion.

4.2 - RS $(n, 1, n)$ codes

The $\lfloor \frac{n-1}{2} \rfloor$-error correcting RS codes are in fact repetition codes, but differently from the binary case they are not perfect. Therefore in spite of the apparent simplicity, the computation of any interesting characteristic requires considerable efforts. To illustrate the reduction coming from the automorphism group we will consider the $(7, 1, 7)$ code.

4.2.1 - RS code $(7, 1, 7)$

The code generating polynomial is:

$$g(x) = \prod_{i=1}^{6} (x - \alpha^i) = 1 + x + x^2 + x^3 + x^4 + x^5 + x^6 \ ,$$

and the set of codewords is simply:

$$\mathcal{C} = \{\gamma \left(1 + x + x^2 + x^3 + x^4 + x^5 + x^6\right) \mid \gamma \in GF(8)\} \ .$$

The $PAut(\mathcal{C})$ group is S_7, so that any permutation performed on a coset leader gives still a coset leader. The number of cosets is $8^6 = 262144$ but exploiting the high symmetry we can consider a considerably smaller number of prototypes.

The ML decoding strategy chooses as coset leaders a word with minimum weight within a coset. Therefore, we classify the coset leader prototypes with increasing weight, starting from weight 0, the results are summarized in Tab. 1. To explain how the table has been constructed, let us consider for example, the weight-4 leader prototype $(000\alpha\alpha\beta\gamma)$ corresponding to row number 10 in Tab. 1. The number of possible such prototypes is

$$\left(\frac{7!}{3! \ 2! \ 1! \ 1!}\right) \ \left(\frac{7 \cdot 6 \cdot 5}{2!}\right) = 44100$$

where

- the first factor accounts for the combinations of the elements $0, \alpha, \beta, \gamma$ taken 3, 2, 1, 1 times respectively;

- the numerator of the second factor accounts for the number of possible choices of α, β, γ regardless of the order;

- the denominator accounts for the number of permutations that lead to a pattern of the same type (in this case, the exchange of β and γ).

Repeating the argument for all the leader prototypes we obtain the coset leader weight distribution. Moreover, we can easily obtain the coset word weight distribution corresponding to each leader prototype. Continuing the example, let us observe that the leader $(000\alpha\alpha\beta\gamma)$ can be summed with any of the possible codewords: (0000000), $(\alpha\alpha\alpha\alpha\alpha\alpha\alpha)$, $(\beta\beta\beta\beta\beta\beta\beta)$, $(\gamma\gamma\gamma\gamma\gamma\gamma\gamma)$, $(\delta\delta\delta\delta\delta\delta\delta$, 4 possible $\delta)$ and the resulting words have weight 4, 5, 6, 6, and 7, respectively, as reported in Tab. 1.

Using the results of Tab. 1, we obtain Tab. 2 which reports the code weight distribution A_i, the coset leader weight distribution L_i and the parameters B_i for SER computation. The coset leader weight distribution for UCL decoding is computed in a similar manner. The results are summarized in table 3, the parameters B_i are taken from [4]. Comparing the results reported in Tab. 2 and Tab. 3, one can notice that B_4 is much smaller in the case of ML decoding than for UCL decoding. This fact implies that the asymptotic error probability after complete decoding *for this code* is far better using ML instead of UCL decoding strategy.

4.3 - RS $(n, n - 2, 3)$ codes

The 1-error correcting RS code $(n, n - 2, 3)$, with $q = 2^m$, is quasi perfect since we can take coset leaders of weight not greater than 2. Therefore, the weight distribution of

Reference number	Number of Cosets	Leader prototype	Coset Weight Distribution							
			0	1	2	3	4	5	6	7
1	1	0000000	1							7
2	49	000000α	1						1	6
3	147	$00000\alpha\alpha$		1				1		6
4	882	$00000\alpha\beta$	1						2	5
5	245	$0000\alpha\alpha\alpha$			1	1				6
6	4410	$0000\alpha\alpha\beta$			1		1	1		5
7	7350	$0000\alpha\beta\gamma$			1			3		4
8	2940	$000\alpha\alpha\alpha\beta$				2		1		5
9	4410	$000\alpha\alpha\beta\beta$				1	2			5
10	44100	$000\alpha\alpha\beta\gamma$				1	1	2		4
11	29400	$000\alpha\beta\gamma\delta$				1		4		3
12	22050	$00\alpha\alpha\beta\beta\gamma$					3	1		4
13	88200	$00\alpha\alpha\beta\gamma\delta$					2	3		3
14	52920	$00\alpha\beta\gamma\delta\epsilon$					1	5		2
15	5040	$0\alpha\beta\gamma\delta\epsilon\eta$						7		1

- greek letters correspond to non-zero elements of $GF(8)$
- coset leader prototypes are listed with increasing weight
  in all possible combinations.

Table 1: RS code (7,1,7) ML decoded - Word weight distribution in Cosets

i	0	1	2	3	4	5	6	7
A_i	1	0	0	0	0	0	0	7
L_i	1	49	1029	12005	80850	163170	5040	0
B_i	0	0	0	0	3185	189777	818503	823542

Table 2: RS code (7,1,7) ML decoded - Coefficients for SER computation

i	0	1	2	3	4	5	6	7
A_i	1	0	0	0	0	0	0	7
L_i	1	49	1029	12005	35910	93240	119910	0
B_i	0	0	0	0	101045	373107	499513	861343

Table 3: RS code (7,1,7) UCL decoded - Coefficients for SER computation

the coset leaders, for either ML or UCL decoding, is

$$L_0 = 1 \qquad L_1 = n^2 \qquad L_2 = 2n \ .$$

To obtain the weight distribution in any coset, we use the transitive property of the code group automorphism. Finally, for the RS code $(7, 5, 3)$, we find the coefficients B_i as defined in Section 4.

First, according to Proposition 4.1, we can choose a RS code $(7, 5, 3)$ that admits a special automorphism. In fact, by taking as generating polynomial

$$g(x) = (x - \alpha^{\frac{n-1}{2}})(x - \alpha^{\frac{n+1}{2}}) \ ,$$

the code admits the automorphism ι defined in Section 3.

All possible 1-weight patterns are coset leaders and the relative cosets share the same weight distribution since two 1-weight cosets can be obtained from each other by suitably applying the automorphisms σ_α and π_α defined in Section 3.

If we consider the general 1-weight coset leader $l(x) = l_j x^{n_j}$ for $j = 1, \ldots, (q-1)^2$, we note that for any codeword $c(x) \in \mathcal{C}$ of weight w:

- there are w leaders ℓ such that $wt(\ell + \mathbf{c}) = w - 1$;

- there are $(n-1)w$ leaders ℓ such that $wt(\ell + \mathbf{c}) = w$;

- there are $n(n-w)$ leaders ℓ such that $wt(\ell + \mathbf{c}) = w - 1$;

where $n = q - 1$.

If we denote by $A_i^{(1)}$ the number of patterns of weight i in a coset with leader of weight 1, according to the above arguments we write out the expression:

$$A_i^{(1)} = \frac{n+1-i}{n} A_{i-1} + \frac{(n-1)i}{n^2} A_i + \frac{i+1}{n^2} A_{i+1} \quad \text{for } i = 0, 1, \ldots, n$$

assuming $A_{-1} = A_{n+1} = 0$.

Now we consider weight-2 coset leaders. Since we have chosen

$$g(x) = (x - \alpha^{\frac{n-1}{2}})(x - \alpha^{\frac{n+1}{2}}) \ ,$$

it is easy to verify that all patterns

$$l(x) = x^j (x - \alpha^{\frac{n-1}{2}}), \qquad x^j (x - \alpha^{\frac{n+1}{2}})$$

have weight 2 and are not codewords. Therefore, they can be chosen as coset leaders according to a ML decoding strategy.

Furthermore, the leaders so defined can be obtained from each other by applying the automorphisms σ_α, γ, and ι defined in Section 3.

Consequently, the cosets share the same distribution and this result is general and independent from the choice of the leaders.

In order to calculate the coset weight distribution, we observe that all the patterns in a weight-2 coset are multiple of either $(x - \alpha^{\frac{n-1}{2}})$ or $(x - \alpha^{\frac{n+1}{2}})$. Therefore, if we denote by $A_i^{(2)}$ the number of i-weight patterns in a coset with 2-weight leader, and by A_i' the number of i-weight codewords in a RS code $(n, n-1, 2)$, we can write the following expression:

$$A_i + n \cdot A_i^{(2)} = A_i'$$

i	0	1	2	3	4	5	6	7
A_i	1	0	0	245	1225	5586	12838	12873
L_i	1	49	14	0	0	0	0	0
B_i	0	0	1995	29225	255150	1297170	3537975	4053525

Table 4: RS code (7,5,3) ML or UCL decoded - Coefficients for SER computation

that gives an expression of $A_i^{(2)}$ in terms of the known quantities A_i *and* A_i' obtained from equation (5) of Section 4. Finally, the following relation holds:

$$A_i + n^2\, A_i^{(1)} + 2n\, A_i^{(2)} = \binom{n}{i} n^i \quad \text{for } i = 0, 1, \ldots, n \ .$$

4.4 - RS $(n, n - 4, 5)$ codes

The 2-error correcting RS code $(n, n - 4, 5)$, with $q = 2^m$, is not a quasi perfect code, because its covering radius is 4, moreover the weight distribution of the coset leaders may depend on the decoding strategy. In particular it is different for ML and UCL decoding. As a consequence of Theorem 4.3, the weight distribution of the coset leaders for ML decoding turns out to be

$$L_0 = 1 \quad L_1 = n^2 \quad L_2 = n^2 \binom{n}{2} \quad L_3 = \frac{n^4 + 9n^3 + 10n^2 + 4n}{2} \quad L_4 = 2n \ ,$$

where L_3 has been evaluated as a complement to the total number $(n + 1)^4$ of cosets. The weight distribution of the coset leaders for UCL decoding must be evaluated by means of a different counting argument. It results:

$$L_0 = 1 \quad L_1 = n^2 \quad L_2 = n^2 \binom{n}{2}$$

$$L_3 = 2n^3 + 14n^2 - 24n \quad L_4 = \frac{n^4 + 5n^3 - 18n^2 + 56n}{2} \ .$$

Here L_4 is obtained as a complement to the total number of cosets, while L_3 is obtained as difference between the total number of patterns of weight 3 filling in the four check positions, that is:

$$n^3 \binom{4}{3} \ ,$$

and the number of patterns of weight 3 with non zero entries falling in check positions, that are in cosets with coset leaders of weight 2. To count the number of such 3-weight words let us first observe that only one, if any, can be present in each coset because we have only one word with all zero in information positions per coset. A second relevant observation is that patterns of weight 3 originate only from codewords of weight 5, it follows that they stem

- either from codewords with 4 non-zero entries in check positions and 1 non-zero entry in information positions, (let us call these words the 4/1 codewords),

ML	i	0	1	2	3	4	5	6	7
	A_i	1	0	0	0	0	147	147	217
	L_i	1	49	1029	3003	14	0	0	0
	B_i	0	0	0	17899	173567	804225	2117461	2391872
UCL	i	0	1	2	3	4	5	6	7
	A_i	1	0	0	0	0	147	147	217
	L_i	1	49	1029	1204	1813	0	0	0
	B_i	0	0	0	16695	153825	771939	2114427	2448138

Table 5: RS code (7,3,5) - Coefficients for SER computation

- or 3 non-zero entries in check positions and 2 non-zero entries in information positions (let us call these words the 3/2 codewords).

The number of 4/1 codewords is

$$\binom{4}{4}\binom{n-4}{1} \quad ,$$

while the number of 3/2 codewords is

$$\binom{4}{3}\binom{n-4}{2} \quad ;$$

finally the 3-weight patterns are obtained by using 2-weight patterns to delete either the information symbol and one information check symbol in 4/1 codewords, the resulting number of patterns is

$$4\binom{4}{4}\binom{n-4}{1} \quad ,$$

or the two information symbols in 3/2 codewords, the resulting number of patterns is

$$\binom{4}{3}\binom{n-4}{2} \quad .$$

Recalling that the automorphism σ_β produces n distinct cosets sharing the same weight distribution, therefore the number of 3-weight words that must be excluded is

$$n\left[4\binom{4}{4}\binom{n-4}{1}+\binom{4}{3}\binom{n-4}{2}\right] \quad .$$

To obtain the weight distribution in any coset, we use the transitive property of the code group automorphism. Finally we find for the RS code $(7,3,5)$, the coefficients B_i as defined in Section 4, for both ML and UCL decoding. The results reported in Tab. 5, except for A_i's and L_i's, were obtained by computer search.

5 - Conclusions

In this paper we have considered the problem of computing the symbol error probability after complete decoding (ML and UCL strategies) for RS codes. Use of the code automorphism group has been made wherever possible. Let us summarize the most significant results:

- a simple proof that the covering radius of a RS code with minimum distance d is $d-1$ (Theorem 4.1);

- the number of cosets with coset leader of minimum $(d-1)$-weight for a RS code;

- the explicit coset weight distribution for the codes $(7,1,7)$, $(7,5,3)$, $(n,n-2,3)$;

- the explicit leader weight distribution for the codes $(7,1,7)$, $(7,5,3)$, $(7,3,5)$, $(n,n-2,3)$, and $(n,n-4,5)$ for either ML and UCL decoding strategies;

- the error probability after ML decoding for the code $(7,1,7)$ and by computer search for $(7,5,3)$, $(7,3,5)$ codes.

The complexity of the subject probably justifies some of the limited results obtained. If we recall that computer search becomes impractical even for small code lengths, such as 15, then the need and the importance for further research work in this area is well motivated. Especially if we want to reach a deeper understanding of the ML decoding process in case of complete decoding, for RS codes.

References

[1] F.J. MacWilliams and N.J.A. Sloane, *The Theory of Error Correcting Codes*, Elsevier, New York, 1976.

[2] A.M. Michelson and A.H. Levesque, *Error-Control Techniques for Digital Communication*, Wiley, New York, 1985.

[3] V. Pless, *The Theory of Error Correcting Codes*, Wiley, New York, 1989.

[4] S. Benedetto, M. Elia and G. Taricco On Symbol Error Probability Computation for Reed-Solomon Codes, *submitted to IEEE Trans. on Inform. Theory*.

[5] G.D. Cohen, M.G. Karpovsky, H.F. Mattson Jr., and J.R. Schatz, Covering radius: A survey and recent results, *IEEE Trans. on Inform. Theory*, vol. IT-31, no. 3, May 1985, pp. 328-343.

[6] A. Dur, The automorphism groups of Reed-Solomon codes, *Journal of Combinatorics Theory Ser. A (USA)*, vol. 44, no.1, Jan. 1987, pp. 69-82.

[7] A. Dur, On the decoding of Reed-Solomon codes, *Inter. Conference AAECC7*, Toulouse, 26-30 June 1989.

[8] M. Elia, A note on the computation of bit error rate for binary block codes, *Journal of Linear Algebra and its Applic.*, vol. 98, Jan. 1988, pp. 199-210.

[9] R.J.McEliece, L.Swanson, On the Decoder Error Probability for Reed-Solomon Codes, *IEEE Trans. on Inform. Theory*, vol. IT-32, Sept. 1986, pp. 701-703.

[10] G.C. Rota, *Finite Operator Calculus*, Academic Press, New York, 1975.

A direct proof for the automorphism group of
Reed Solomon codes

Thierry BERGER

Département de Mathématiques, Faculté des Sciences de Limoges

123,Av. Albert Thomas, 87060 Limoges Cedex.

France.

Abstract: We introduce a special basis for the description of the primitive extended cyclic codes, considered as subspaces of the modular algebra $A=GF(p^m)[GF(p^m)]$. Using properties of this basis, we determine the automorphism group of some extended cyclic codes, among the extended Reed Solomon codes.

1.Introduction. Let p be a prime. A cyclic code of length $n = p^m -1$ over a field of characteristic p is called primitive. We denote by G the Galois field $GF(p^m)$, and by K a subfield of G, $K = GF(q)$, $q = p^r$, $m = m'r$.

The algebra $A=K[G]$ is the set of formal polynomials

$$x=\sum_{g\in G} x_g X^g, \; x_g \in K$$

with the usual operations:

$$a\sum_{g\in G} x_g X^g + b\sum_{g\in G} y_g X^g = \sum_{g\in G} (ax_g+by_g)X^g$$
$$aX^g bX^h = abX^{g+h}, \; 1=X^0$$

where $a \in K$, $b \in K$, $x_g \in K$, $y_g \in K$, $g \in G$, $h \in G$.

By definition, a K-subspace of A is a code of A. An automorphism of a code is a permutation of the p^m places of the coordinates which transforms codewords into codewords. More precisely, let us identifie a permutation of G with a transformation of A of the type:

$$\sigma : \sum_{g\in G} x_g X^g \rightarrow \sum_{g\in G} x_g X^{\sigma(g)} = \sum_{g\in G} x_{\sigma^{-1}(g)}X^g$$

We denote by Aut(C) the automorphism group of a code C. A permutation σ is an element of Aut(C) if and only if $\sigma(C) = C$.

Let T be a subset of [0,n], $0 \in T$, invariant under the multiplication by q mod n. It is known (see [2],[3],[4]) that a primitive extended cyclic code of K can be represented as a code of A by this way:

$$C=\{x \in A \ / \ \varphi_s(x)=0 \ , \ \forall s \in T\}$$

with $x= \sum_{g \in G} x_g X^g$, $\varphi_s(x)=\sum_{g \in G} x_g g^s$ - conventionally $\varphi_0(x)=\sum_{g \in G} x_g$ -

We say that T is the definition set of C.

Since K is a subfield of G, if T is the definition set of an extended cyclic code C on the algebra A = K[G], it is also the definition set of an extended cyclic code C' on the algebra A' = G[G].

Lemma 1. The codes C and C' have the same automorphism group.

Proof. It is a direct consequence of the two properties:

  i,   C = C'∩A.

  ii, a basis of the K-subspace C of A is a basis of the G-subspace C' of A' (see[2]).

The lemma 1 means that in the study of the automorphism group of the extended cyclic codes in A, we can suppose that K = G. Without lost of generality, from now on let us assume A = G[G].

Now, we will describe a basis $\theta_0, ..., \theta_n$ which verifies

$$\varphi_s(\theta_k) = 0 \text{ if } s \neq k, \text{ and } \varphi_k(\theta_k) = -1.$$

Thus, this basis is very well adapted to describe the extended cyclic codes. It has good properties for the operations of the algebra A, and for the action of certain permutations of the support G.

The automorphism group of the Reed-Solomon codes has been determined by A.Dür [5] in a very complex and difficult way. Using this basis, we can prove the result very easily.

This method has more general applications: we give another example of determination of the automorphism group of extended cyclic codes, in particular of certain Bose-Chaudhuri-Hocquenghem codes.

2.A special basis describing the extended cyclic codes.

For $k = 0,.....,n-1,$ we define
$$\theta_k = \sum_{g\in G} g^{n-k} X^g \text{ , and }\ \theta_n = \sum_{g\in G} X^g$$

Notice that for $k<n$, $\theta_k = \sum_{g\in G*} g^{n-k}X^g$; in particular $\theta_0 = \theta_n - X^0$.

Theorem 1. The set $B = \{\ \theta_k\ /\ 0 \leq k \leq n\ \}$ is a basis of A and

if $x = \sum_{g\in G} x_g X^g$, then $x = \sum_{s=0}^{n-1} -\varphi_s(x)\theta_s + x_0\theta_n$

Proof:

If $W =$
$$\begin{bmatrix} 1 & 1 & 1 & . & . & 1 \\ 0 & \alpha & \alpha^2 & . & . & \alpha^n \\ 0 & \alpha^2 & \alpha^4 & . & . & \alpha^{2n} \\ 0 & . & . & . & . & . \\ 0 & . & . & . & . & . \\ 0 & \alpha^n & \alpha^{2n} & . & . & \alpha^{n^2} \end{bmatrix}$$

then
$$\begin{bmatrix} \theta_n \\ \theta_{n-1} \\ \theta_{n-2} \\ . \\ . \\ \theta_0 \end{bmatrix} = W \ . \ \begin{bmatrix} X^0 \\ X^\alpha \\ X^{\alpha^2} \\ . \\ . \\ X^{\alpha^n} \end{bmatrix}$$

It is well known (see [7] p.323) that W is invertible, hence B is a basis of A.

For $0\leq s\leq n$,
$$\varphi_s(\theta_k) = \sum_{g\in G} g^{n-k}g^s = \sum_{g\in G} g^{n-k+s}, \text{ then}$$

$$\varphi_s(\theta_k)=0 \text{ if } s\neq k, \text{ and } \varphi_k(\theta_k)=-1$$

For $0\leq s<n,$ if $x =\sum \lambda_k\theta_k,$ then $\varphi_s(x) =\sum \lambda_k\varphi_k(\theta_k)=-\lambda_s.$

The element θ_n is the only one to have a non-zero component on X^0, which implies $\lambda_n = x_0$. ■

Remarks:

1, The polynomial $P(Z) = \sum_{j=0}^{n-1} \varphi_{n-j}(x) Z^j$ is in fact the Mattson Solomon polynomial of x. (See [7]).

2, If C is an extended cyclic code of A, and T its definition set, then $\{\theta_k / k \notin T\}$ is a basis of C.

Properties 1.

1, Let σ_a, ψ_h ($a \in G^*$, $h \in (\mathbb{Z}/n\mathbb{Z})^*$) be the permutations on G defined by:

$$\sigma_a: g \to ag, \qquad \psi_h: g \to g^h,$$

then $\quad \sigma_a(\theta_k) = a^k \theta_k \quad$ and $\quad \psi_h(\theta_k) = \theta_{hk \bmod n}$.

2, For all permutation σ on G, $\sigma(\theta_n) = \theta_n$.

The proofs are straightforward.

Let $S = \{0,...,n\}$; each element s of S is identified with $(s_0,........,s_{m-1})$, the coefficients of the p-ary expansion of s. We denote by $<<$ the partial order relation:

$$h \in S, k \in S, \quad h << k \Leftrightarrow h_i \leq k_i \quad \forall i \in \{0,.....,n-1\}$$

Theorem 2. Let $h \in S$, $k \in S$.

If $k = h = 1$, $\theta_0 \theta_0 = -\theta_0 - \theta_n$, else
If $k << n-h$ then $\theta_k \theta_h = -\binom{k+h}{h} \theta_{k+h}$, otherwise $\theta_k \theta_h = 0$.

Proof: For x,y $\in$ A, $\varphi_s(xy) = \sum_{t << s} \binom{s}{t} \varphi_t(x) \varphi_{s-t}(y)$ (see [4])

Consequently, $\varphi_s(\theta_h \theta_k) = \sum_{t << s} \binom{s}{t} \varphi_t(\theta_h) \varphi_{s-t}(\theta_k)$.

So we have:

if $k+h>n$, $\varphi_s(\theta_h \theta_k) = 0$ for all $s \in [0,n[$

if $k+h \leq n$, $\varphi_s(\theta_h \theta_k) = 0$ for $s \neq k+h$, and $\varphi_{h+k}(\theta_h \theta_k) = \binom{h+k}{h}$.
Applying Lucas' theorem, $\binom{h+k}{h} \neq o \bmod p \Leftrightarrow h << k+h$,

(the conditions $h+k \leq n$ and $h << h+k$ are equivalent to $k << n-h$), using theorem 1: $x = \sum\limits_{s=0}^{n-1} -\varphi_s(x)\theta_s + x_0\theta_n$,

and property $x_0 = \sum\limits_{g\in G} x_g - \sum\limits_{g\in G^*} x_g = \varphi_0(x) - \varphi_n(x)$, we obtain the result. ∎

We denote by $*$ the componantwise product of two elements of A :

$$x*y = \sum\limits_{g\in G} x_g y_g X^g, \text{ where } x = \sum\limits_{g\in G} x_g X^g, \text{ and } y = \sum\limits_{g\in G} y_g X^g.$$

Properties 2.

1, For all x of A, $x*\theta_n = x$.

2, For all x, y of A, and for all permutation σ on G, $\sigma(x*y) = \sigma(x)* \sigma(y)$.

3, For $k \neq n$ or $h \neq n$, $\theta_k * \theta_h = \theta_{k+h \bmod n}$.

4, For $k \neq n$, $\theta_k = \theta_1 *^k$, and $\sigma(\theta_k) = \sigma(\theta_1)*^k$, where $x*^k = x*....*x$, k times.

5, $(\sum \lambda_i \theta_i)*^{p^j} = \sum \lambda_i^{p^j} \theta_{ip^j}$, $\lambda_i \in G$.

The proofs are straightforward.

Lemma 2. If $x = x_h\theta_h + \sum\limits_{i>h} x_i\theta_i$ and $y = y_k\theta_k + \sum\limits_{j>k} y_j\theta_j$ with $x_h\neq 0$, $y_k\neq 0$, $h\geq\frac{n}{2}$ $k\geq\frac{n}{2}$,

$$\text{then } x*y = x_h y_k \theta_{k+h-n} + \sum\limits_{i>h+k-n} \lambda_i\theta_i.$$

Proof:

With the conditions $h\geq\frac{n}{2}$ and $k\geq\frac{n}{2}$, $x*y = \sum\limits_{i\geq h, j\geq k} x_i y_j\, \theta_{i+j \bmod n} = \sum\limits_{i\geq h, j\geq k} x_i y_j\, \theta_{i+j-n}$.

Moreover for $i \neq h$ or $j \neq k$, $i+j-n > h+k-n$. ∎

3. The automorphism group of Reed Solomon codes.

The extended Reed Solomon code of designed distance d+1 over G is the code RS(d) of A generated by the basis $B_d = \{\theta_k \,/\, d \leq k \leq n \}$. Its definition set is

$T_d = \{0,.....,d-1\}$.

We denote by GA the affine group of G on G, whose elements are the permutation on G of the form $\sigma : g \to ag+b$, $a\in G^*$, $b\in G$.

Lemma 3. The automorphism group of the RS(n-1) code is the affine group GA.

Proof: the inclusion GA$\subset$Aut(RS(d)) is well known [6]. Let σ be an element of

Aut(RS(n-1)); this code is generated by θ_n and θ_{n-1}, it follows that

$\sigma(\theta_{n-1}) = a\,\theta_{n-1} + b\,\theta_n$,   a,b$\in$ G.

Then $\sigma(\theta_{n-1}) = a \sum_{g\in G} g\,X^g + b \sum_{g\in G} X^g = \sum_{g\in G} (ag+b)X^g$.

But we have $\sigma(\theta_{n-1}) = \sum_{g\in G} \sigma^{-1}(g)X^g$,

then $\sigma^{-1}(g) = ag+b$ for all g on G, $\sigma \in$ GA,   Aut(RS(n-1)) = GA. ■

Lemma 4. For n/2$\leq$d<n-1, Aut(RS(d))$\subset$Aut(RS(d+1)).

Proof: Suppose that $\sigma \in$ Aut(RS(d)) and $\sigma \notin$ Aut(RS(d+1)),   then there exists

k $\in$ [d+1,n-1] such that $\sigma(\theta_k) \in$ RS(d) but $\sigma(\theta_k) \notin$ RS(d+1).

Thus $\sigma(\theta_k) = \sum_{i=d}^{n} \lambda_i\theta_i$, $\lambda_d \neq 0$.

Since d<n, then $\theta_{n-1} \in$ RS(d), $\sigma(\theta_{n-1}) \in$ RS(d), and $\sigma(\theta_{n-1}) = \sum_{j=h}^{n} \mu_j\theta_j$

with $\mu_h \neq 0$,   n-1 $\geq$ h $\geq$ d $\geq$ n/2   (the condition h $\neq$ n is obvious: θ_n and θ_{n-1} have

not the same weight).

       Since k > d, $\theta_{k-1} = \theta_k * \theta_{n-1} \in$ RS(d), $\sigma(\theta_{k-1}) = \sigma(\theta_k)*\sigma(\theta_{n-1})$,   (property

2.2). In accordance with lemma 2 we have:

$\sigma(\theta_{k-1}) = \lambda_d\mu_h\theta_{d+h-n} + \sum_{i>d+h-n} v_i\theta_i$, the condition $n > d \geq \frac{n}{2}$ implie d > d+h-n ≥ 0;

then $\theta_{d+h-n} \notin$ RS(d) , $\sigma(\theta_{k-1}) \notin$ RS(d), that is a contradiction. ■

Theorem 3. For 1 < d < n, the automorphism group of the RS(d) code is the

affine group GA.   (see[5])

Proof: For n/2 $\leq$ d < n, it's a direct consequence of lemmas 3 and 4, and of the

property   GA $\subset$ Aut(RS(d)).

For 1 < d < n/2, the dual of the RS(d) code is the RS(n-d+1) code (see[6]),

then Aut(RS(d)) = Aut(RS(n-d+1)) = GA.   ■

4. Another example of determination of the automorphism group of an extended cyclic code.

Let C be the extended cyclic code of A generated by the basis

      B=$\{\theta_k \,/\, k=n-\varepsilon_i p^i-\varepsilon_j p^j$, $\varepsilon_i, \varepsilon_j \in \{0,1\}$,   i,j$\in$ [0,m[, i>j $\}$

i.e.   B=$\{\theta_n,\theta_{n-1},...,\theta_{n-p^{m-1}},...,\theta_{n-1-p},...,\theta_{n-p^{m-2}-p^{m-1}}\}$

Remarks :

1, If $p = 2$, this code is in an extended Reed-Muller code of second order, and its automorphism group is the affine group of G on GF(2) (see[6]).

2, If $m = 1$, C is the extended Reed-Solomon code RS(n-1) on GF(p).

     We suppose now $p > 2$ and $m > 1$.

3, If $m = 2$, i.e. $G = GF(p^2)$, C is an extended Bose-Chaudhuri-Hocquenghem code of designed distance $d = n-p-1 = p^2-p-2$ (see[4],[6]).

4, If $m = 3$, i.e. $G = GF(p^3)$, C is an extended Bose-Chaudhuri-Hocquenghem code of designed distance $d = n-p-p^2 = p^3-p^2-p-1$ (see[4],[6]).

     We denote by GSA the semi-affine group of G on G, whose elements are the permutations on G of the type

$$\sigma : \quad g \to ag^{p^i}+b, \quad a \in G^*, \; b \in G, \; i \in \{0,....,m-1\}.$$

Lemma 5. The automorphism group of C contains the semi-affine group GSA.

Proof: The code C is invariant under the affine group (see[6])

Let $a \in G^*$, then $\sigma_a: g \to ag$ is an element of Aut(C):

$$\sigma_a(\theta_k) = a^k \, \theta_k \quad \text{(property 1.i,)}$$

Let $b \in G$, then $\tau_b: g \to g+b$ is an element of Aut(C): $\quad \tau_b(\theta_n) = \theta_n,$

$$\tau_b(\theta_{n-p^i}) = \theta_{n-p^i} - b^{p^i}\theta_n , \quad \tau_b(\theta_{n-p^i-p^j}) = \theta_{n-p^i-p^j} - b^{p^j}\theta_{n-p^i} - b^{p^i}\theta_{n-p^j} + b^{p^i+p^j}\theta_n$$

The definition-set of C is invariant under the multiplication by p mod n, C is invariant by $\psi_p: g \to g^p$.

    These permutations generate GSA. ■

Let C' be the extended cyclic code generated by the basis
$$B'=\{\theta_n, \theta_{n-p^i} / i \in [0,m[\}$$

This code is the extended generalized Reed Muller code of first order, whose automorphism group is the affine group of G on GF(p) (see[1]), but this result is not necessary here.

Lemma 6. The automorphism group of C' contains the automorphism group of C.

Proof: For all $\theta_k \in C$, with the conditions $p > 2$ and $m > 1$, we have

$$k \geq n-p^{m-1}-p^{m-2} \geq n/2.$$

Suppose that $\sigma \in$ Aut(C), but $\sigma \notin$ Aut(C'), then there exists $i \in [0,m-1[$ such that

$\sigma(\theta_{n-p^i}) \in$ Aut(C) but $\sigma(\theta_{n-p^i}) \notin$ Aut(C'),   then $\sigma(\theta_{n-p^i}) = \lambda_s \theta_s + \sum \lambda_i \theta_i$

with $\lambda_s \neq 0$ and $\theta_s \in$ C-C'.

Using relation $\sigma(\theta_{n-p^i}) = \sigma(\theta_{n-1}) * p^i$ (property 2.4), we can suppose $i = 0$ without lost of generality.

Moreover, eventually composing σ with ψ_{p^j}, we can suppose $\sigma(\theta_{n-1}) = \lambda_s \theta_s + \sum\limits_{i>s} \lambda_i \theta_i$

$\theta_i \in$ C, $\lambda_s \neq 0$ and $\theta_s \in$ C-C', i.e. $s = n - p^{i_1} - p^{i_2}$, $i_1 \neq i_2$.

Let $\sigma(\theta_{n-p}) = \sum\limits_{j=h}^{n} \mu_j \theta_j$, $\mu_h \neq 0$, $h \neq n$, $\theta_j \in$ C for $\mu_j \neq 0$.

Then $\theta_{n-1-p} = \theta_{n-1} * \theta_{n-p}$, $\sigma(\theta_{n-1-p}) = \sigma(\theta_{n-1}) * \sigma(\theta_{n-p})$, and, using lemma 2, we have:

$$\sigma(\theta_{n-1-p}) = \lambda_s \mu_h \theta_{s+h-n} + \sum\limits_{i>s+h-n} \nu_i \theta_i$$

where $\theta_{s+h-n} \notin$ C since $s = n - p^{i_1} - p^{i_2}$ and $h = n - p^i - \varepsilon_j p^j$, $\varepsilon_j \in \{0,1\}$.

That means $\sigma \notin$ Per(C), which contradicts the hypothesis ∎

Theorem 4. The automorphism group of the code C is the semi-affine group GSA.

Proof: Let $\sigma \in$ Aut(C), composing eventually σ with a translation, we suppose $\sigma(0) = 0$. This implies that, for $k \neq n$, $\sigma(\theta_k)$ have a zero component on θ_n.

Using lemma 6, $\sigma \in$ Aut(C'), $\sigma(\theta_{n-1}) = \sum\limits_{i=0}^{m-1} \lambda_i \theta_{n-p^i}$.

Composing eventually σ with ψ_{p^j}, we can suppose without lost of generality that $\lambda_0 \neq 0$.

Suppose that there exists another index r, $r \neq 0 \mod m$, with $\lambda_r \neq 0$,

Then $\sigma(\theta_{n-p^{m-r}}) = \sigma(\theta_{n-1}) * p^{m-r}$, $\sigma(\theta_{n-p^{m-r}}) = \sum\limits_{i=0}^{m-1} \lambda_i^{p^{m-r}} \theta_{n-p^{i-r}}$

$\theta_{n-1-p^{m-r}} = \theta_{n-1} * \theta_{n-p^{m-r}} \in$ C, $\sigma(\theta_{n-1-p^{m-r}}) = \sigma(\theta_{n-1}) * \sigma(\theta_{n-p^{m-r}})$,

$\sigma(\theta_{n-1-p^{m-r}}) = \sum\limits_{i,j \in [0,m[} \lambda_i \lambda_j^{p^{m-r}} \theta_{n-p^i-p^{j-s}}$.

The expression $\theta_{n-2} = \theta_{n-1-1}$ is obtained only one time, for $i = 0$ and $j = r$, with a non-zero coefficient.

Since $\theta_{n-2} \notin$ C, that implies $\sigma(\theta_{n-1-p^{m-r}}) \notin$ C, that is a contradiction with the hypothesis.

The only one non-zero coefficient is λ_0, $\sigma(\theta_{n-1}) = \lambda_0 \theta_{n-1}$, $\sigma: g \rightarrow \lambda_0 g$, $\sigma \in$ GSA, hence Aut(C) = GSA. ∎

Conclusion: This method to determine an automorphism group has more applications; the reader would be referred to [2].

References:

[1] T.Berger & P.Charpin *The automorphism group of the generalized Reed Muller codes* Rapport INRIA, to appear.

[2] T.Berger *Sur le groupe d'automorphismes des codes cycliques étendus primitifs affine-invariants* Thèse de l'Université de Limoges, in preparation.

[3] P.Charpin *The extended Reed Solomon codes considered as ideals of a modular algebra* Annals of Discrete Mathematics.17(1983)171.176.

[4] P.Charpin *Codes cycliques étendus invariants sous le groupe affine* Thèse de Doctorat d'Etat, Université Paris VII, LITP (1987).

[5] A.Dür *The automorphism group of Reed Solomon codes* J. of Combinatorial Theory, serie A, vol.44, n°1 (1987).

[6] T.Kasami, S.Lin &W.W.Peterson *Some results on cyclic codes which are invariant under the affine group and their applications* Info. and Control, vol 11, p475-496 (1967).

[7] F.J.Macwilliams & N.J.A.Sloane *The theory of error correcting codes* North-Holland (1986).

A FAMILY OF BINARY CODES WITH ASYMPTOTICALLY GOOD DISTANCE DISTRIBUTION

T. BETH, D. E. LAZIĆ, AND V. ŠENK

T. Beth is with Universität Karlsruhe, Fakultät für Informatik, Institut für Algorithmen und Kognitive Systeme, Am Fasanengarten 5 (Geb. 5034), D-7500 Karlsruhe, FR Germany.

D. E. Lazić is with Universität Karlsruhe, Fakultät für Informatik, Institut für Algorithmen und Kognitive Systeme, Am Fasanengarten 5 (Geb. 5034), D-7500 Karlsruhe, FR Germany, Alexander von Humboldt Fellow, on leave from Faculty of Technical Sciences, Computer Science, Control and Measurements Institute, V. Vlahovica 3, 21000 Novi Sad, Yugoslavia.

V. Šenk is with Faculty of Technical Sciences, Computer Science, Control and Measurements Institute, V. Vlahovica 3, 21000 Novi Sad, Yugoslavia.

ABSTRACT: After proving that long binary block codes having the same error exponent as optimum codes (those that attain the minimum possible probability of error) have binomial distance distribution, an infinite sequence of even self-dual codes based on Hadamard matrices is constructed that is conjectured to satisfy the requirements. The first two codes in the sequence are the extended Hamming [8,4,4] and Golay [24,12,8] codes.

I INTRODUCTION

Ever since Claude Shannon published his famous 1948 papers [1], information theorists used and developed his random coding technique of obtaining an asymptotic (in code length) upper bound on the probability of block decoding error for the optimal channel block code. Formally, it consists in calculation of the average probability of block decoding error over the ensemble of all possible sets of M codewords with dimensionality (length) N that can be constructed over a given encoding space. This technique was enthusiastically welcomed as the "way out of the impasse" and frequently designated as the central technique of information theory. For a large number of coding channels it has really succeeded in obtaining lower bounds on channel error exponents (reliability functions) defined as

$$E(R) = \lim_{N \to \infty} \sup \left\{ -\frac{1}{N} \, ld \left[P_{eopt}(R, N) \right] \right\}, \quad R = \frac{ld(M)}{N}, \quad ld(x) = \log_2(x), \quad (1a)$$

where R is the bit code rate, and $P_{eopt}(R,N)$ is the smallest possible probability of block decoding error for codes of bit code rate R and length N used on the given channel. Thus, the upper bound on $P_{eopt}(R,N)$ is given by

$$P_{eopt}(R, N) \leq \tilde{P}_{eopt}(R, N) = \exp2[-N \cdot E(R) + o(N)], \quad \exp2(x) = 2^x, \quad (1b)$$

where o(N) is a term of order less than N; as N→∞ it becomes small relative to N·E(R).

Two aspects of this technique were exceptionally satisfactory:
1) the obtained lower bound was positive for all bit code rates below the channel capacity, and equal to zero at the channel capacity, and
2) it was tight in the most important range of bit code rates, i.e. it agreed with the upper bound on E(R) for all bit code rates above R_{crit} (critical rate), and below the channel capacity, R_c. R_{crit} is typically well below R_c.

The most unsatisfactory aspect of the random coding technique is that it can not determine the requirements that a specific family of channel block codes should meet in order to attain a positive error exponent, the maximum possible being the channel error exponent.

It is the aim of this paper to present a new technique for derivation of a tight lower bound on the error exponent for specific families of binary block codes with known Hamming distance distribution, used on the binary symmetric channel (BSC), to establish the criterion for choosing good binary codes, and to construct a family of even self-dual codes that is conjectured to satisfy the requirements.

II GENERAL METHOD FOR BOUNDING THE ERROR EXPONENT FOR BINARY CODES WITH KNOWN HAMMING DISTANCE DISTRIBUTION USED ON THE BINARY SYMMETRIC CHANNEL

The overall probability of block decoding error for an N-dimensional binary block code B used on the BSC has the form

$$P_e = \sum_{m=1}^{M} P[\mathbf{x}] P_{em} \tag{2}$$

where P_{em} is the probability of erroneous decoding of the codeword $\mathbf{x}_m$ (one of M codewords from B). This latter probability can be stated as

$$P_{em} = P[\mathbf{y} \notin V(\mathbf{x}_m)|\mathbf{x}_m] \tag{3}$$

where $\mathbf{y} = \mathbf{x}_m + \mathbf{n}$ is the channel response and $V(\mathbf{x}_m)$ is the Voronoi region (decision region of the maximum likelihood decoder, identical to the minimum Hamming distance region) of the codeword $\mathbf{x}_m \in B$. Here, $\mathbf{n}$ is the noise vector. The Voronoi regions of all codewords from B form a complete partitioning of the channel output space b^N ($b = \{0,1\}$). Introducing the notion of the error effect of the codeword $\mathbf{x}_j$ on the probability of erroneous decoding when $\mathbf{x}_m$ is sent over the BSC as

$$e_m(\mathbf{x}_j) = P[\mathbf{y} \in V(\mathbf{x}_j)|\mathbf{x}_m, j \neq m], \quad \mathbf{x}_m, \mathbf{x}_j \in B, \quad m = 1, \ldots, M, \tag{4}$$

(3) can be restated as

$$P_{em} = \sum_{\substack{m=1 \\ j \neq m}}^{M} e_m(\mathbf{x}_j) . \tag{5}$$

The channel output space b^N can be divided into two disjoint parts,

$\Omega^{int}(x_m,d^*)$ and $\Omega^{ext}(x_m,d^*)$, defiined by

$$\Omega^{int}(x_m,d^*) = \{y: d_H(y,x_m) \le d^*\}, \qquad \Omega^{int}(x_m,d^*) \cup \Omega^{int}(x_m,d^*) = b^N, \qquad (6)$$

where d^* is the radius of the confining sphere $\Omega(x_m,d^*)$. Using this partitioning of b^N, the error effect $e_m(x_j)$ can be divided intoo two parts, $e_m'(x_j,d^*)$ and $e_m''(x_j,d^*)$, defined as

$$e_m'(x_j,d^*) = P[y \in V(x_j) \cap \Omega^{int}(x_m,d^*) \mid x_m] , \qquad (7a)$$

$$e_m''(x_j,d^*) = P[y \in V(x_j) \cap \Omega^{ext}(x_m,d^*) \mid x_m] , \qquad (7b)$$

so that

$$e_m(x_j) = e_m'(x_j,d^*) + e_m''(x_j,d^*). \qquad (7c)$$

It is obvious that the confined error effect $e_m'(x_j,d^*) \le e_m(x_j)$, and that $e_m'(x_j,d^*) = 0$ if $d_H(x_m,x_j) > 2d^*$. Introducing (7) into (5), and then into (2), one obtains that

$$P_e = \sum_{m=1}^{M} P[x_m] \sum_{\substack{m=1 \\ j \ne m}}^{M} e_m(x_j) =$$

$$= \sum_{\substack{m=1 \\ j \ne m}}^{M} \sum_{m=1}^{M} P[x_m] e_m'(x_j,d^*) + \sum_{\substack{m=1 \\ j \ne m}}^{M} \sum_{m=1}^{M} P[x_m] e_m''(x_j,d^*) = P_e' + P_e'' . \qquad (8)$$

The exterior error probability P_e'' may be upper-bounded by

$$P_e'' = \sum_{\substack{m=1 \\ j \ne m}}^{M} \sum_{m=1}^{M} P[x_m] \, P[y \in V(x_j) \cap \Omega^{ext}(x_m,d^*) \mid x_m] =$$

$$= \sum_{m=1}^{M} P[x_m] \, P[y \in (\bigcup_{\substack{j=1 \\ j \ne m}}^{M} V(x_j)) \cap \Omega^{ext}(x_m,d^*) \mid x_m] \le$$

$$\leq \sum_{m=1}^{M} P[x_m] \, P[y \in \Omega^{ext}(x_m, d^*) \mid x_m] = P[y \in \Omega^{ext}(0, d^*) \mid 0], \qquad (9)$$

i.e. by the probability that the Hamming weight of the noise vector n is greater than d^*, namely

$$\tilde{P}_e^{\,''} = P[w_H(n) > d^*]. \qquad (10)$$

The Voronoi region $V(x_m)$ can be expanded into $V(x_m \mid x_j)$, defined as

$$V(x_m \mid x_j) = \{y: d_H(y, x_m) \geq d_H(y, x_j)\}. \qquad (11)$$

The confined error effect $e_m'(x_j, d^*)$ can now be upper-bounded by

$$e_m'(x_j, d^*) = P[y \in V(x_m \mid x_j) \cap \Omega^{int}(x_m, d^*) \mid x_m] . \qquad (12)$$

The confined error probability P_e' can now be upper-bounded by

$$\tilde{P}_e^{\,'} = \sum_{m=1}^{M} \sum_{\substack{m=1 \\ j \neq m}}^{M} P[x_m] \, \tilde{e}_m'(x_j, d^*) . \qquad (13)$$

Sorting the $M(M-1)$ upper bounds on the confined error effects in (13) by descending order, L $(1 \leq L \leq N))$ different values (levels) of upper bounds on the confined error effects are obtained so that $1 \geq \tilde{e}_1'(d^*) > \ldots > \tilde{e}_L'(d^*) \geq 0$, i.e. $\tilde{e}_l'(d^*) > \tilde{e}_{l+1}'(d^*)$, $l=1,\ldots,L-1$ where there are M_l $(1 \leq M_l \leq M(M-1))$ upper bounds on the confined error effects that have the value $\tilde{e}_l'(d^*)$, $1 \leq l \leq L$, out of which M_{ml} $(0 \leq M_{ml} \leq M-1)$ have the sole influence on the determination of $\tilde{P}_{em}'$. Now,

$$\tilde{P}_e^{\,'} = \sum_{m=1}^{M} P[x_m] \sum_{l=1}^{L} M_{ml} \tilde{e}_l'(d^*) = \sum_{l=1}^{L} \langle M_l \rangle \tilde{e}_l'(d^*) , \qquad \langle M_l \rangle = \sum_{m=1}^{M} P[x_m] M_{ml} . \qquad (14)$$

Introducing (14) and (10) into (8), the upper bound on the overall probability of error, $\tilde{P}_e$, attains the form

$$\tilde{P}_e = \sum_{l=1}^{L} <M_l> \tilde{e}_l'(d^*) + P[w_H(n) > d^*] \ . \tag{15}$$

The error exponent of a family $\mathcal{B}$ with known expected Hamming distance distribution $\{<M>\}$ can be lower bounded introducing (15) into (1a) as

$$E_{\mathcal{B}}(R) = \max_{d} \min \left\{ E_{\mathcal{B}}'(R,d^*), \ E_{\mathcal{B}}''(R,d^*) \right\} \tag{16}$$

where

$$E_{\mathcal{B}}'(R,d^*) = \min_{1 \le l \le L} \left\{ -\frac{1}{N} \ ld(<M_l> \tilde{e}_l'(d^*)) \right\} \tag{17}$$

and

$$E_{\mathcal{B}}''(R,d^*) = -\frac{1}{N} \ ld \ P[w_H(n) > d^*] \ . \tag{18}$$

The upper bound on the confined error effect,

$$\tilde{e}_l'(d^*) = P[d_H(y,x_m) \ge d_H(y,x_j) \wedge w_X(n=y+x_m) \le d^* \ | \ d_H(x_m,x_j)=l \ , \ x_m] \ , \tag{19}$$

can be evaluated as

$$\tilde{e}_l'(d^*) = \sum_{n=l/2}^{d^*} \sum_{i=\lceil l/2 \rceil}^{l} \binom{l}{i} \cdot \binom{N-l}{n-i} \ p^n(1-p)^{N-n}, \tag{20}$$

since at least $i=\lceil l/2 \rceil$ errors need exist on the l places where the two codewords (x_m and x_j) differ, and the total ammount of errors (nonzero positions in n) has to be at least equal to i, but not greater than d^*. Using the well-known inequality [2, p. 309],

$$\binom{N}{n} \le 2^{NH(n/N)}, \quad H(x)=-xld(x)-(1-x)ld(1-x), \quad ld(x)=\log_2(x), \tag{21}$$

(20) can be upper-bounded by

$$N^2 \cdot \max_{l/2 \le n \le d^*} \left\{ 2^{l+(N-l)H\left[\frac{2n-l}{2(N-l)}\right]} + n \cdot ld(p) + (N-n) \cdot ld(1-p) \right\} \ . \tag{22}$$

Introducing (22) into (17) one obtains that

$$E'_{\mathcal{B}} (R, d^*) = \min_{1 \leq l \leq L} \min_{1/2 \leq \underline{n} \leq \underline{d}^*} \left\{ - \frac{1}{N} \; \mathrm{ld}(<\mathcal{M}_l> S_1(l, n, N, p) \right\} \tag{23}$$

where

$$S_1(l, n, N, p) = \underline{l} + (1 - \underline{l}) H \left[\frac{2\underline{n} - \underline{l}}{2(1 - \underline{l})} \right] + \underline{n}\,\mathrm{ld}(p) + (1 - \underline{n})\,\mathrm{ld}(1-p), \quad \underline{l} = \frac{l}{N}, \quad \underline{n} = \frac{n}{N} \;. \tag{24}$$

The upper bound on P''_e, (9), can be evaluated as:

$$\tilde{P}''_e = \sum_{n=d^*+1}^{N} \binom{N}{n} p^n (1-p)^{N-n}\;, \tag{25}$$

and upper bounded by

$$N \cdot \max_{d^* < n \leq N} \left\{ 2^{NH(n/N) + n\,\mathrm{ld}(p) + (N-n)\,\mathrm{ld}(1-p)} \right\} \tag{26}$$

Introducing (26) into (18) one obtains that

$$E''_{\mathcal{B}} (R, d^*) = \min_{d^* < n \leq N} \{ -T(n, N, p) \}\;, \tag{27}$$

where

$$T(n, N, p) = H(\underline{n}) + \underline{n}\,\mathrm{ld}(p) + (1 - \underline{n})\,\mathrm{ld}(1-p)\;. \tag{28}$$

III   THE BSC ERROR EXPONENT FOR BINARY CODES UNIFORMLY DISTRIBUTED OVER THE N-DIMENSIONAL HYPRCUBE

Suppose that a binary code B of length N was obtained using a uniform expurgation of an N-dimensional hypercube b^N, so that the expected number of codewords on Hamming distance l from any codeword in the code is proportional to $\binom{N}{l}$. For instance, the so-called "random code" has the feature when N grows beyond all bounds. Such a code with M codewords has thus

$$<\mathcal{M}_l> = M \cdot \frac{1}{2^N} \binom{N}{l} \leq 2^{N \cdot [H(l/N) - (1-R)]}, \quad l = 1, \ldots, N. \tag{29}$$

There is a certain bound, d_{GV} , under which $<M_1> \le P(N)$, where $P(N)$ is a polynomial in N. This bound is easily obtained equating the exponent in (29) with zero, so that

$$d_{GV} = N \cdot H^{-1}(1-R) . \tag{30}$$

It is obvious that this is in fact the Gilbert-Varshamov lower bound on the maximum minimum distance in a binary code. Expurgating one codeword from any pair of codewords in B that is on Hamming distance below d_{GV} does not change neither the exponent in (29) nor the bit code rate R. Such an expurgated family B^* thus attains (and does not exceed) the Gilbert-Varshamov bound, and has the binomial Hamming distance distribution (29).

It is easily seen that the optimum value of d^* is that one for which $\Omega^{int}(0,d^*)$ contains exactly $2^N/M$ vertices from b^N (i.e. has the same number of vertices from b^N as the average Voronoi region). This condition means that $\sum_{n=0}^{d^*} \binom{N}{n} = 2^{N(1-R)}$, i.e. that $d^* = d_{GV}$. Assuming this value of d^*, and introducing (29) into (23), after a straightforward manipulation one obtains that

$$E'^*_B (R) = \min_{\underline{d}_{GV} \le \underline{l} \le 1} \quad \min_{1/2 \le \underline{n} \le \underline{d}_{GV}} \left\{ -H(\underline{l}) + (1-R) - S_1(\underline{l}, \underline{n}, p) \right\} =$$

$$= \begin{cases} Z \cdot \underline{d}_{GV} , & R \le R_{crit1} , \\ R_0 - R , & R_{crit1} \le R \le R_{crit2} , \\ (1-R) + U(R,p) , & R_{crit2} \le R , \end{cases} \quad \underline{d}_{GV} = d_{GV}/N , \tag{31}$$

where

$$Z = ld(V), \quad V = \sqrt{4p(1-p)} , \tag{32a}$$
$$R_0 = 1 - ld(1+V) , \tag{32b}$$
$$U(R,p) = \underline{d}_{GV} ld(p) + (1-\underline{d}_{GV}) ld(1-p) , \tag{32c}$$
$$R_{crit1} = 1 - H[V/(1-V)] , \tag{32d}$$
$$R_{crit2} = 1 - H[\sqrt{p} / (\sqrt{p} + \sqrt{1-p})] . \tag{32e}$$

Assuming $d^* = d_{GV}$, one also obtains that

$$E_{\mathcal{B}}''^* (R) = \begin{cases} (1-R) + U(R,p) & R \le R_c , \\ 0 , & R \ge R_c , \end{cases} \tag{33}$$

where

$$R_c = 1 - H(p) . \tag{34}$$

The lower bound on the error exponent of expurgated family of codes uniformly distributed over b^N is, introducing (33) and (31) into (16), obtained as

$$E_{\mathcal{B}}^* (R) = \begin{cases} Z \cdot \underline{d}_{GV} , & R \le R_{crit1} , \\ R_o - R , & R_{crit1} \le R \le R_{crit2} , \\ (1-R)+U(R,p) , & R_{crit2} \le R \le R_c , \\ 0 , & R_c \le R . \end{cases} \tag{35}$$

This function is identical to the one obtained for the ensemble of all possible binary codes used over the BSC. Furthermore, since this expurgated family has the minimum Hamming distance equal to $\underline{d}_{GV}$, (35) is also the upper bound on the error exponent of $\mathcal{B}^*$, obtained using the sphere packing, space partitioning and straight-line bounds [3,ch.10]. The expression (35) thus represents the tight solution to the error exponent of expurgated family of codes uniformly distributed over b^N at all bit code rates.

IV   A FAMILY OF BINARY CODES WITH BINOMIAL HAMMING DISTANCE DISTRIBUTION

Linear codes having the binomial Hamming distance distribution (29) have to meet the Gilbert-Varshamov bound, and may not exceed it, since they are distance invariant. A question that could be raised is: do we know any constructive family of linear binary block codes that have the Hamming distance distribution (29)? It has been proved that a randomly chosen long linear code meets, with probability 1, both the Shannon bound on the probability of error of optimum codes [4,pp.206-209],and the Gilbert-Varshamov bound [5], and, with probability 1, exceeds neither. Thus, it may be expected that any linear binary block code is arbitrarily close to the optimal one. But, this does not tell that any known constructive family of linear binary block codes meets the bounds.

It is well-known (see [2, pp 130-132 and 282-284]) that when the dual of a linear binary block code has minimum distance d', then the first d'-1 central and binomial moments of the normalized weight (and thus the distance) distribution of the code are identical to that of the binomial distribution obtained for the [N, N, 1] linear binary block code. This means that the duals of linear codes meeting the Shannon bound should be asymptotically good, and hopefully those meeting the Gilbert-Varshamov bound. In the random coding case, it is an easy task to find such a pair of linear codes, but is there a constructive family of such codes (that, together with their duals meet the Gilbert-Varshamov bound)?

The task of searching for such a twin family is obviously easiest in the case of self-dual codes. It is known [2, pp. 556-558 and 629-633] that long self-dual codes meet the Gilbert-Varshamov bound with probability one, so that any constructive family of such codes is conjectured to have the property, and thus meet the Shannon bound.

A very simple construction of an infinite sequence of binary self-dual codes can be obtained using Hadamard matrices for all dimensions $N=8 \cdot (2l-1), l=1, 2, \ldots$ (provided Hadamard matrices exist in each case). The first few codes of this type have lengths $8, 24, 40, 56, 72, \ldots$. A normalized Hadamard matrix H_n, $n=4 \cdot (2l-1)$, $l=1.2, \ldots$, is changed into the binary Hadamard matrix A_n replacing +1's in H_n by 0's and -1's by 1's, and deleting the first row and column (that now contain only zeros). Any row and column of such a matrix has exactly $n/2$ 1's, and any two rows and columns of it have exactly $n/4$ 1's in the same places (see theorem 5. on p.44 of [2]). Construct a generator matrix G_n as

$$G_n = \left[I_n \left| \begin{array}{c|c} A_n & \mathbf{1}^T \\ \hline \mathbf{1} & 0 \end{array} \right. \right] \tag{36}$$

where $\mathbf{1}$ is a row vector of all 1's and I_n is the $n \times n$ unit matrix. It is readily seen that for any $n=4 \cdot (2l-1)$ the first $n-1$ rows of this matrix contain exactly $1+2(2l-1)+1=4l$ 1's, while the last row contains $1+4(2l-1)-1=4(2l-1)$ 1's. Furthermore $G_n G_n^T = 0 \pmod 2$ since the last row has 1's in all the $2(2l-1)$ places where the rows of A_n do, and other pair of rows of G_n have exactly $0+(2l-1)+1=2l$ common 1's. Thus, G_n defines a self-dual code of length $N=2n=8(2l-1)$. Furthermore, since the Hamming weight of any row of G_n is divisible by 4, Hamming weight of any codeword in the code defined by the matrix is also divisible by 4 [2, p.27].

The first two codes of this family are the extended Hamming [8,4,4] and Golay [24,12,8] codes. The weight generators of these two codes form an integrity basis for the weight generatos of all doubly even self-dual codes [6]. Thus, any code in the proposed family has the weight generator (weight distribution polynomial) in the form

$$D_1(Z) = \sum_{r=0}^{\lceil(2l-1)/3\rceil} a_r D_1(Z)^{2l-1-3r} D_2'(Z)^r , \qquad 1 = 3,4,\ldots \qquad (37)$$

where $D_1(Z) = 1 + 14Z^4 + Z^8$ is the weight generator of the extended Hamming [8,4,4] code, and $D_2'(Z) = [D_1(Z)^3 - D_2(Z)]/42 = Z^4(1-Z^4)$, $(D_2(Z) = 1 + 759Z^8 + 2576Z^{12} + 7592Z^{16} + Z^{24}$ is the weight generator of the extended Golay code). The coefficients a_r in (37) are integers, while $a_0 = 1$ (Gleason's theorem, see [2,ch.19]). The ratio of the i'th coefficient of $D_1(z)$ for the first four codes in the sequence to the optimum value of $2^{-N/2}\binom{N}{1}$ (that corresponds to (29), differs percentually from 4 as follows (K=N/2):

N	D_K	$D_{K\pm4}$	$D_{K\pm8}$	$D_{K\pm12}$	$D_{K\pm16}$
8	-20.00				
24	-2.453	+5.676			
40	-0.064	+0.087	-0.151	-2.852	
56	-0.017	+0.023	-0.059	+0.273	-1.569

The obvious conjecture follows that the sequence $D_1(z)$, $1=1,2,\ldots$ converges to:

$$D_1(Z)= 1 + 4 \cdot \sum_{r=1}^{2(21-1)-1} \left\lfloor 2^{-4(21-1)} \binom{8(21-1)}{4r} \right\rfloor Z^{4r} + Z^{8(21-1)} \qquad (38)$$

as $1 \to \infty$. This is simply the Hamming distance distribution (29) with every 4'th term retained and multiplied by 4. It is easily recognized that this sequence of codes has the same error exponent as the sequence of codes uniformly distributed over b^N.

This family of codes, that includes some quadratic residue and double circulant codes is the first one, to our knowledge, to be conjectured to meet the Shannon bound.

It is important to note that inequivalent Hadamard matrices of the same size give rise to different self dual codes in our construction, so that the Hamming distance distribution of these codes need not be equal.

The criterion of choosing good codes presented in this paper is derived in general case for all channels in [7]. This criterion was already conjectured for the binary case in [8], but without other evidence than simulation results.

REFERENCES

[1] C. E. Shannon: "A Mathematical Theory of Communications", BSTJ, 27, pp. 379-423 and 623-656, 1948.
[2] F. J. MacWilliams, N. J. A. Sloane: "The Theory of Error Correcting Codes", Part I and II, North-Holland, 1977.
[3] R. E. Blahut: "Principles and Practice of Information Theory" , Addison-Wesley, 1987.
[4] R. G. Gallager: "Information Theory and Reliable Communication", John Willey and Sons, 1968.
[5] J. N. Pierce: "Limit Distributions of the Minimum Distance of Random Linear Codes", IEEE Transactions on Information Theory, Vol IT-13, No. 4, pp. 595-599, October 1967.
[6] T. Beth : "Codes und Invarianten" in "Mathematishes Codierung Theorie", Erlangen 1978.
[7] D. E. Lazić, V. Šenk: "A Simple Geometrical Method for Determination of the Error Exponent for Deterministic Channel Block Codes – Part I: Cutoff Rate Bound, and Part II: the Tight Solution", submitted to IEEE Transactions on Information Theory, 1990.
[8] G. Battail: "Construction explicite de Bons Codes Longs", Ann. Telecommun., 44, No. 7-8, pp. 392-404, 1989.

A TRANSFORMATION ON BOOLEAN FUNCTIONS, ITS CONSEQUENCES ON SOME PROBLEMS RELATED TO REED - MULLER CODES

Claude Carlet[*]

Université de Picardie, France

ABSTRACT

We introduce a transformation defined on the set of all boolean functions defined on Galois fields $GF(2^m)$, $m \in \mathbf{N}^*$, (or on F_2^m), which changes their weights in a way easy to be followed, and which, when we iterate it, reduces their degrees down to 2 or 3. We deduce that it is as difficult to find a general characterization of the weights in the Reed-Muller codes of order 3 as it is to obtain one in the Reed-Muller codes of any orders. We also use this transformation to characterize the existence of some affine sets of bent functions, and to obtain bent functions of degree 4 from bent functions of degree 3.

INTRODUCTION

For any positive integers m and r with $r \leq m$, the Reed-Muller code of order r, $R(r, m)$, is the set of all boolean functions on the Galois Field G of order 2^m, (or on F_2^m), whose expression by means of coordinate functions is a polynomial of degree at most r.

The weight distribution of $R(2, m)$ is well known since the weight of an element $f \in R(2, m)$ (ie the cardinal of its support) is 2^{m-1} (we say then that f is balanced) if f is non-constant on the kernel of its associated symplectic form $\varphi_f : (u, v) \rightarrow f(0) + f(u) + f(v) + f(u+v)$, and is otherwise (simply) related to the dimension of the kernel (cf [1] for instance).

Unfortunately, for $r \geq 3$, the weight distribution of $R(r, m)$ is unknown. This open problem is related to other very difficult open problems in discrete mathematics. Some people have tried to work on the case $r = 3$, which seemed to be easier than the general case .

We introduce (proposition 1) a mapping from the set of boolean functions on G to the set of boolean functions on $G \times F^2$ (with $F = F_2 = \{0, 1\}$), which changes their weights in a very simple way , and we use it to prove that the weight of a general boolean function of any degree

---

[*] INRIA, Bat 10, Domaine de Voluceau, BP 105, 78153 Le Chesnay, Cedex, France

defined on G is related to the weight of a boolean function of degree at most 3, defined on a Galois field of a greater order (corollary 1).

That means that to characterize the weight distribution of R(3, m), we will have either to find a way to calculate the weight of any boolean function of any given degree or to obtain this weight distribution globally.

We use the same idea to prove (proposition 2) that an affine set of boolean functions on $GF(2^m)$ (m even) of dimension n (n even) contains only bent functions if and only if some function on $GF(2^m) \times F^n$ is bent, and to obtain bent functions of degree 4 from bent functions of degree 3.

1. WEIGHTS OF BOOLEAN FUNCTIONS

Let us recall that the weight w(f) of a boolean function f on the Galois field G =

$GF(2^m)$, is related to the sum : $\sum_{u \in G} (-1)^{f(u)}$ by the relation :

$$w(f) = 2^{m-1} - \frac{1}{2} \sum_{u \in G} (-1)^{f(u)} \quad (1).$$

PROPOSITION 1

Let m be any integer, and G the Galois field $GF(2^m)$.

Let f_1, f_2 and g be boolean functions on G, and f the function defined on G by :

$\forall u \in G, f(u) = f_1(u) f_2(u) + g(u).$

Then f, f_1, f_2 and g satisfy the following relation :

$$\sum_{u \in G} (-1)^{f(u)} = \frac{1}{2} \sum_{u \in G ; a,b \in F} (-1)^{ab + a f_1(u) + b f_2(u) + g(u)} \quad \text{and therefore :}$$

$w(f) = \frac{1}{2} w \left(f^{[1]} \right) - 2^{m-1}$, *where $f^{[1]}$ is the function on $G \times F^2$ defined by:*

$\forall u \in G, \forall a, b \in F, f^{[1]}(u, a, b) = ab + a f_1(u) + b f_2(u) + g(u).$

Proof :

We have :

$$\sum_{u \in G} (-1)^{f(u)} = \frac{1}{2} \sum_{u \in G; a,b \in F} (-1)^{ab + f(u)}$$

(since ab takes 3 times the value 0 and once the value 1)

$$= \frac{1}{2} \sum_{u \in G; \, a,b \in F} (-1)^{\left(a + f_2(u)\right)\left(b + f_1(u)\right) + f(u)}$$

(since F is invariant under any translation, and in particular under the translations $a \rightarrow a+f_2(u)$ and $b \rightarrow b+f_1(u)$)

$$= \frac{1}{2} \sum_{u \in G; \, a,b \in F} (-1)^{ab + af_1(u) + bf_2(u) + g(u)}.$$

Using equality (1) applied to f and to $f^{[1]}$(where GxF^2 is identified with $GF(2^{m+2})$, we obtain

$$w(f) = 2^{m-1} - \frac{1}{2} \sum_{u \in G} (-1)^{f(u)} = 2^{m-1} - \frac{1}{4}\left(\sum_{u \in G; \, a,b \in F} (-1)^{f^{[1]}(u,\, a,\, b)} \right)$$

$$= 2^{m-1} - \frac{1}{4}\left(2^{m+2} - 2w\left(f^{[1]}\right)\right) = \frac{1}{2} w\left(f^{[1]}\right) - 2^{m-1}.$$

Remark

the degree $d^\circ f^{[1]}$ of $f^{[1]}$ (when we express it by means of coordinate functions on GxF^2) is equal to : $\max(2,\, d^\circ f_1 + 1,\, d^\circ f_2 + 1,\, d^\circ g)$.

COROLLARY 1

For any boolean function f on G, there exists an integer p and a boolean function $f^{[p]}$ on GxF^{2p} such that $d^\circ f^{[p]} \leq 3$ and :

$$\sum_{u \in G} (-1)^{f(u)} = \frac{1}{2^p} \sum_{u \in G; \, v \in F^{2p}} (-1)^{f^{[p]}(u,\, v)}$$

$$ie, \; w(f) = 2^{m-1} - 2^{m+p-1} + \frac{1}{2^p} w\left(f^{[p]}\right).$$

Proof :

If $d^\circ f \leq 3$, then we take : $p = 0$ and $f^{[p]} = f$.

Otherwise, let us choose a base of the F-space G. Any boolean function f on G can be expressed by means of the coordinate functions as the sum of some (say q) monomials (cf [4] ch 13) :

$$f = \sum_{i=1}^{q} f_i , \quad d^\circ f_i \leq d^\circ f.$$

Each monomial f_i can be written as the product of two monomials $f_{i,1}$ and $f_{i,2}$ of degrees at most $\left\lceil \frac{d^\circ f}{2} \right\rceil$ (the smallest integer greater than $d^\circ f / 2$).

By applying Proposition 1 to f with $g = f_{2,1} f_{2,2} + \ldots + f_{q,1} f_{q,2}$, we obtain :

$$\sum_{u \in G} (-1)^{f(u)} = \frac{1}{2} \sum_{u \in G;\, a_{1,1},\, a_{1,2} \in F} (-1)^{f^{[1]}(u,\, a_{1,1},\, a_{1,2})}$$

where $f^{[1]}(u, a_{1,1}, a_{1,2}) = a_{1,1} a_{1,2} + a_{1,1} f_{1,1} + a_{1,2} f_{1,2} + \sum_{i=2}^{q} f_{i,1} f_{i,2}$.

By applying Proposition 1 to $f^{[1]}$ (where GxF^2 is identified as a linear space with $GF(2^{m+2})$
we obtain :

$$\sum_{u \in G;\, a_{1,1},\, a_{1,2} \in F} (-1)^{f^{[1]}(u,\, a_{1,1}, a_{1,2})} =$$

$$\frac{1}{2} \sum_{u \in G} \sum_{a_{1,1},\, a_{1,2},\, a_{2,1},\, a_{2,2} \in F} (-1)^{f^{[2]}(u,\, a_{1,1},\, a_{1,2},\, a_{2,1},\, a_{2,2})} \qquad \text{where :}$$

$f^{[2]}(u, a_{1,1}, a_{1,2}, a_{2,1}, a_{2,2}) = a_{1,1} a_{1,2} + a_{2,1} a_{2,2} + a_{1,1} f_{1,1} + a_{1,2} f_{1,2} + a_{2,1} f_{2,1} + a_{2,2} f_{2,2} + \sum_{i=3}^{q} f_{i,1} f_{i,2}$.

By applying again (q-2) times Proposition 1 , we see that there exists a boolean function $f^{[q]}$ on
GxF^{2q} such that :

$$d^{o}f^{[q]} \le 1 + \left\lceil \frac{d^{o}f}{2} \right\rceil \text{ and }$$

$$\sum_{u \in G} (-1)^{f(u)} = \frac{1}{2^{q}} \sum_{u \in G;\, v \in F^{2q}} (-1)^{f^{[q]}(u,\, v)}.$$

Let us iterate this process, the degree of $f^{[q]}$ stops decreasing when $d^{o}f^{[q]} = 1 + \left\lceil \dfrac{d^{o}f^{[q]}}{2} \right\rceil$,

ie, $d^{o}f^{[q]} = 3$. So, the existence of p is proved and $f^{[p]}$ satisfies :

$$w(f) = 2^{m-1} - \frac{1}{2} \sum_{u \in G} (-1)^{f(u)} = 2^{m-1} - \frac{1}{2^{p+1}} \sum_{u \in G;\, v \in F^{2p}} (-1)^{f^{[p]}(u,\, v)} =$$

$$2^{m-1} - \frac{1}{2^{p+1}} \left(2^{m+2p} - 2w(f^{[p]}) \right) = 2^{m-1} - 2^{m+p-1} + \frac{1}{2^{p}} w\left(f^{[p]} \right).$$

So, it is as hard to find a way to characterize the weight of any function of degree 3, as it is for
any other degree.

COROLLARY 2

For any k∈ Z, there exists integers p and q, and an element h of R(3, q) such that
$$k = \frac{1}{2^{p}} \left(2^{q} - 2w\,(h) \right).$$

Proof :

Let m be an integer such that : $-2^{m-1} \leq k \leq 2^{m-1}$.

Since we have : $0 \leq 2^{m-1} - k \leq 2^m$, there exists a boolean function f on $G = GF(2^m)$

such that $w(f) = 2^{m-1} - k$, ie $k = \frac{1}{2} \sum_{u \in G} (-1)^{f(u)}$.

Corollary 1 (with $q = m + 2(p-1)$ and $h = f^{[p-1]}$) completes the proof.

We see here the great difference between the weight distribution of the Reed-Muller codes of orders 2 and 3, since for any h belonging to R(2, q), $2^q - 2w(h)$ is equal to $\pm 2^i$, $q/2 \leq i \leq q$, or to 0.

Remark

Proposition 1 can be generalized to functions from any Galois field towards any Galois field.

Indeed, let G be a Galois field of charateristic p and order q, G' a Galois field of characteristic p' and order q', and f a function from G to G'.

Let $\omega = e^{\frac{2i\pi}{p'}}$, and for any u in G', ψ_u the character on G' defined by : $\psi_u(x) = \omega^{tr(ux)}$ where tr denotes the trace function from G' to GF(p').

It is well known that for any b in G', we have :
$$\sum_{u \in G'} \psi_u(b) = q' \text{ if } b = 0, \text{ and } 0 \text{ if } b \neq 0 \quad (2)$$

and therefore, if N(f = b) denotes the cardinal of $f^{-1}(b)$:
$$N(f = 0) = \frac{1}{q'} \sum_{u \in G'} \sum_{x \in G} \psi_u(f(x)) .$$

We have :
$$\sum_{u, a, be G'} \left(\sum_{x \in G} \psi_u(ab + f(x)) \right) = (q'-1) \sum_{u \in G', a \in G'-\{0\}} \left(\sum_{x \in G} \psi_u(a + f(x)) \right) + (2q'-1) \sum_{u \in G'} \left(\sum_{x \in G} \psi_u(f(x)) \right)$$

since when a and b range over G', ab takes q'-1 times any value different from 0, and 2q'-1 times the value 0, and therefore :
$$\sum_{u, a, be G'} \left(\sum_{x \in G} \psi_u(ab + f(x)) \right) = (q'-1) \sum_{u \in G', a \in G'} \left(\sum_{x \in G} \psi_u(a + f(x)) \right) + q' \sum_{u \in G'} \left(\sum_{x \in G} \psi_u(f(x)) \right)$$

$$= (q'-1)q \sum_{u \in G', a \in G'} \psi_u(a) + q' \sum_{u \in G'} \left(\sum_{x \in G} \psi_u(f(x)) \right)$$

$$= (q'-1)q'q + q' \sum_{u \in G'} \left(\sum_{x \in G} \psi_u(f(x)) \right) \text{ (according to (2))} .$$

Therefore :
$$N(f = 0) = \frac{1}{q'^2} \sum_{u, a, be G'} \left(\sum_{x \in G} \psi_u(ab + f(x)) \right) - \frac{q(q'-1)}{q'} .$$

So, if f is such that $f(x) = f_1(x) f_2(x) + g(x)$, where the product between $f_1(x)$ and $f_2(x)$ is calculated in the Galois field G', then for any c in G':

$$N(f = c) = \frac{1}{q'^2} \sum_{u,\,a,\,b \in G'} \left(\sum_{x \in G} \psi_u \big(ab + f(x) - c\big) \right) - \frac{q(q'-1)}{q'} =$$

$$\frac{1}{q'^2} \sum_{u,\,a,\,b \in G'} \left(\sum_{x \in G} \psi_u \big(ab + f_1(x)f_2(x) + g(x) - c\big) \right) - \frac{q(q'-1)}{q'} =$$

$$\frac{1}{q'^2} \sum_{u,\,a,\,b \in G'} \left(\sum_{x \in G} \psi_u \big((a + f_1(x))(b - f_2(x)) + f_1(x)f_2(x) + g(x) - c\big) \right) - \frac{q(q'-1)}{q'} =$$

$$\frac{1}{q'^2} \sum_{u,\,a,\,b \in G'} \left(\sum_{x \in G} \psi_u \big(ab - a\,f_2(x) + b\,f_1(x) + g(x) - c\big) \right) - \frac{q(q'-1)}{q'}$$

and if $f^{[1]}$ is the function defined on $G \times G'^2$ by $f^{[1]}(x,a,b) = ab - af_2(x) + bf_1(x) + g(x)$, we have :

$$N(f = c) = \frac{1}{q'} N(f^{[1]} = c) - \frac{q(q'-1)}{q'} .$$

We can also generalize corollary 1 to functions from any Galois field towards any Galois field of same characteristic.

2. AFFINE SETS OF BENT FUNCTIONS

PROPOSITION 2

Let m and n be even positive integers.
Let $g, g_1, \dots, g_n$ be boolean functions on $G = GF(2^m)$, and E the F-linear space
generated by $g_1, \dots, g_n$. Then all the elements of the affine set $g + E$ are bent
functions if and only if the following boolean function on $G \times F^n$:

$$f : (u, (a_1, \dots, a_n)) \to g(u) + \sum_{i=1}^{n/2} \big(a_{2i-1} + g_{2i-1}(u)\big)\big(a_{2i} + g_{2i}(u)\big)$$

is a bent function on $G \times F^n \simeq GF(2^{m+n})$.

proof:

Let us recall that a boolean function g on G is bent if (cf [2], [5]) for any linear form l on G, we have :

$$\sum_{u \in G} (-1)^{g(u)+l(u)} = \pm 2^{m/2}.$$

Therefore, f is bent if and only if for any $(\lambda_1, \ldots, \lambda_n) \in F^n$ and any linear form l on G :

$$\sum_{(u,a) \in G \times F^n} (-1)^{f(u, a_1, \ldots, a_n) + \sum_{j=1}^{n} a_j \lambda_j + l(u)} = \pm 2^{(m+n)/2} \quad (\text{where } a = (a_1, \ldots, a_n)).$$

We have :

$$\sum_{(u,a) \in G \times F^n} (-1)^{f(u, a_1, \ldots, a_n) + \sum_{j=1}^{n} a_j \lambda_j + l(u)}$$

$$= \sum_{(u,a) \in G \times F^n} (-1)^{\sum_{i=1}^{n/2}[(a_{2i-1}+g_{2i-1}(u)+\lambda_{2i})(a_{2i}+g_{2i}(u)+\lambda_{2i-1})+g_{2i-1}(u)\lambda_{2i-1}+g_{2i}(u)\lambda_{2i}+\lambda_{2i}\lambda_{2i-1}] + g(u) + l(u)}$$

$$= (-1)^{\sum_{i=1}^{n/2} \lambda_{2i}\lambda_{2i-1}} \sum_{(u,a) \in G \times F^n} (-1)^{\sum_{i=1}^{n/2}[a_{2i-1}a_{2i}+g_{2i-1}(u)\lambda_{2i-1}+g_{2i}(u)\lambda_{2i}] + g(u) + l(u)}$$

(since F is invariant under any translation and in particular under the translations
$a \to a + g_{2i-1}(u) + \lambda_{2i}$ and $a \to a + g_{2i}(u) + \lambda_{2i-1}$)

$$= \pm \left(\sum_{(a_1, \ldots, a_n) \in F^n} (-1)^{\sum_{i=1}^{n/2} a_{2i-1}a_{2i}} \right) \sum_{u \in G} (-1)^{\sum_{i=1}^{n/2}[g_{2i-1}(u)\lambda_{2i-1}+g_{2i}(u)\lambda_{2i}] + g(u) + l(u)}$$

$$= \pm 2^{n/2} \sum_{u \in G} (-1)^{\sum_{i=1}^{n/2}[g_{2i-1}(u)\lambda_{2i-1}+g_{2i}(u)\lambda_{2i}] + g(u) + l(u)}$$

since $\displaystyle \sum_{(a_1, \ldots, a_n) \in F^n} (-1)^{\sum_{i=1}^{n/2} a_{2i}a_{2i-1}} = \left(\sum_{(a_1, a_2) \in F^2} (-1)^{a_1 a_2} \right)^{\frac{n}{2}} = 2^{\frac{n}{2}}$

and therefore, f is bent if and only if for any $(\lambda_1, \ldots, \lambda_n) \in F^n$, the boolean function on G equal to :

$$g + \sum_{i=1}^{n/2} \left(\lambda_{2i-1} g_{2i-1} + \lambda_{2i} g_{2i} \right) = g + \sum_{j=1}^{n} \lambda_j g_j$$

is bent.

Remark

If p is an arbitrary positive integer, and $g, g_1, \dots, g_p$ are boolean functions on G, then we can apply this proposition to g and to the functions $0, g_1, 0, g_2, \dots, 0, g_p$ (with n = 2p).

So, if E is the linear space generated by $g_1, \dots, g_p$, the affine space g + E contains only bent functions if and only if the function :

$$(u, (a_1, \dots, a_{2p})) \rightarrow g(u) + \sum_{i=1}^{p} (a_{2i-1})(a_{2i} + g_i(u))$$

is a bent function on $GF(2^{m+2p}) \approx GxF^{2p}$.

An example of application of proposition 2

We know (cf [2]) that a boolean function f on G is bent if and only if for any $u \in G$, $u \neq 0$, the function $f + f \circ t_u : v \rightarrow f(v) + f(u+v)$ is balanced on G (ie its support has cardinal 2^{m-1}).

If f belongs to R(3, m), then for any u, the function $f + f \circ t_u$ belongs to R(2, m), and according to the result recalled in the introduction, it is balanced if and only if there exists $v \in G$ such that :

$\forall w \in G, (f + f \circ t_u)(v+w) = (f + f \circ t_u)(w) + 1$.

Therefore, f is bent if and only if for any $u \neq 0$, there exists v such that :
$\forall w \in G, f(u+v+w) = f(w) + f(u+w) + f(v+w) + 1$, or, equivalently :
$\forall w \in G, \forall a, b \in F, f(au+bv+w) = f(w) + a[f(w) + f(u+w)] + b[f(w) + f(v+w)] + ab$.

In this last relation, we recognize the function f of proposition 2 with n = 2 :
let G' be a subspace of G of dimension m-2, which does not contain u, v, and u+v (we will identify G' with $GF(2^{m-2})$ as a F-linear space) and h, g_1, g_2 the restrictions of f, f + f ∘ t_v and f + f ∘ t_u to G'(respectively).

We can apply proposition 2 to f (where n=2, G' stands for G, G for GxF^2, a for a_1, b for a_2, g_1 for f_1, g_2 for f_2, and $g_1 g_2 + h$ for g).
We deduce that the following functions are bent on $GF(2^{m-2})$:
$g_1 g_2 + h$, $g_1 g_2 + h + g_1$, $g_1 g_2 + h + g_2$, and $g_1 g_2 + h + g_1 + g_2$.

Notice that the degree of these functions is 4 (in general). So, *any bent function of degree 3 on $GF(2^m)$ leads to $(2^{m+2} - 4)$ (perhaps not all distinct) functions of degree 4 (in general) on $GF(2^{m-2})$*.

REFERENCES

(1) C. Carlet, thèse, publication du LITP, Institut Blaise Pascal, Université Paris 6, n° 90.59 (1990).

(2) J. F. Dillon, "Elementary Hadamard Difference sets", Ph. D. Thesis, Univ. of Mariland (1984).

(3) T. Kasami and N. Tokura "On the weight structure of the Reed Muller codes", IEEE Trans. Info. Theory 16 (1970) 752-759.

(4) F. J. Mac Williams and N. J. Sloane, "The theory of error-correcting codes", Amsterdam, North Holland.

(5) O. S. Rothaus, "On bent functions", J. Comb. Theory, 20A (1976) 300-305

(6) J. H. Van Lint, "Coding Theory", Springer Verlag 201.

(7) H.C.A. Van Tilborg "On weights in codes" Report 71-WSK03, Dep.of math. Technological University of Eindhoven, Netherlands (1971).

COVERING RADIUS OF RM(1,9) IN RM(3,9)

Philippe LANGEVIN
G.E.C.T.
Université de Toulon
83130 La Garde, France.

ABSTRACT. We give new properties about Fourier coefficients and we prove that the distance of the first order Reed-Muller code of length 512 to any cubic is at most 240.

INTRODUCTION.

Let m be a positive integer, and E the vector space F_2^m. The ring of the mappings from E into F_2 (we shall simply say "functions") is identified with the ring of reduced polynomials $F_2[X_1,X_2,...,X_m]/(X_1^2 - X_1, X_2^2 - X_2,...,X_m^2 - X_m)$. The Reed-Muller code of order k and length 2^m, denoted by RM(k,m), is the subspace of functions with degree at most k (see [7]). RM(m,m) is the entire space of functions. The weight of a function f is $wt(f)=\#\{\ x \in E\ /\ f(x)=1\}$ and the mapping d from RM(m,m)×RM(m,m) into N, $(f,g) \to wt(f+g)$ is a metric. If A and B are subsets of a metric space (M,d) we define the covering radius of A in B to be the least integer $\rho(A,B)$ such that the spheres of radius $\rho(A,B)$ around the points of A cover B, $\rho(A,B) = \sup_{y \in B} d(y,A)$, where d(y,A) is the distance of y to A, i.e. $d(y,A) = \inf_{x \in A} d(y,x)$. For example, if C is a binary code of length n the covering radius of C in F_2^n is the classical covering radius of C (see [7]). The covering radius of RM(k,m) in RM(1,m) will be denoted by $\rho_m(k,l)$. The following survey is given in [8].

$\rho_m(-\infty,m)$	2^m	
$\rho_m(0,m)$	2^{m-1}	
$\rho_m(1,m)$	$2^{m-1} - 2^{(m-2)/2}$	m even
	$2^{m-1}- 2^{(m-1)/2} \leq ... < 2^{m-1}-2^{(m-2)/2}$	m odd
$\rho_m(k,m)$	$... \geq 2^{m-k-3}(k+4)$	k even
	$... \geq 2^{m-k}$	$m \geq 6$
$2 \leq k \leq m-3$	$... \geq 2^{m-k-3}(k+5)$	k odd
$\rho_m(m-3,m)$	$m+2$	m even
	$m+1$	m odd
$\rho_m(m-2,m)$	2	
$\rho_m(m-1,m)$	1	
$\rho_m(m,m)$	0	

For m even, we have $\rho_m(1,m)=\rho_m(1,2)$ (see [7]). A function f such that $d(f,RM(1,m))=\rho_m(1,m)$ is said to be "bent". There exist bent functions of degree exactly k iff $2 \le k \le m/2$ (see [1]).

For m odd, The lower bound of the inequality on $\rho_m(1,m)$ is $\rho_m(1,2)$ and when m=1, 3, 5 or 7, $\rho_m(1,m)$ is equal to this bound (see [2,3,4]), but Patterson and Wiedemann have found $\rho_{15}(1,15) \ge 16276 > \rho_{15}(1,2)$ and more generally $\rho_m(1,m) > \rho_m(1,2)$ for all odd m such that m≥15 (see [5,6]). A function f such that $d(f,RM(1,m)) = \rho_m(1,m)$ will also be called a "bent" function.

To complete this survey, we recall two conjectures:
$$\rho_m(1,m) = \rho_m(1,m-1) \quad \text{for } m \ge 1 \text{ (see [9,10])}$$
$$\text{and } \lim_{m \to +\infty} \frac{2^m - 2\rho_m(1,m)}{2^{m/2}} = 1 \text{ (see [5])}.$$

At present time, nothing is known about the intermediate values m=9, 11 or 13. In this paper, we prove that $\rho_9(1,3)=240$, therefore if the covering radius of RM(1,9) is greater than 241 then the bent functions have degree at least 4.

1 FOURIER COEFFICIENTS.

Definition.
Let <,> denote a non degenerate symmetric bilinear form (for example the usual inner product of E), χ the non trivial additive character over F_2, χ is the mapping from F_2 into $\{1,-1\}$ such that $\chi(0) = 1$ and $\chi(1) = -1$. The Fourier coefficient of a function f at the point a is defined by $\hat{f}_a = \sum_{x \in E} \chi(f(x)+<a,x>)$.

Proposition 1.1.
If f is a function then $d(f,RM(m,m)) = 2^{m-1} - \frac{1}{2}\sup_{a \in E} |\hat{f}_a|$, and

$$\rho_m(1,m) = 2^{m-1} - \frac{1}{2}\inf_f \sup_{a \in E} |\hat{f}_a|$$

Formulas.

The following formulas are not difficult to prove.

$$\mathbf{F_1} \quad \chi(f(x))= 2^{-m} \sum_{a \in E} \hat{f}_a \chi(<a,x>)$$

Denoting the function $x \to f(x+a)+g(x)$ by $\Delta_a(f,g)$, then for every subspace F of E:

F2.1 $\qquad \sum_{a \in F} \hat{f}_a \hat{g}_{a+b} = |F| \sum_{a \in F^\perp} \widehat{\Delta_a(f,g)}_b \, \chi(<a,b>)$

Denoting the function $x \to f(x+a)+f(x)$ by $\Delta_a f$, then for every subspace F of E:

F2.2 $\qquad \sum_{a \in F} \hat{f}_a \hat{f}_{a+b} = |F| \sum_{a \in F^\perp} \widehat{\Delta_a f}_b \, \chi(<a,b>)$

In particular

F2.3 $\qquad \sum_{a \in E} \hat{f}_a \hat{f}_{a+b} = 2^{2m} \delta_{b,0}$

F3.1 $\qquad |F^\perp| \sum_{a \in F} \chi(f(a)) = \sum_{a \in F^\perp} \hat{f}_a$

If F is a subspace of dimension k then

F3.2 $\qquad \sum_{a \in F} f(a) = 2^{k-1} - 2^{k-m-1} \sum_{a \in F^\perp} \hat{f}_a$

where the left hand side is calculated in the ring Z of integers.

Proposition 1.2.

The Fourier coefficients of the constants are $\hat{0}_a = 2^m \delta_{a,0}$ and $\hat{1}_a = -2^m \delta_{a,0}$.
Let q be a quadratic form over E whose kernel N is of dimension k.

(a) If q has no default then $\begin{cases} \hat{q}_a = 0 & \text{iff } a \not\perp N \\ \hat{q}_a = \pm 2^{(m+k)/2} & \text{iff } a \perp N \end{cases}$

(b) If q has a default d, set $N=M \oplus \{0,d\}$, where $M=\{x \in N \ / \ q(x)=0\}$.

Then $\begin{cases} \hat{q}_a = 0 & \text{iff } a \perp d \text{ or } a \not\perp M \\ \hat{q}_a = \pm 2^{(m+k)/2} & \text{iff } a \not\perp d \text{ and } a \perp M \end{cases}$

proof

Let φ be the bilinear form associated to q. $N=\{x \in E \ / \ \forall \ y \in E, \ \varphi(x,y)=0 \}$.

$(\hat{q}_a)^2 = \sum_{x \in E} \sum_{y \in E} \chi(q(x)+q(y)+<a,x+y>) = \sum_{x \in E} \sum_{z \in E} \chi(q(z)+\varphi(x,z)+<a,z>)$

$(\hat{q}_a)^2 = 2^m \sum_{z \in N} \chi(q(z)+<a,z>)$

(a) If q is without default then we get $(\hat{q}_a)^2 = 2^m \sum_{z \in N} \chi(<a,z>)$

(b) If q has a default d, $(\hat{q}_a)^2 = 2^m \sum_{z \in M} \chi(q(z)+<a,z>)+\chi(q(z+d)+<a,z+d>)$

$(\hat{q}_a)^2 = 2^m \sum_{z \in M} \chi(<a,z>)-\chi(<a,z+d>) = 2^m \sum_{z \in M} \chi(<a,z>)(1-\chi(<a,d>))$

$\blacklozenge$

Remarks: $\hat{q}_0 = 0$ iff q has a default.

One can calculate the signs distribution, using $F_{3.1}$.

Proposition 1.3

If a sequence of complex numbers $(n_a)_{a \in E}$ satisfies $F_{2.3}$ then the n_a's are integers and there exists a function f such that $\hat{f}_a = n_a$, for all $a \in E$. In particular, if F is a subspace of dimension k, the 2^{m-k} complex numbers n_b indexed by the quotient space E/F, and defined by $n_b = 2^{-k} \sum_{a \in F} \hat{f}_{a+b}$ are the Fourier coefficients of a function f of $RM(m-k,m-k)$.

proof

The first part is easy, and the n_b are defined over E/F. We now show that the n_b's satisfy $F_{2.3}$. Set $S(b') = \sum_{b \in E/F} n_b\, n_{b+b'}$

$2^k S(b') = \sum_{c \in F} \sum_{b \in E/F} n_{b+c}\, n_{b+c+b'} = 2^{-2k} \sum_{c \in F} \sum_{b \in E/F} \sum_{a \in F} \sum_{a' \in F} \hat{f}_{a+b+c}\, \hat{f}_{a'+b+c+b'}$

$2^k S(b') = 2^{-2k} \sum_{a \in F} \sum_{a' \in F} \sum_{c \in F} \sum_{b \in E/F} \hat{f}_{a+b+c}\, \hat{f}_{a'+b+c+b'} = 2^{2(m-k)} \sum_{a \in F} \sum_{a' \in F} \delta_{a,\, a'+b'}$

Hence $S(b') = 2^{2(m-k)} \delta_{b',0}$. $\blacklozenge$

The following proposition precises this result.

Proposition 1.4.

Let f a function, F a subspace of dimension k, and S a complementary of F. If $S \cap S^\perp = \{0\}$ then the integers $n_b = 2^{-k} \sum_{a \in F} \hat{f}_{a+b}$ $b \in S$ are the Fourier coefficients of the function $f \circ p$, where p is the projection on $F^\perp$ along $S^\perp$.

proof

Let $x \in S$,

$2^{-k} \sum_{s \in S} \sum_{a \in F} \hat{f}_{a+s} \chi(<s,x>) = 2^{-k} \sum_{s \in S} \sum_{a \in F} \sum_{y \in E} \chi(f(y)+<a+s,y>+<s,x>)$

$$= 2^{m-2k} \sum_{a\in F} \sum_{y\in x+S^\perp} \chi(f(y)+<a,y>)$$

$$= 2^{m-k} \sum_{y\in (x+S^\perp)\cap F^\perp} \chi(f(y))$$

Since $E=S \oplus F=S^\perp \oplus F^\perp$ then x decomposes in one and only one way as $x=p(x)+(x-p(x))$, whence $2^{-k} \sum_{s\in S} \sum_{a\in F} \hat{f}_{a+s} \chi(<s,x>) = 2^{m-k} \chi(f\circ p(x))$ ♦

Remark. If $F\cap F^\perp = \{0\}$, we can choose $S=F^\perp$ and then $f\circ p$ is the restriction of f to $F^\perp$.

Proposition 1.5.

If all the Fourier coefficients of a function f are divisible by 2^s (resp: have diadic valuation $s \geq 2$, that means all the coefficients are divisible by 2^s and not by 2^{s+1}) then f is a function of RM(m-s+1,m) (resp: RM(m-s,m)).

proof

Let $u\in F_2^m$ and $F=\{v\in F_2^m / \text{ supp}(u)\supset\text{supp}(v)\}$, where $\text{supp}(v)=\#\{i / v_i = 1\}$. The weight k of u is the dimension of F. The coefficient a_u of $x_1^{u_1} x_2^{u_2}... x_m^{u_m}$ in f is $\sum_F f(v)$ (see [7]).

By $F_{3.2}$, $a_u = 2^{k-1} - 2^{k-m-1} \sum_{a\in F^\perp} \hat{f}_a = 2^{k-1} - 2^{k-m-1+s} N$. Since $N= 2^{-s} \sum_{a\in F^\perp} \hat{f}_a$ is an integer, this implies $a_v = 0$ when k>m-s+1. In the second case, f is not of degree m, necessarily s<m and for k≤m-1, the integer N is even ♦

Notation.

If f is a function and F a subspace of E, we write $\Delta_F f$ the function defined by $x\rightarrow \sum_{u\in F} f(x+u)$. If $F=\{0\}$ then $\Delta_F f=f$ and if $a\neq 0$ then $\Delta_{(0,a)} f$ is the $\Delta_a f$ of $F_{2.2}$.

Proposition 1.6.

Let F be a subspace of dimension d and p an odd prime which is not a divisor of $2^d - 2i$, for all integer i such that $0 \leq i< 2^{d-1}$. If p divides $\hat{f}_a$ for all $a\in F^\perp$ then $\Delta_F f$ is a constant. More precisely,

a) dim F > 0, b) if dim F = 1 then $\Delta_F f = 1$, c) if dim F ≥ 2 then $\Delta_F f = 0$

proof

$$2^m \sum_{u\in F} \chi(f(x+u)) = \sum_{a\in E} \hat{f}_a \chi(<a,x>) \sum_{u\in F} \chi(<a,u>) = 2^d \sum_{a\in F^\perp} \hat{f}_a \chi(<a,x>)$$

Since p divides the RHS, $\sum_{u\in F}\chi(f(x+u))=0$ which is impossible if dim F= 0 and

for each x the function u→f(x+u) is balanced over F (the number of u such f(x+u)=0 is equal to the number of u such f(x+u)=1), whence $\Delta_F f(x) = 0$ or $\Delta_F f(x) = 1$ according to whether d>1 or d=1 ◆

Proposition 1.7.

Let H be an affine hyperplane of E and p an odd prime. If p divides $\hat{f}_a$ for all a∈ H then $\hat{f}_a = 0$ for all a∈ H.

proof

We may assume H is a subspace of E defined by the equation <u,x>=0. The proposition 1.6 implies $\Delta_u f =1$ and using F$_{2.2}$, we find $\sum_{a\in H}\hat{f}_a{}^2 = 0$ ◆

2 FUNCTIONS WITH FOURIER 'S COEFFICIENTS EQUAL TO ±α OR ±β

In this part, we study the functions with two kinds of Fourier coefficients and more specially those that have their Fourier coefficients equal to $\pm 2^{t-1}$ or $\pm 3\ 2^{t-1}$, such a function f would exceed the quadratic bound (that is d(f,RM(1,m)) > $\rho_m(1,2)$) and the proposition 1.5 implies that its degree is at most t+2. One can see that the examples of Patterson and Wieddemann are of this type; one of them is extremal since its degree is 9.

Proposition 2.1.

Assume f has Fourier coefficients equal to ±α or ±β (α≠β). Let g the indicator of the set {a∈ E / $\hat{f}_a$ =±β}. Then the Fourier coefficients of g are

$\hat{g}_a = 2m\ A\ \Delta_a \hat{f}_0$ for a≠0 and $\hat{g}_0 = 2^{2m}A+2^mB$, where $A=\dfrac{2}{\alpha^2-\beta^2}$ and $B= -\dfrac{\alpha^2+\beta^2}{\alpha^2-\beta^2}$

proof

We have $A\alpha^2+B = 1$ and $A\beta^2+B = -1$, whence $\chi(g(u))= A\ \hat{f}_u{}^2 +B$.

$\hat{g}_a = \sum_{u\in E}\chi(g(u)+<a,u>) = \sum_{u\in E}(A\ \hat{f}_u{}^2 +B)\ \chi(<a,u>)$

$\hat{g}_a = A\ \sum_{u\in E}\hat{f}_u{}^2\chi(<a,u>)+B\ \sum_{u\in E}\chi(<a,u>)$

If a=0 then we get $2^{2m}\ A + 2^m\ B$ else

$\hat{g}_a = A\ \sum_{u\in E}\sum_{x\in E}\sum_{y\in E}\chi(f(x)+f(y)+<u,x+y+a>) = 2^m\ A\ \sum_{x\in E}\chi(f(x)+f(x+a))$

$$\hat{g}_a = 2^m \, A \, \hat{\Delta_a f_0} \quad \blacklozenge$$

Proposition 2.2.

If all the Fourier coefficients of a function f are equal to $\pm 2^{t-1}$ or $\pm 3 \, 2^{t-1}$, and if g is the indicator of the $\pm 2^{t-1}$ then $\hat{g}_a = 2 \, \hat{\Delta_a f_0}$ for $a \neq 0$ and $\hat{g}_0 = 3 \, 2^{m-2}$.

Proposition 2.3.

If f is a function of RM(3,9) with Fourier coefficients equal to $\pm 2^3$ et $\pm 3 \, 2^3$ then there exists $a \neq 0$ such $| \hat{\Delta_a f_0} | > 2^5$.

proof

Let g be the indicator of $\{ a \in E \, / \, \hat{f}_a = \pm 2^{t-1} \}$, we may assume $\chi(g(0)) = 1$.

$\Delta_a f$ is a function of degree at most 2 and by the proposition 1.2, its Fourier coefficients are in the set $\{0\} \cup \{ \pm 2^{t+i} \, / \, i > 1 \}$.

Now suppose $\hat{g}_a = 0$ or $\hat{g}_a = \pm 2^{t+2}$ for all $a \neq 0$. Let x,y et z be the number of Fourier coefficients of g equal to -2^{t+2}, 0 and 2^{t+2}. x,y et z satisfy the equations :

$$\begin{pmatrix} 1 & 1 & 1 \\ -2^{t+2} & 0 & 2^{t+2} \\ 2^{2t+4} & 0 & 2^{2t+4} \end{pmatrix} \begin{pmatrix} x \\ y \\ z \end{pmatrix} = \begin{pmatrix} 2^m - 1 \\ 2^m \chi(g(0)) - 3 \, 2^{m-2} \\ 2^{2m} - 9 \, 2^{2m-4} \end{pmatrix}$$

There is only one solution: x=13, y=447 and z=15. Set $P = \{ a \, / \, \hat{\Delta_a f_0} = 2^5 \}$ and $N = \{ a \, / \, \hat{\Delta_a f}_{,0} = -2^5 \}$.

We now prove that if a, b $\in$ P and a$\neq$b then a+b$\in$ N.

Suppose a$\neq$b and let F=$\{0,a,b,a+b\}$.
If $\hat{\Delta_a f_0} = \hat{\Delta_b f_0} = \hat{\Delta_{a+b} f_0} = 2^5$ then the average of the $\hat{f}_u^2$ over $F^{\perp}$ is $2^9 + 3 \, 2^5 > 24^2$ and there exists $u \in F^{\perp}$ such $| \hat{f}_u | > 24$.
If $\hat{\Delta_a f_0} = \hat{\Delta_b f_0} = 2^5$ et $\hat{\Delta_{a+b} f_0} = 0$ then the average of the $\hat{f}_u^2$ over $F^{\perp}$ is 24^2 and all the Fourier coefficients are equal to ± 24 over $F^{\perp}$, and proposition 1.6 implies $\Delta_F f(x)=0$. The vector a is always in the kernel of the quadratic form $q_a(x) = f(x+a) + f(x) + f(a) + f(0)$, the bilinear form associated to q_a is $\varphi_a(x,y) = f(x+y+a) + f(x+y) + f(x+a) + f(x) + f(y+a) + f(y) + f(a) + f(0)$. Since $\Delta_F f = 0$, b is also in the kernel of q_a.

Hence because the dimension of E is odd the dimension of the kernel of q_a is at least 3, using proposition 1.2 we get $2^5 = \hat{\Delta_a f_0} = \chi \ (f(a) + f(0)) \ \hat{q_{a_0}} > 2^5$, that is absurd.

Let a denote a fixed point of P, the map from P-{a}→N defined by x→x+a would be one to one, which is absurd because $|P-\{a\}| = 14$ and $|N| = 13$.

Finally, there exists a≠0 such that $| \ \hat{\Delta_a f_0} \ | > 2^5$ ◆

Proposition 2.4.
The cubics of RM(3,9) don't exceed the quadratic bound.

proof
Ax's theorem (see [11]) shows that the cubics and quartic have weight divisible by 4. Suppose f to be a cubic of RM(3,9) exceeding the quadratic bound then its Fourier coefficients are 0, ±8, ±16 or ±24.

But $\hat{f_b} - \hat{f_a} = 2 \ w(a+b) - 4 \ w(\ (a+b)(b+f))$ implies that if one of the Fourier coefficients is divisible by 16 then all are divisible by 16, so the Fourier coefficients of f are equal to ±8 or ± 24.

Applying the proposition 2.3, there exists an hyperplane H such that

$$\sum_{a \in H} \hat{f_a}^2 \geq 2^8 \ (2^9 + 2^6) = 2^8 \ 24^2,$$ which implies $| \ \hat{f_a} \ | = 24$ over H and the

proposition 1.7 concludes the proof ◆

CONCLUSION.
Our contribution about the covering radius of RM(1,9) holds in the equality $\rho_9(1,3) = \rho_9(1,2)$ and now the question is to know if this holds for all first order Reed-Muller codes ?.

NOTATIONS.

N, Z	the set of positive and rational integers
F_2	the field with two elements
RM(k,m)	the set of polynomials of degree at most k in
	$F_2[X_1,X_2,...X_m]/(X_1^2-X_1,X_2^2-X_2,...,X_m^2-X_m)$
wt(f)	the weight of f, wt(f)= Number of x such that f(x)=1
d(f,A)	the distance of f to A, $d(f,A) = \inf_{x \in A} d(f,x)$.
$\rho_m(k,l)$	the covering radius of RM(k,m) in RM(l,m)
	$\rho_m(k,l)=\sup_{f \in RM(l,m)} d(f,RM(k,m))$
χ	the additive character of F_2, $\chi(0) = 1$ and $\chi(1) = -1$
<,>	a non degenerate bilinear symetric form
$\hat{f}_a$	the Fourier coefficient of f, $\hat{f}_a = \sum_{x \in E} \chi(f(x)+<a,x>)$
$F^\perp$	the orthogonal of F, $F^\perp=\{x \in E \ / \ <x,y>=0 \ \forall \ y \in E\}$
supp(v)	the support of v, supp(v)=$\{i \ / \ v_i=1\}$.
#A or $|A|$	the cardinal of the set
$\Delta_F f$	the function $x \rightarrow \sum_{u \in F} f(x+u)$
$\Delta_a f$	the function $x \rightarrow f(x)+f(x+a)$
$\blacklozenge$	End of proof

BIBLIOGRAPHY.

[1] O. S. Rothaus, On "bent" functions, J.C.T. 20 (1976), 300-305.
[2] E. Berlekamp and L.R. Welch, Weigth distributions of the cosets of the (32,6) Reed-Muller code, IEEE T. I. T., vol. IT-18 (1972), 203-207.
[3] T. Helleseth, T. Klove, and J. Mykkelveit, On the covering radius of binary codes, IEEE T.I.T. vol. IT-24 (1978), 627-628.
[4] J. Mykkelveit, The covering radius of the (128,8) Reed-Muller code is 56. IEEE T.I.T. vol-26 (1980) 359-362.
[5] N.J. Patterson and D.H. Wiedemann, The covering radius of the $(2^{15},16)$ Reed-Muller code is at least 16276, IEEE T.I.T. vol IT-29 (1983) 354-356.
[6] J. Constantin, B. Courteau and J. Wolfmann, Numerical experiments related to the covering radius of some first order Reed-Muller codes.
[7] F.J.S. Mac Williams and N.J.A Sloane, The theory of error correcting codes, North Holland.
[8] G.D. Cohen, M.G. Karpovsky, H.F. Mattson, Jr. and J.R. Schatz Covering radius – Survey and Recent Results IEEE T.I.T. vol-31 (1985), (328-344).
[9] R.A. Brualdi and V.S. Pless, Orphans of the first order Reed-Muller codes. IEEE T.I.T., vol I.T-18 (1972), (203-207).
[10] R.A. Brualdi,N. Cai, and V.S. Pless, Orphan Structure of the first order Reed-Muller codes.
[11] J.R Joly, Equations et variétés algébriques sur un corps fini in L'enseignement des mathématiques, t.XIX, fasc 1-2.

THE WEIGHTS OF THE DUALS OF BINARY BCH CODES

OF DESIGNED DISTANCE δ = 9

François RODIER

Equipe C.N.R.S. Arithmétique et théorie de l'information
U.F.R. de Mathématiques
Université Paris 7
75231 PARIS CEDEX 05

1. Statement of the results

Let C_m be a binary BCH code of length $2^m - 1 = q - 1$ with designed distance $\delta = 2t + 1$. Then the Carlitz-Uchiyama bound (cf. MacWilliams and Sloane [2], p. 280) says that the weight w of any nonzero codeword in the dual of C_m is such that

$$| w - 2^{m-1} | \le (t - 1)2^{m/2}.$$

In their book *The Theory of Error-Correcting Codes* , MacWilliams and Sloane suggest that there should be a slightly stronger result (cf. loc. cit., Research Problem (9.5) p. 282):

$$| w - 2^{m-1} | \le (t - 1)2^{[m/2]}, \qquad (*)$$

where [.] stands for the integer part. For an odd m it means

$$| w - 2^{m-1} | \le (t - 1)2^{(m-1)/2} .$$

It is easy to see that this inequality (*) is true for codes with designed distance 3, 5 or 7. Indeed the dual of the extended code $\hat{C}_m$ embeds itself in a Reed-Muller code $R(2,m)$ (cf. loc. cit. p. 385), and the weights of this one are multiples of $2^{[(m-1)/2]}$, i.e. of $2^{(m-1)/2}$ if m is odd. The inequality (*) is therefore a simple consequence of this fact and the Carlitz-Uchiyama bound.

For $\delta = 9$, we will show that this inequality (*) does not hold for infinitely many m. More precisely we will prove the following theorem.

THEOREM. *For infinitely many odd m the code C_m contains a nonzero word c_m whose weight w_m is such that*

$$|w_m - 2^{m-1}| > 3.12 \times 2^{(m-1)/2} > 3 \times 2^{(m-1)/2}.$$

2. Proof

a) The extended binary BCH codes

Let $\widehat{C}_m$ be the extended binary BCH code of length $2^m = q$ with designed distance $\delta = 9$. We will prove first that the dual of $\widehat{C}_m$ is the code associated to the space $F_q[x]_7$ of polynomials $f \in F_q[x]$ such that $d°f \leq 7$; the codewords can be written

$$c_f = (tr\ f(a_1),\ tr\ f(a_2),...,\ tr\ f(a_q))$$

where the a_i are the elements of F_q and tr is the trace from F_q to F_2.

Indeed let C be the code given by the image of the map

$$c: F_q[x]_7 \rightarrow (F_q)^q$$

defined by $c(f) = (f(a_1),\ ...\ ,\ f(a_q))$. A basis of $F_q[x]_7$ is given by the family of monomials $1, x, ..., x^7$. The dual $C^\perp$ has therefore a parity check matrix given by

$$H = \begin{pmatrix} 1 & \cdots & 1 \\ a_1 & \cdots & a_q \\ \cdots & \cdots & \cdots \\ a_1^7 & \cdots & a_q^7 \end{pmatrix}.$$

It shows that the code $(tr\ C)^\perp = (C^\perp)|\,F_2$ (cf. Delsarte's theorem, [2] Chap. 7,

Theorem 11) is the extended binary BCH code $\hat{C}_m$, of length 2^m and designed distance $\delta = 9$ (cf. [1], example (7.2), and [2], example 5, p. 345).

Let f be a polynomial in $F_q[x]_7$, and r an integer. We will call c_r the associate codeword in the dual of $\hat{C}_{mr}$, and w_r the weight of the codeword c_r .

b) The sums S_r

Let $S_r = \sum (-1)^{tr_r \ f(x)}$ where the sum is over F_q , and tr_r is the trace from F_q to F_2 . By the preceding section, one has $S_r = q^r - 2w_r$. By [3] or [4] one knows that there exists π_i and $\omega_i = q^{-1/2}\pi_i$ such that

$$-S_r = \sum_{1 \leq i \leq 6} \pi_i^r \qquad \text{and } |\pi_i| = q^{1/2}$$
$$= q^{r/2} \sum_{1 \leq i \leq 6} \omega_i^r \qquad \text{and } |\omega_i| = 1.$$

Moreover the numbers π_i (and hence the ω_i) are algebraic integers and pairwise complex conjugate. One can suppose that $\pi_i = \bar{\pi}_{i+3}$ for $i = 1$, 2 and 3.

To compute the π_i it is enough to compute first the $b_i = \pi_i + \bar{\pi}_i$ and then use the fact that $\pi_i\bar{\pi}_i = q$. The b_i are the roots of an equation of degree 3 whose coefficients can be computed easily from S_1 , S_2 and S_3 .

c) An example

Let $m = 1$, and $f(x) = x^7 + x^3$. One checks that:

$$S_1 = 2, S_2 = 0, S_3 = 2.$$

Hence the b_i are the roots of the equation:

$$X^3 + 2X^2 - 4X - 6 = 0.$$

The roots are

$$b_1 \cong -2.655$$
$$b_2 \cong -1.210$$
$$b_3 \cong 1.866 .$$

Hence the π_i are given by

$$\pi_1 \cong \sqrt{2}\exp(2.790 \ \sqrt{-1})$$

$$\pi_2 \cong \sqrt{2}\exp(2.013 \ \sqrt{-1})$$
$$\pi_3 \cong \sqrt{2}\exp(0.850 \ \sqrt{-1})$$

and their complex conjugates. From this it is easy to see that

$$S_{17} = 1\ 600.$$

Thus one has

$$2\ |\ w - 2^{16}\ | = |\ S_{17}\ | = 1\ 600\ > 1\ 536 = 6.2^8.$$

Hence $|\ w - 2^{16}\ | > 3.12 \times 2^8 > 3 \times 2^{(17-1)/2}$ which proves the conclusion of the theorem for $m = 17$.

d) The number of r such that the theorem is true is infinite

Let M_r be the point $(\omega_1{}^r, \omega_2{}^r, \omega_3{}^r)$ in the space T^3 where T is the set of complex numbers of absolute value equal to 1. As T^3 is a compact group, the theory of Fourier series tells us that the points M_r for $r \geq N$ where N is any given integer, span a set whose closure is a subgroup G of T^3 and is independant of N. In the same way, the closed span of the points M_{2r} for $2r \geq N$ is a subgroup G_1 of G. Thus the closed span Γ of the points M_{2r+1} for $2r + 1 \geq N$ is a coset of G_1, hence it is either $G - G_1$ if $G \neq G_1$, or the whole of G if $G = G_1$. Let Ω be a neighbourhood of M_{17}. For any given integer N there is an odd r such that $r \geq N$ and M_r is in Ω. One has

$$|\ 2w_r - 2^r\ | = |\ S_r\ | = 2^{r/2}\ |\Sigma_{1 \leq i \leq 6}\ \omega_i{}^r\ |$$
$$= 2 \times 2^{r/2}\ |\ \text{Re}\ \Sigma_{1 \leq i \leq 3}\ \omega_i{}^r\ |.$$

In our case, for $r = 17$, this gives

$$1\ 600 = |\ S_{17}\ | = 2 \times 2^{17/2}\ |\ \text{Re}\ \Sigma_{1 \leq i \leq 3}\ \omega_i{}^{17}\ |,$$

whence

$$|\ \text{Re}\ \Sigma_{1 \leq i \leq 3}\ \omega_i{}^{17}\ | = 1\ 600/(2 \times 2^{17/2}) > 2.209 .$$

Let us take for Ω the set of points $M = (m_1, m_2, m_3)$ in T^3 such that

$$|\ \text{Re}\ \Sigma_{1 \leq i \leq 3}\ m_i\ | > 2.209 .$$

Then there are infinitely many odd integers r such that

$$|\ \text{Re}\ \Sigma_{1 \leq i \leq 3}\ \omega_i{}^r\ | > 2.209 ,$$

that is

$$|\ w_r - 2^{r-1}\ | = |\ S_r/2\ |\ >\ 2^{r/2} \times 2.209 > 2^{(r-1)/2} \times 3.12 .$$

The theorem is proved.

3. References

[1]   G. Lachaud: Exponential sums, algebraic curves and linear codes, preprint, 1989.

[2]   F.J. MacWilliams and N.J.A. Sloane: The Theory of Error-Correcting Codes, North-Holland, Amsterdam, 1977.

[3]   J.P. Serre: Majoration de sommes exponentielles, Astérisque 41-42 (1977), p. 111-126.

[4]   A. Weil: One some exponential sums, Proc. Nat. Ac. Sc. 34 (1948), p. 204-207.

The minimum distance of some binary codes via the Newton's identities

D. Augot [*]   P. Charpin [†]   N. Sendrier [†]

Abstract

In this paper, we give a natural way of deciding whether a given cyclic code contains a word of given weight. The method is based on the manipulation of the locators and of the locator polynomial of a codeword x.

Because of the dimensions of the problem, we need to use a symbolic computation software, like Maple or Scratchpad II. The method can be ineffective when the length is too large.

The paper contains two parts :

In the first part we will present the main definitions and properties we need.

In the second part, we will explain how to use these properties, and, as illustration, we will prove the three following facts :

The dual of the BCH code of length 63 and designed distance 9 has true minimum distance 14 (which was already known).
The BCH code of length 1023 and designed distance 253 has minimum distance 253.
The cyclic codes of length $2^{11}, 2^{13}, 2^{17}$, with generator polynomial $m_1(x)$ and $m_7(x)$ have minimum distance 4 (see [5]).

1 Notation and properties.

1.1 Cyclic codes.

We use the following definition for cyclic codes :

Definition 1 *A primitive cyclic code C of length $n = 2^m - 1$ over $GF(2)$ is an ideal of the polynomial algebra $GF(2)[x]/(x^n - 1)$. The roots of its generator polynomial are called the zeros of C. The code C is characterized by its defining set $I(C)$:*

$$I(C) = \{i \in [0..n-1] \mid \alpha^i \text{ is a zero of } C\} \tag{1}$$

where α is a nth root of unity in $GF(2^m)$.

---

[*]Université Paris 6, UFR d'Informatique, LITP, 2 pl. Jussieu, 75251 Paris CEDEX 05, FRANCE
[†]INRIA, Domaine de Voluceau, Rocquencourt, BP 105, 78153 Le Chesnay Cedex, FRANCE

We denote by $cl(s)$ the cyclotomic class of 2 mod n, containing s :

$$cl(s) = \{s, 2s, 2^2s \ldots, 2^{m-1}s \bmod n\} \tag{2}$$

If α^i is a zero of the code C, then $(\alpha^i)^2$ is also a zero of the code C. So we can see that $I(C)$ is the union of cyclotomic classes mod n.

As an example of a cyclic code, let us recall the definition of the BCH codes :

Definition 2 *The narrow-sense BCH code of designed distance δ is the cyclic code whose defining set is the union of the cyclotomic classes $cl(1), cl(2) \ldots cl(\delta - 1)$.*

It is well-known that the true minimum distance of a BCH code of designed distance δ is greater or equal than δ. This is the BCH bound :

Theorem 1 *Let C be a cyclic code with defining set $I(C)$. If $I(C)$ contains a sequence of $\delta - 1$ consecutive integers (0 is treated consecutive to $n - 1$), then the minimum distance of C is greater or equal to δ.*

There exist many ways to find lower bounds for the minimum distance of a cyclic code. Van Lint and Wilson give an overview of these methods in [6]. In this paper we will also need the following bound, called the Hartmann-Tzeng bound ([4]) :

Theorem 2 *Let C be a cyclic code of length n over $GF(q)$ with defining set $I(C)$. Let β be a primitive n^{th} root of unity. If the defining set $I(C)$ contains the following roots :*

$$\{\beta^{l+i_1c_1+i_2c_2}\} \ i_1 = 0, 1 \ldots d_0 - 2, \ i_2 = 0, 1 \ldots s, \ where \ gcd(n, c_1) = 1, \ gcd(n, c_2) < d_0, \tag{3}$$

then the minimum distance d of C satisfies $d \geq d_0 + s$.

In general, no method is known for finding the true minimum distance of a given cyclic code. Our approach is to find properties of a given cyclic code by using equations in $GF(2^m)$, and by doing effective computations in $GF(2^m)$.

1.2 Locators, locator polynomial of a word

All the notions introduced here can be found in [3].

Definition 3 *Let α be a nth root of unity in $GF(2^m)$, let x be a word of length n and weight w, $\mathbf{x} = (x_0, x_1 \ldots, x_{n-1})$. The locators of x are the $X_1, X_2 \ldots, X_w$, defined as follows :*

$$X_i = \alpha^{k_i} \quad where \quad x_{k_i} \neq 0 \tag{4}$$

We also define the locator polynomial of x :
The locator polynomial of x is the following polynomial $\sigma_x(Z)$:

$$\sigma_x(Z) = \prod_{i=1}^{w}(1 - X_iZ) = \sum_{i=0}^{i=w} \sigma_i Z^i \ , \tag{5}$$

where the σ_i are the elementary symmetric functions of the X_i :

$$\sigma_i = \sum_{1 \leq k_1 < k_2 \ldots < k_i \leq w} X_{k_1} X_{k_2} \cdots X_{k_i} \tag{6}$$

We also recall the *power sum symmetric functions* :

Definition 4 *The power sum symmetric functions A_k of the locators of $\mathbf{x}$ are* :

$$A_k(X_1, X_2, \ldots, X_w) = A_k = \sum_{i=1}^{w} X_i^k \ , \forall k \in [0..n].$$

In accordance with the definition of cyclic codes, we have the following property :

Proposition 1 *Let C be a cyclic code with defining set $I(C)$ and $\mathbf{x}$ a word of length n. Then $\mathbf{x} \in C$ if and only if* :

$$A_k = 0 \ for \ k \in I(C). \tag{7}$$

The following identities show how the A_i's are related to the σ_i's.

Proposition 2 *Let $X_1 \ldots X_w$ be w indeterminates in a ring K, σ_i their elementary symmetric functions, A_i their power sum symmetric functions. Then the following identities hold* :

$$\begin{cases} r \leq w, & (eq_r): \quad A_r + A_{r-1}\sigma_1 + \cdots + A_1\sigma_{r-1} + r\sigma_r = 0 \\[2mm] r > w, & (eq_r): \quad A_r + A_{r-1}\sigma_1 + \cdots + A_{r-w}\sigma_w = 0 \end{cases}$$

These identities are called the Newton's identities.

2   Presentation of the method

Let C be a cyclic code of length $n = 2^m - 1$ with defining set $I(C)$, let $\mathbf{x}$ be a codeword of C, let w be the weight of $\mathbf{x}$.

From the preceding part, we can state the following facts :

Let $X_1, X_2 \ldots, X_w$ be the locators of $\mathbf{x}$, $A_1, A_2 \ldots, A_n$ be the power sum symmetric functions of the X_i's. Then $A_i = 0$ for $i \in I(C)$.

The locator polynomial of $\mathbf{x}$, $\sigma_x(Z)$ splits over $GF(2^m)$.

The A_i's and the σ_i's are related by the Newton's identities.

So to find the codeword of weight w, we do the following :

1. Write the Newton's identities for a weight w with generic A_i's and σ_i's, using the fact that $A_i = 0$ for $i \in I(C)$.

2. Solve the system given by these equations, in order to obtain the σ_i's in terms of the A_i's.

3. Construct the locator polynomial corresponding to these σ_i's and test if it splits over $GF(2^m)$.

If we can find such a locator polynomial, then with the locators we can construct a word of weight w.

If there is a failure at any step of the method then we know that C contains no word of weight w.

We will show three ways of using this method.

2.1 Finding a contradiction in the Newton's Identities.

We want to find the minimum distance of the dual of the BCH code of designed distance 9 and length 63.

The defining set of this code C is $I(C)$:

$$0, cl(1), cl(3), cl(5), cl(9), cl(11), cl(13), cl(21), cl(27). \tag{8}$$

So the BCH bound gives $\delta \geq 8$ (As well as the Carlitz-Uchiyama bound).

Let us suppose there exists a codeword x of weight 8. Then its power sum symmetric functions satisfy :

$$A_0 = A_1 = A_3 = A_5 = A_9 = A_{11} = A_{13} = A_{21} = A_{27} = 0, \tag{9}$$

and we must have $A_7 \neq 0$ (If not then, by the BCH bound, the weight of x would be greater than 16.).

Now we can write down the Newton's identities, one by one :

$$
\begin{aligned}
(eq_7) & & A_7 + \sigma_7 &= 0 \\
(eq_8) & & 0 &= 0 \\
(eq_9) & & A_7\sigma_2 &= 0 \\
(eq_{10}) & & 0 &= 0 \\
(eq_{11}) & & A_7\sigma_4 &= 0 \\
(eq_{12}) & & 0 &= 0 \\
(eq_{13}) & & A_7\sigma_6 &= 0 \\
(eq_{14}) & & A_7{}^2 + A_7\sigma_7 &= 0 \\
(eq_{15}) & & A_{15} + A_7\sigma_8 &= 0 \\
(eq_{16}) & & A_7{}^2\sigma_2 &= 0 \\
(eq_{17}) & & A_{15}\sigma_2 &= 0 \\
(eq_{18}) & & A_7{}^2\sigma_4 &= 0 \\
(eq_{19}) & & A_{15}\sigma_4 &= 0 \\
(eq_{20}) & & A_7{}^2\sigma_6 &= 0 \\
(eq_{21}) & & A_{15}\sigma_6 + A_7{}^2\sigma_7 &= 0
\end{aligned}
$$

The contradiction appears when we see that A_7 supposed to be non zero has to be null :

$$
\begin{aligned}
(eq_7) & \Rightarrow & \sigma_7 = A_7 \neq 0 \\
(eq_{13}) & \Rightarrow & \sigma_6 = 0 \\
(eq_{21}) & \Rightarrow & (A_7)^3 = 0
\end{aligned}
$$

Because there are no σ_i solutions of the Newton's identities, this proves that there can be no locator polynomial of a codeword of weight 8 in C. So there is no codeword of weight 8. The bound for C is raised, and we can try the method for a word of weight 10.

Using several times this method we proved that C contains no codewords of weight less than 14. Then we found a word of weight 14, which is an idempotent of C. So the true minimum distance of C is 14.

How to find an idempotent is presented in the next subsection :

2.2 Finding idempotents in a cyclic code.

We want to find the minimum distance of the BCH code of length 1023 and designed distance 253. Let us suppose there exists a codeword x of weight 253. For the power sum symmetric functions of the locators of x, we have :

$$A_1 = A_2 = \cdots = A_{252} = 0, \tag{10}$$

and :

$$A_{253} \neq 0 \tag{11}$$

(If A_{253} is equal to zero, then the weight of x is greater than 253 because of the BCH bound.)

Because the length is large, in order to symplify the problem, we first try to find a codeword x which is an idempotent. This implies (see [3]) :

$$\Rightarrow \quad \forall k \, , \, A_k \in GF(2).$$
$$\tag{12}$$
$$\Rightarrow \quad A_{253} = 1.$$

So the computations become much simplier. Here we give a part of a Maple session, where we compute the A_i :

Initialization of the parameter for the programm :
length=1023, designed distance=253, weight=253.
> init(1023,253,253);
The representants of the cyclotomic classes which are not in the defining set of C are :

$$\{511, 495, 479, 447, 439, 383, 379, 375, 367, 363, 351, 347, 343, 341, 255\}$$

We try to find a solution for the particular case $A_{255} = 0$:
> $A_{255} := 0;$

$$A_{255} := 0$$

After computing the values of the σ_i in terms of the A_i, we have equations in terms of the A_i. With these equations, we can find the values of the A_i :

$$
\begin{array}{lll}
eq_{507}: & A_{447} + A_{375} + A_{347} = 0 & \Rightarrow A_{447} := A_{347} + A_{375} \\
eq_{509}: & A_{383} + A_{351} = 0 & \Rightarrow A_{351} := A_{383} \\
eq_{511}: & A_{511} + A_{379} + A_{343} = 0 & \Rightarrow A_{343} := A_{511} + A_{379} \\
eq_{517}: & A_{341} + 1 = 0 & \Rightarrow A_{341} := 1 \\
eq_{519}: & A_{511} + A_{379} = 0 & \Rightarrow A_{511} := A_{379} \\
eq_{523}: & A_{347} = 0 & \Rightarrow A_{347} := 0 \\
eq_{527}: & A_{383} = 0 & \Rightarrow A_{383} := 0 \\
eq_{543}: & A_{367} = 0 & \Rightarrow A_{367} := 0 \\
eq_{551}: & A_{375} = 0 & \Rightarrow A_{375} := 0 \\
eq_{555}: & A_{379} = 0 & \Rightarrow A_{379} := 0 \\
eq_{571}: & A_{439} = 0 & \Rightarrow A_{439} := 0 \\
eq_{583}: & A_{363} + 1 = 0 & \Rightarrow A_{363} := 1 \\
eq_{611}: & A_{479} = 0 & \Rightarrow A_{479} := 0 \\
eq_{627}: & A_{495} + 1 = 0 & \Rightarrow A_{495} := 1 \\
\end{array}
$$

In a similar way we obtain a solution for the Newton's identities in the case $A_{255} = 1$. These two solutions give us the following locator polynomials :

$$\boxed{A_{255} = 0}$$

$$1 + z^{66} + z^{88} + z^{110} + z^{132} + z^{198} + z^{242} + z^{253}$$

$$\boxed{A_{255} = 1}$$

$$
\begin{aligned}
&1 + z^2 + z^4 + z^6 + z^8 + z^{10} + z^{12} + z^{14} + z^{16} + z^{18} \\
&+ z^{20} + z^{22} + z^{24} + z^{26} + z^{28} + z^{30} + z^{32} + z^{34} \\
&+ z^{36} + z^{38} + z^{40} + z^{42} + z^{44} + z^{46} + z^{48} + z^{50} \\
&+ z^{52} + z^{54} + z^{56} + z^{58} + z^{60} + z^{62} + z^{64} + z^{68} \\
&+ z^{72} + z^{76} + z^{80} + z^{84} + z^{90} + z^{94} + z^{96} + z^{100} \\
&+ z^{104} + z^{108} + z^{110} + z^{112} + z^{116} + z^{122} + z^{128} \\
&+ z^{130} + z^{136} + z^{138} + z^{144} + z^{146} + z^{152} + z^{154} \\
&+ z^{160} + z^{162} + z^{168} + z^{170} + z^{176} + z^{178} + z^{184} \\
&+ z^{186} + z^{192} + z^{200} + z^{208} + z^{218} + z^{220} + z^{222} \\
&+ z^{224} + z^{232} + z^{238} + z^{240} + z^{242} + z^{246} + z^{252} \\
&+ z^{253}
\end{aligned}
$$

(Performed with Maple symbolic computation software.)

These polynomials factor over $GF(2)$ in distinct polynomials of degree 2, 5 or 10, which split over $GF(2^{10})$. So these locator polynomials have 253 distinct roots in $GF(2^{10})$. (Of course the factorization of these polynomials is also accomplished with Maple software.) There exist codewords of weight 253 in C ; in conclusion, the code C has true minimum distance 253.

2.3 Finding a locator polynomial by doing an exhaustive research.

We want to find the true minimum distance of the following codes :

let C_m be the binary primitive cyclic code of length $2^m - 1$ and zeros α^1 and α^7.

VAN LINT and WILSON showed that C has minimum distance $\delta < 5$, when $m \neq 5, 11, 13, 17$ ([5]). Their demonstration is based on algebraic geometry theory, for they need to know deep algebraic properties of $GF(2^m)$.

In the case $m = 5$, they proved that the minimum distance is five. So the remaining unsolved cases are : $m = 11, 13, 17$.

We use our method to find a minimum weight codeword :

The defining set contains : 1, 2, 7, 8. So, the Hartmann-Tzeng bound shows that $d \geq 4$. ($d_0 = 3$, $s = 1$, $l = 1$, $c_1 = 1$, $c_2 = 7$.)

Let us suppose there exists a codeword x of C of weight 4. We associate to x its locator polynomial $\sigma(Z)$:

$$\sigma_{\mathbf{x}}(Z) = 1 + \sigma_1 Z + \sigma_2 Z^2 + \sigma_3 Z^3 + \sigma_4 Z^4 .$$

For the power sum symmetric functions, we have $A_1 = A_7 = 0$.
The Newton's Identities give :

$$
\begin{array}{llll}
(eq_1) & : & A_1 + \sigma_1 = 0 & \implies \sigma_1 = \quad\; 0 \\
(eq_3) & : & A_3 + \sigma_3 = 0 & \implies \sigma_3 = \quad A_3 \\
(eq_5) & : & A_5 + A_3\sigma_2 = 0 & \implies \sigma_2 = A_5/A_3 \\
(eq_7) & : & A_5\sigma_2 + A_3\sigma_4 = 0 & \implies \sigma_4 = \quad \sigma_2^2
\end{array}
$$

Shifting the word by one position corresponds to multiplication of the locators by α. Because 3 and n are coprime we can shift the word in such a way that $A_3 = 1$. So we want to find a polynomial with the following form :

$$\sigma_x(Z) = 1 + A_5 Z^2 + Z^3 + A_5^2 Z^4 , \tag{13}$$

which satisfies the following condition :

$$\sigma_x(Z) \text{ has four different roots in } GF(2^m).$$

Because there is no explicit condition on the coefficients of a polynomial for its decomposition, our only solution is to make an exhaustive research on all the possible values of A_5, testing each time if $\sigma_x(Z)$ splits over $GF(2^m)$.

The test is the following : compute $Z^{2^m} \bmod \sigma_x(Z)$ and test if it is equal to Z. It only requires m repeated squaring operations of a polynomial modulus an other polynomial, which is of small degree (i.e. 4).

Using Scratchpad II computation software which is able to compute with polynomials with coefficients in an extension of $GF(2^m)$, we found a word of weight 4 in the three cases $m = 11, 13, 17$, at relative low cost. After doing the change of indeterminates $Z^2 \leftarrow A_5 Z^2$, we search for polynomials of the form :

$$\sigma_x(Z) = 1 + Z^2 + Y Z^3 + Z^4, \tag{14}$$

where :

$$Y = A_5^{-\frac{3}{2}}$$

because less operations in finite field are needed for computing these polynomials.

Here are the locators we found in the three cases $m = 11$, $m = 13$, $m = 17$:

$$\boxed{m = 11}$$

$$GF(2^{11}) = GF(2)[X]/(X^{11} + X^2 + 1) , \; u = \overline{X} \bmod (X^{11} + X^2 + 1)$$

(In the three cases we have $\alpha = \overline{X}$.)

$$
\begin{aligned}
u^{10} + u^8 + u^7 + u^6 + u^2 + u + 1 &= \alpha^{660} \\
u^9 + u^3 + u &= \alpha^{487} \\
u^9 + u^3 + 1 &= \alpha^{1769} \\
u^{10} + u^8 + u^7 + u^6 + u^2 &= \alpha^{1178}
\end{aligned}
$$

So the polynomial x in $GF(2)[X]/(X^{11} - 1)$ is :

$$X^{487} + X^{660} + X^{1178} + X^{1769}$$

Then we can verify that α^1 and α^7 are zeros of x :

$$(\alpha^1)^{660} + (\alpha^1)^{487} + (\alpha^1)^{1769} + (\alpha^1)^{1178} = 0 \qquad (15)$$

$$(\alpha^7)^{660} + (\alpha^7)^{487} + (\alpha^7)^{1769} + (\alpha^7)^{1178} = 0 \qquad (16)$$

$$\boxed{m = 13}$$

$$GF(2^{13}) = GF(2)[X]/(X^{13} + X^4 + X^3 + X + 1), \; u = \overline{X} \bmod (X^{13} + X^4 + X^3 + X + 1)$$

$$1 = \alpha^0$$
$$u^{12} + u^6 + u^5 + u^4 + 1 = \alpha^{6399}$$
$$u^{12} + u^{10} + u^8 + u^7 + u^6 + u^3 + u = \alpha^{2735}$$
$$u^{10} + u^8 + u^7 + u^5 + u^4 + u^3 + u = \alpha^{6454}$$

$$x = X^0 + X^{6399} + X^{2735} + X^{6454}$$

And we can verify that x is in C :

$$(\alpha^1)^0 + (\alpha^1)^{6399} + (\alpha^1)^{2735} + (\alpha^1)^{6454} = 0 \qquad (17)$$

$$(\alpha^7)^0 + (\alpha^7)^{6399} + (\alpha^7)^{2735} + (\alpha^7)^{6454} = 0 \qquad (18)$$

$$\boxed{m = 17}$$

$$GF(2^{17}) = GF(2)[X]/(X^{17} + X^3 + 1), \; u = \overline{X} \bmod (X^{13} + X^3 + 1)$$

$$u^{16} + u^{14} + u^{13} + u^{10} + u^8 + u^7 + u^3 + u = \alpha^{124733}$$
$$u^{16} + u^{15} + u^{13} + u^{12} + u^{11} + u^{10} + u^8 + u^6 + u^5 + u^3 = \alpha^{58930}$$
$$u^{16} + u^{12} + u^9 + u^5 + u^4 + u^2 = \alpha^{88726}$$
$$u^{16} + u^{15} + u^{14} + u^{11} + u^9 + u^7 + u^6 + u^4 + u^2 + u = \alpha^{120824}$$

$$x = X^{124733} + X^{58930} + X^{88726} + X^{120824}$$

And we can again verify :

$$(\alpha^1)^{124733} + (\alpha^1)^{58930} + (\alpha^1)^{88726} + (\alpha^1)^{120824} = 0 \qquad (19)$$

$$(\alpha^7)^{124733} + (\alpha^7)^{58930} + (\alpha^7)^{88726} + (\alpha^7)^{120824} = 0 \qquad (20)$$

So in these three cases, the minimum distance is 4.

3 Conclusion

We saw how the use of the Newton's identities is helpful for studying the minimum weight codewords of cyclic codes. Of course, we need to use a symbolic computation software for the manipulation of these identities.

With this method, we proved that the minimum distances of the two BCH codes of length 255 and designed distance 59 and 61 (See [2]) are 61 and 63 respectively. We also found a general property of the minimum weight codewords of the BCH codes of length $2^m - 1$ and designed distance $2^{m-2} - 1$ (See [1]).

Acknowledgment

The authors wish to thank G. NORTON for his careful and patient reading which improved the paper.

References

[1] D. Augot, P. Charpin, and N. Sendrier. Studying the locator polynomials of minimum weight codewords of BCH codes. *submitted.*

[2] G. Cohen. On the minimum distance of some BCH codes. *IEEE Transaction on Information Theory*, 26, 1980.

[3] F.J. MacWilliams and N.J.A. Sloane. *The Theory of Error Correcting Codes.* North-Holland, 1986.

[4] J.H. van Lint. *Coding Theory.* Springer-Verlag, 1971.

[5] J.H. van Lint and R.M. Wilson. Binary cyclic codes generated by $m_1 m_7$. *IEEE Transaction on Information Theory*, 32(2):283, March 1986.

[6] J.H. van Lint and R.M. Wilson. On the minimum distance of cyclic codes. *IEEE Transaction on Information Theory*, 32(1):23, January 1986.

SECTION 2

COMBINATORIAL CODES

MINIMUM-CHANGE BINARY BLOCK-CODES WHICH ARE WELL BALANCED

J.Burger(*), H.Chabanne(**), M.Girault(*)

(*) Service d'Etudes communes des Postes et Télécommunications
42, rue des Coutures
B.P. 6243
14066 CAEN CEDEX (FRANCE)

(**) Université de Limoges
123, rue A.Thomas
87000 LIMOGES (FRANCE)

Abstract

In this paper, we deal with the following problem : how to code a counter (that is an interval of integers such as [0,n-1]) by the means of an adequate binary block-code so that each bit changes as few as possible when the counter runs the whole cycle and comes back to the first state ? This problem can be divided in two sub-ones. First, how to minimize the Hamming distance between two consecutives states of the counter ? Solutions of this problem have been known for a long time, as Gray codes. Second, how to spread the changed bits over the whole counter in a well balanced way ? The goal of this paper is to provide some solutions to the latter problem.

Such codes are useful in the case of wearing parts such as particular memory counters or input/output devices. In these cases it is requested to avoid that some places often change in the binary data, while some other places would rarely change (as in the case of natural binary coding). As a single example, we can think of some E^2PROM set up in the most recent smart cards.

I. Notations and definitions.

Let n and k be two non-null positive integers. A *binary code* $C_{n,k}$, also denoted by C, is an injective function from the interval of integers X=[0,n-1] over the set $\{0,1\}^k$. If k=$\log_2$ n, such a code is called *exact* and denoted by C_k (from now on, we will write log n instead of $\log_2$ n). Similarly, a code is called *almost-exact* if k=[log n], where [z] denotes the smallest integer greater than or equal to the real number z. An element y of $\{0,1\}^k$ is denoted by $y_k y_{k-1} \cdots y_1$ where y_j=0 or 1. We call *bit j* of y the jth bit of y, that is y_j.

We can now define a *change function* B as follows : for any x in [0,n-1], for any integer j in [1,k] we have : if $C(x-1)_j$ is equal to $C(x)_j$ then B(x,j)=0 else B(x,j)=1, where, as a convention, C(-1) means C(n-1). We write B(x) the summation from j=1 to k of B(x,j). So B(x)=d[C(x-1),C(x)] where d is the Hamming distance in $\{0,1\}^k$. We also denote by B_j the summation from x=0 to n-1 of B(x,j). So B_j is equal to the number of changes supported by the bit j when the counter has run any whole cycle.

I.1 Minimum-change codes.

Definition : The *change rate* of $C_{n,k}$ is $t = \frac{1}{n} \sum_{x=0}^{n-1} B(x)$.

Remarks:

1) t is greater than or equal to 1 . In case $C_{n,k}$ is the natural binary code, $t=2-2^{1-k}$.

2) If n is odd, we cannot expect t=1 because the total number of changes in a whole cycle is always even. In this case, we should deal with non-cyclic runs of the counter and have a different definition of the change rate. From here, n will always be even in order to simplify the paper.

3) t is less than or equal to $k-\frac{1}{2}$ if n is greater than or equal to 4.

A natural question is : can the above bounds be reached ? The answer, known for a long time, is positive. However, in this paper, we are only interested in the lower limit, and a code $C_{n,k}$ will be called a *minimum-change code* if t=1.

For example, Gray codes are a well-known class of exact or almost-exact minimum-change codes. They are defined by the following recursion : let G_k be the Gray code over $\{0,1\}^k$, let x be an element of the integers in the range $[0,2^{k+1}-1]$. If the most significant bit x_k is equal to 0 then $G_{k+1}(x) = 0 \| G_k(x)$ else $G_{k+1}(x) = 1 \| G_k(2^{k+1}-x-1)$, where $G_1(0)=0$, $G_1(1)=1$ and $\|$ is the symbol for concatenation (see illustration in section III.5).

I.2 Well balanced codes.

Definition : A code is called a *well balanced code* if for any i and any j in [1,k], $|B_i-B_j|$ is less than or equal to 2.

Neither natural binary codes nor Gray codes belong to well balanced codes for n greater than or equal to 5, because in both cases $B_k = \frac{n}{2}$ and $B_1 = 2$.

Remark :

That definition is justified by the fact that for any even integer m, and for any integer k smaller than m, m can be expressed as the summation of k even integers m_i where for any i and any j in [1,k], $|m_i - m_j|$ is 0 or 2. Note that if k|m, this condition simply implies that $m_i = m_j$ for all i,j. Conversely, all these equalities can hold only if k|m. A well balanced code $C_{n,k}$ such that k|n will be called *perfectly well balanced*.

I.3 Minimum-change well balanced codes.

Definition : A code is a *minimum-change (perfectly) well balanced code* if it is both a minimum-change code and a (perfectly) well balanced code. Such a code is briefly called *MCWB-code*.

Remark :

If there is no condition on k, building a well balanced code becomes an evidence for any n, since unary code gives birth to a convenient one ! (Flip the bits from 0 to 1 from the right to the left, then flip them from 1 to 0, still from the right to the left). In fact we search well balanced codes for reasonable values of k, say k=O(log n). Nevertheless, unary coding leads to a method to build asymptotical minimum-change well balanced codes, as will be described later.

First, we examine the case of exact and almost-exact codes. Recall that $C_{n,k}$ is exact iff $n=2^k$, almost-exact iff n is in $[2^{k-1}+1, 2^k]$.

Proposition :

1) Let $C_{n,k}$ be an exact MCWB-code. $C_{n,k}$ is perfectly well balanced iff $k=2^r$, for some r.

2) Let $C_{n,k}$ be an almost-exact MCWB-code. $C_{n,k}$ is perfectly well balanced iff 2k divides n.

Examples :   1) n=2,4,16,256,16384.
             2) n=6,20,30,48,60.

Our aim is to find (almost-)exact MCWB-codes. But, as it is rather hard to find such codes, we will also give our attention to asymptotical MCWB-codes, that is codes for which the average number of changes of any bit tends to $\frac{n}{k}$ (this ratio being assumed to be constant) when k tends to the infinity. Our main results are the following : First, there exist exact MCWB-codes for k less than or equal to 6 but the coding and decoding procedures are not trivial. Second, there exist (almost-)exact asymptotical MCWB-codes easy to construct for any k and n. Moreover, the coding and decoding procedures are very easy. So are the increment and decrement procedures.

II. Particular MCWB-codes.

II.1 Representation of $C_{n,k}$ minimum-change codes.

As stated by Gilbert[G], the search of a $C_{n,k}$ minimum-change code, is equivalent to that of a cycle of length n on the k-cube Q_k. The geometrical picture Q_k consists of a k-regular graph of 2^k vertices and $k2^{k-1}$ edges. Each vertex is labelled by its own k-tuple $(y_1,...,y_k)$ where $y_i=0$ or 1 is called the ith coordinate. Two vertices of Q_k are adjacent if and only if their coordinates agree on all but one of the k places.

A path on Q_k can be represented by the list of the labels of the successive vertices. Better is to list the place of the changing coordinate of the successives vertices on the path on Q_k. This last representation, called *sequence*, is more compact and is independent of the first label of the path. As the symmetry group of Q_k is known as the hyperoctohedral group O_k of order $2^k k!$, two paths on Q_k are called *equivalent* if one can be changed into the other by applying an operation of O_k. On the contrary, they are called of *different type*. So we are interested in cycles of different type and, in case $n=2^k$, in hamiltonian circuits of Q_k.

II.2 Building exact $C_{n,k}$ MCWB-codes.

We denote by K_2, the graph composed of two connected points. As stated by Harary[H1][H2], Q_k can be defined recursively in terms of the cartesian sum of two graphs $Q_k = K_2 + Q_{k-1}$, where $Q_1 = K_2$. So we say that Q_k results from the *displacement* or the *duplication of Q_{k-1} along the kth direction*. Two nodes joined by the kth direction are called *twin nodes*.

This point of view enabled us to construct k-bit MCWB-codes for k less than or equal to 6 in the following manner :
- Assume we know a well balanced hamiltonian circuit of Q_{k-1}, say H_{k-1}.
- We draw Q_k as the duplication of Q_{k-1}. We choose a starting node S on the left Q_{k-1} and trace H_{k-1} starting at S. Then we follow the left H_{k-1} on half of its nodes and arrive at the node T.
- Now we jump from T to the right Q_{k-1}, and trace H_{k-1} starting at T', the twin of T.
- Now we consider only the figure formed by the two H_{k-1} and the edges joining them so that each node is of degree 3. We have $2*2^{k-1}+2^{k-1}$ edges and have to break 2^{k-1} of them in order to decrease the degree of all the nodes down to 2 and get a hamiltonian circuit of Q_k.
- We start the H_k at the node S and walk on the left H_{k-1} up to T. We jump onto the right H_{k-1}, and walk on it up to S', the twin of S. Here half of the nodes of Q_k have been visited.
- Then we try to go on walking as much as possible on the H_{k-1}, jumping from left to right and from right to left in order to obtain a well balanced circuit H_k.

The minimum-change perfectly well balanced sequence H_4 is :
0123210132310302 (four changes in each direction). This sequence can also be found in

Gilbert's paper, in which all the hamiltonian circuits of Q_4 are listed. There are nine such (non-equivalent) circuits, and only one of them is well balanced. Furthermore, this circuit is the only one which is not composed or ultra-composed, that is, roughly speaking, which is not a natural extension of hamiltonian circuits of hypercubes of lower dimension.

The minimum-change well balanced sequence H_5 is :
0123210401232102030403014342434 (six or eight changes in each direction).

The minimum-change well balanced sequence H_6 is:
01232104012321050123210401232102505350545053505143424345342 43421 (ten or twelve changes in each direction).

II.3 Building $C_{n,k}$ asymptotical MCWB-codes (see more details in [C]).

Assume $n=2^k$. Let us call Q_m-twins of Q_k two nodes of Q_k labels of which agree on m places. In case $m=k-1$ and they differ on place j, they are called twins of Q_k along the jth direction. The set of all the Q_m-twins of Q_k agreeing in the same m places is a sub-cube Q_{m-k} of Q_k. Once set the m places in the labels of the nodes, Q_k contains 2^m different sub-cubes Q_{k-m}.

We know trivial well balanced circuits on Q_k (see remark in section I.3) and call them chenilles. They are specified by a starting node on Q_k and the sequence $(0,1,...,k-1,0,1,...,k-1)$. Two chenilles are different if and only if they do not share any node of Q_k. These cycles are 2^*k long. In order to cover all the nodes of Q_k with such different chenilles, we need and suppose that $k=2^r$ so the number of chenilles is an integer.

We first need to find 2^{2^r-r-1} different chenilles. Let us call diagonal an edge joining two nodes of Hamming distance greater than or equal to two. A diagonal is denoted by $(i+j+...)$ where $i,j,...$ are the changed bits. Two chenilles may be joined in one single cycle of length 2^*2k by breaking two parallel edges, one on each circuit, and joining the vertices with at most two diagonals. By cleverly repeating this process, we can obtain an asymptotical MCWB-code.

Here Q_k is seen as containing 2^{2^r-r} sub-cubes Q_r indexed from 0 to $2^{2^r-r}-1$. The forward walk of the first chenille takes r edges on the sub-cube of index 0, and one edge on each sub-cube of index i iff i is a Mersenne's number, for i less than or equal to 2^{2^r-r}. Then the backward walk on the chenille 0 takes r edges on the sub-cube 2^{2^r-r} and one on each sub-cube of index i', for $i'=2^{2^r-r}-i$, where i is the ith Mersenne's number. The jth chenille is built in the same way as the first one except that the indexes of the sub-cubes are added 2j modulo (2^{2^r-r}). In the end, each sub-cube participates in $r+1 + 2^r-r-1$ points spread on 2^{2^r-r-1} chenilles.

This only provides a necessary condition of the existence of a partition of Q_k. We need now to find the starting node of each chenille. The search becomes easy in looking at the sequence $(012..k-1,012..k-1)$. Assume $a\|b$ is the label of the starting node of a chenille, were b is r-bit long. Then $(a+i)\|b'$ is in the same chenille, where + stands for exclusive-or, b' is the complement to 1 of b and i is 1,2,4,...2^{k-r}. a'||(b+j) is in the same chenille, where j is 1,2,...2^r.

Last, we need to link the chenilles together. As there are many nodes which labels are of weigth $\frac{k}{2}$, the best way seems to join the different chenilles at these places.

Example on Q_8:
The complete list of starting nodes is in hexadecimal notation : 00,90,21,B1,12,82,33,A3,74,E4,55,C5,66,F6,47,D7. The sixteen chenilles are linked together to form an asymptotical MCWB-code. Here follows an example of sequence containing only one diagonal and where the number of changes are 34,32,32,32,28,36,32,32 :

$$\begin{array}{l}
0123456701230432 \\
1076543210762654 \\
3210765432103701 \\
2345670123457567 \\
0123456701230432 \\
1076543210761567 \\
0123456701230567 \\
0123456701235345 \\
6701234567016210 \\
7654321076545345 \\
6701243567015123 \\
4567012345675701 \\
2345670123451567 \\
0123456701236076 \\
5432107654325076 \\
543210765432(2+4+5)567.
\end{array}$$

III. A class of asymptotical MCWB-codes easy to construct   (Patented by SEPT).

III.1 Some more notations.

The interval of integers X=[0,n-1], also called here *counter*, is supposed to be run in a sequential way once or many times (in the latter case, 0 is the successor of n-1). For any x in X, x is coded in a k-bit register denoted by C(x). C(x) is divided in two fields : the *reference field* $C_r(x)$ and the *rotation* field $G_s(x)$, where C_r denotes any r-bit coding function and G_s denotes the s-bit Gray code. $Incr_s$ will denote the incremental function of G_s, so that $Incr_s(G_s(x))=G_s(x+1)$, for any x in $\{0,1,...,2^s-1\}$. Similarly, $Decr_s$ will denote the decremental function of G_s, so that $Decr_s(G_s(x))=G_s(x-1)$. $Incr_s$ and $Decr_s$ are more precisely defined in annex.

---

* We describe only the code for which coding, decoding, increasing and decreasing procedures are very simple. It is obvious that some well selected permutations instead of rotations will work as well.

Here are the two advantages of the code we are going to define. First, when the counter runs from $C(x)$ to $C(x+1)$, only one bit changes in the rotation field. Rarely, one bit changes in the reference field too. Second, when the counter runs one or more whole cycles, the difference between the number of changes of any two bits of the rotation field is at most two.

III.2 The basic code.

In this section we suppose that $n=2^k$ (else take the smallest k such that n is less than or equal to 2^k). The reference field is r-bit long, and the rotation field is q-bit long, where $k=r+q$. The best situation occurs when $q=2^r$, $k=2^r+r$ and $n=2^k$ so if $r=1,2,3,4,5$ then $k=3,6,11,20,37$ and $n=8,64,2048,2^{20},2^{37}$.

The change-rate in the rotation field is equal to $\frac{1}{q}$, whilst the changes are not frequent in the reference field.

The coding procedure
The counter is divided in 2^r intervals I_i of size 2^q. For any i in $\{0,..,2^r-1\}$ and for x in $I_i=[i*2^q,(i+1)*2^q-1]$, $C(x)$ is composed of $C_r(i)$ in the reference field. The rotation field of $C(x)$ results from the left rotation of magnitude i applied to $G_q(x-i*2^q)$. Assume $G_q(x-i*2^q)$ is $g_q g_{q-1} \dots g_1$ before the rotation, it changes into $g_{q-i} \dots g_1 g_q \dots g_{q-i+1}$ after the rotation. In case i is greater than q, i is reduced modulo q.

In case C_r is the r-bit natural binary code then the reference field is simply composed of the r most significant bits of the natural binary code of x.

The decoding procedure
First extract the data $C_r(i)$ from the reference field and decode i. Then apply a right rotation of magnitude i on the rotation field and use the standard decoding procedure of the Gray code G_q to get the decoded less significant q bits of x.

The incremental procedure
First extract the data $C_r(i)$ from the reference field and decode i. Then apply a right rotation of magnitude i on the rotation field and use the $Incr_q$ procedure. If bit q changed from 1 to 0 during the increment function, increment the reference field too and decode its new value i. Last, carry out the left rotation of magnitude i on the rotation field.

Remark: The incremental operation on the reference field depends on the C_r code and is not described here. It is very simple when C_r is the Gray code G_r.

The decremental procedure
As for the incremental procedure, it is recommended to take the following method. First extract the data $C_r(i)$ from the reference field and decode i. Then apply a right rotation of magnitude i on the rotation field and use the $Decr_q$ procedure. If a change 0 to 1 occurred at place q during the decrement function, decrement the reference field too and decode its new value i. Last, carry out the left rotation of magnitude i on the rotation field.

III.3 A first variant.

In order to decrease the change number in the rotation field, we can add some more bits in it. Here the reference field is r-bit long, and the rotation field $(p+q)$-bit long, where $k=r+p+q$. The best situation occurs when $p+q=2^r$, $k=2^r+r$ and $n=2^{k-p}$ so if $r=3$ and $p=1$, then $k=11$ and $n=1024$. The case $p=0$ is described in the previous section.

The change-rate of the rotation field is equal to $\dfrac{1}{p+q}$ whilst in the reference field the changes are not frequent.

The coding an decoding procedure
They are the same as in the case of the basic code except that the rotation field results from the left rotation of magnitude i applied to $O_p \parallel G_q(x-i*2^q)$, where O_p denotes a null p-bit field. In case i is greater than $p+q$, i is reduced modulo $p+q$.

The incremental and decremental procedure
They are the same as in the case of the basic code except that $Incr_s$ and $Decr_s$ are applied to the q less significant bits of the rotation field.

III.4 A second variant.

This concerns particularly counters intended to be run many times and which might be erased. The reference field is r-bit long, and the rotation field q-bit long. The best situation occurs when $q=2^r$, $k=2^r+r$, $n=2^q$ so if $r=1,2,3,4,5$ then $k=3,6,11,20,37$ and $n=4,16,256,65536,2^{32}$.

The change-rate of the rotation field is equal to $\dfrac{1}{q}$, whilst in the reference field the changes are not frequent.

The coding procedure
When the counter is run for the first time, the reference field is set to $C_r(0)$ and the rotation field to $G_q(x)$. When the counter is run for the ith time, the reference field is set to $C_r(i)$ and the rotation field results from the left rotation of magnitude i of $G_q(x)$.

The decoding procedure
First extract the data $C_r(i)$ from the reference field and decode i. Then apply a right rotation of magnitude i on the rotation field and use the standard decoding procedure of the Gray code G_q applied to the q less significant bits of the rotated field to get the decoded q bits of x.

The incremental procedure and the decremental procedure are the same as for the basic code.

The erase procedure
To erase the counter, increment the reference field and set the rotation field to $G_q(0)$.

III.5 Examples.

Let us recall the Gray code :

x	Gray code
0	0000
1	0001
2	0011
3	0010
4	0110
5	0111
6	0101
7	0100
8	1100
9	1101
10	1111
11	1110
12	1010
13	1011
14	1001
15	1000

Here is an illustration of the basic asymptotical MCWB-code where n=64, k=6, r=2. For coding, we use the natural binary code C_2 for the reference field and the Gray code G_4 for the rotation field.

x	binary code	MCWB-code
0	000000	000000
1	000001	000001
2	000010	000011
...	...	...
15	001111	001000
16	010000	010000
17	010001	010010
18	010010	010110
...	...	...
31	011111	010001
32	100000	100000
33	100001	100100
34	100010	101100
...	...	...
...	...	...
62	111110	111100
63	111111	110100

Now follows an illustration of the first variant of the asymptotical MCWB-code where n=16, k=6, r=2 and p=2. For coding, we use the natural binary code C_2 for the reference field and the Gray code G_2 for the rotation field.

x	binary code	MCWB-code
0	0000	000000
1	0001	000001
2	0010	000011
3	0011	000010
4	0100	010000
5	0101	010010
6	0110	010110
7	0111	010100
8	1000	100000
9	1001	100100
10	1010	101100
11	1011	101000
12	1100	110000
13	1101	111000
14	1110	111001
15	1111	110001

Last follows a particular illustration of the second variant of the MCWB-code where n=16, k=6, r=2, in which occur some resets. For coding, we use the natural binary code C_2 for the reference field and the Gray code G_4 for the rotation field.

x	binary code	MCWB-code
0	0000	000000
1	0001	000001
2	0010	000011
0	0000	010000
1	0001	010010
2	0010	010110
3	0011	010100
0	0000	100000
1	0001	100100
2	0010	101100
3	0011	101000
4	0100	101001
5	0101	101101
...	...	...
14	1110	100110
0	0000	110000

IV. References.

[C] Chabanne H., "Etude de codes uniformes pour mémoires E^2PROM", UFR des sciences de Limoges, 1990.
[F] Flores I., "Reflected number systems", IRE Trans. on Elect. Comp., Vol. 5, 1956.
[G] Gilbert E.N., "Gray codes and paths on the n-cube", Bell Syst. Tech. J., Vol.37, 1958.
[H1] Harary F., "Graph theory", Addison-Wesley, 1969.
[H2] Harary F., Hayes J.P. and Wu H.J., "A survey of the theory of hypercube graphs", Comput. Math. Applic., Vol.15, 1988.

Annex

Here we describe a practical way, stated by Flores[F], of transcoding from natural binary code to Gray code and conversely from Gray code to natural binary code. (+ stands for bitwise exclusive or)

The coding function from natural binary code to Gray code :
$$g_s = h_s,$$
$$g_{s-1} = h_{s-1} + h_s,$$
$$g_{s-2} = h_{s-2} + h_{s-1},$$
$$...$$
$$g_{s-1} = h_1 + h_2.$$

The decoding function from Gray code to natural binary code :
$$h_s = g_s,$$
$$h_{s-1} = g_{s-1} + g_s,$$
$$h_{s-2} = g_{s-2} + g_{s-1} + g_s,$$
$$...$$
$$h_1 = g_1 + g_2 + ... + g_s.$$

We also describe a practical way to increment or decrement Gray codes. Assume $G_s(x)$ is $g_s...g_1$, and $h_s...h_1$ the natural binary code of x where x is in $\{0,1\}^s$. In case $x=2^s-1$, x+1 denotes 0; conversely, in case x=0, x-1 denotes 2^s-1.

The $Incr_s$ function connects $G_s(x)$ to $G_s(x+1)$ as follows :
If $g_1 + g_2 + ... + g_s = 0$ then change g_1.
Else If $g_1 = 1$ then change g_2.
...
Else if $g_{s-2} = 1$ then change g_{s-1}.
Else change g_s.

Finally, the $Decr_s$ function connects $G_s(x)$ to $G_s(x-1)$ as follows :
If $g_1 + g_2 + ... + g_s = 1$ then change g_1.
Else If $g_1 = 1$ then change g_2.
...
Else If $g_{s-2} = 1$ then change g_{s-1}.
Else change g_s.

CONSTRUCTION OF UNEQUAL ERROR PROTECTION CODES

A. O. Mabogunje and P. G. Farrell
Communications Research Group
Department of Electrical Engineering
University of Manchester
Manchester M13 9PL

Introduction

The study of unequal error protection(uep) codes has been prompted by an increase in the volume of digital data being transmitted across communication channels, where an increasing percentage of this data has reliability requirements that vary within a block of data. Classical error control codes, also referred to in this text as equal error protection (eep) codes, provide the same level of protection for all parts of an information block. This makes them non–optimal for protecting data with uep requirements. It is possible to optimize the redundancy in an error control code by giving different parts of an information block different levels of protection. This can be done by using uep codes. The concept of uep codes was first introduced by Masnick and Wolf (1967).

The objective in this paper is to draw attention to construction methods by which array codes can be modified to provide unequal error protection, and to highlight some of the advantages of using uep codes. Simulation results are shown for some array codes with unequal error protection.

Relative minimum distance

The minimum distance of a code relative to any information position is always the same in an eep code and is equal to the true minimum distance of the code. The error correction capability of such random error correction codes is based on the true minimum distance of the code. In a uep code, the situation is different because the relative minimum distance of the code differs from one information position to another and is not always the same as the true minimum distance of the code. In other words, information symbols in different positions see the code as having different minimum distances. The error correction capability of a uep code is therefore based on a more complicated distance parameter known as the separation vector, which was introduced by Dunning and Robbins (1978). Unlike the minimum distance of a code, which is independent of the generator matrix used in encoding, the separation vector is dependent on the generator matrix. It is therefore defined with respect to a particular generator matrix. The individual components of the separation vector refer to the minimum distance of a code relative to a particular position in the information stream being encoded.

Preliminary definitions.

Let the symbols in the information stream to be encoded be taken from GF(q) and let

$I = \{ I_j : j=0, 1,...,k-1 \}$ be the information stream with k positions per word,

$C = \{ C_j : j=0, 1,...,n-1 \}$ be an n-symbol codeword of a code V,

and $\quad C^{(i,j)}$ be a codeword formed from an information stream with symbol 'i' in position I_j.

The *separation* of position I_j in the information stream I with respect to a code with generator matrix G and weight function w is the minimum distance of the code relative to position I_j and is defined as

$$ s_w(G)_j = \min_i w\left(C^{(i,j)} - C^{(i',j)}\right) \qquad i \neq i' $$

The *separation vector* is defined along the same lines as

$$ S_w(G) = \{ s_w(G)_i : i = 0,1,\ldots, k-1 \} $$

When no ambiguity will result, the separation $s_w(G)_i$ will be represented as s_i and the separation vector $S_w(G)$ as S. For large codes, the above form of the separation vector is cumbersome and it becomes convenient to use the polynomial form of the separation vector which is introduced below:

$$ S = \sum_{b=0}^{z} ax^b $$

where a is the number of information bits that have a separation of b and z is the number of protection levels.

When reference needs to be made to a particular generator matrix and weight function, subscripts may be used in the usual manner.

The *protection level* t_j of an information position j with separation s_j is defined as

$$ t_j = \left\lfloor (s_j - 1)/2 \right\rfloor $$

where $\lfloor x \rfloor$ is the integer part of x.

The *protection spread* of a code is defined as: $p = 1 + (s_{max} - s_{min})$ where s_{max} and s_{min} are the maximum and minimum components of the separation vector. By this definition, eep codes have a unit protection spread. Note that this is a slightly different from the protection bandwidth introduced by Pingzhi et al. (1988).

Construction of unequal error protection codes from array codes.

Two information arrays (square or rectangular, say) are said to be orthogonal in the jth order if and only if no more and no less than j positions are common to any of their rows and columns. There is a finite number of orthogonal arrays of the jth order that can be formed on any particular array of information positions. In this study, orthogonal array codes of the first order are of primary interest because they are easy to understand and are easily decoded by one step majority logic decoding algorithms. Most of the arguments for the first order orthogonal arrays hold for higher order arrays. The three possible first order orthogonal 5x5 arrays that can be formed with twenty-five information positions are shown in figure 1. Let each position hold a binary information symbol. Each row and column can be made a codeword of an even

1	2	3	4	5
6	7	8	9	10
11	12	13	14	15
16	17	18	19	20
21	22	23	24	25

1	10	14	18	22
7	11	20	24	3
13	17	21	5	9
19	23	2	6	15
25	4	8	12	16

1	8	15	17	24
9	11	18	25	2
12	19	21	3	10
20	22	4	6	13
23	5	7	14	16

Figure 1: Orthogonal arrays possible with 25 positions

parity check code by adding a 1 or a 0 to make the number of non-zeros in the row an even number. The result is a 6x6 array without the check on checks (figure 2). There are two sets of parity information on each array. One set corresponds to symbols a-e and the other to symbols f-k in figure 2. Due to the fact that the parity information is formed from orthogonal information arrays, each set of parity is orthogonal to the others and can be used independently to increase the minimum distance of the code by one. All six sets of parity information formed on the above arrays can be combined to give a triple error correcting (55,25,7) code. By combining different numbers of parity sets, the following eep (n,k,d) array codes can be formed:(30,25,2), (35,25,3), (40,25,4), (45,25,5), (50,25,6) and (55,25,7). The simulated bit error rates of the (55,25,7), (45,25,5) and (35,25,3) codes when used for random error correction are shown in figure 3, from which it is seen that there is little difference in performance below E_b/N_o values of 10dB. The decoding algorithm used in the simulation is a one step majority logic decoding algorithm. It should be noted from table 2 that the asymptotic coding gains of these codes differ.

Table 2

code	asymptotic coding gain (dB) 10log(rate x min. dist)
(55,25,7)	5.027
(45,25,5)	4.437
(35,25,3)	3.3099

1	2	3	4	5	a
6	7	8	9	10	b
11	12	13	14	15	c
16	17	18	19	20	d
21	22	23	24	25	e
f	g	h	j	k	

Figure 2: 6x6 array without check on checks.

If all the information in a parity set is not used simultaneously, a code with non-unity protection spread will result. For example, if only those parity symbols that check the information symbol in one particular position are included in the code, only one parity check is included from each of the parity sets and the separation vector of the resulting (31,25) code will be

$$S = x^7 + 25x^2$$

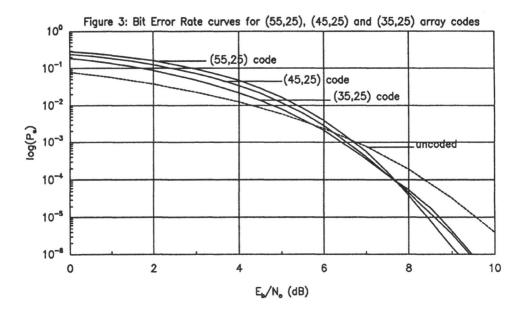

Figure 3: Bit Error Rate curves for (55,25), (45,25) and (35,25) array codes

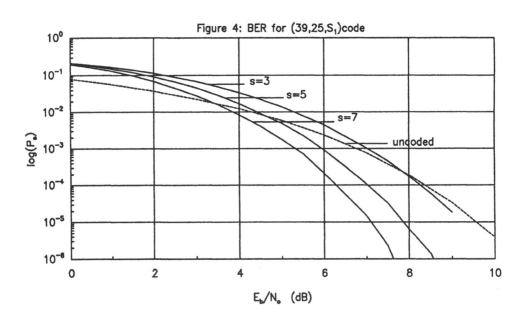

Figure 4: BER for (39,25,S_1)code

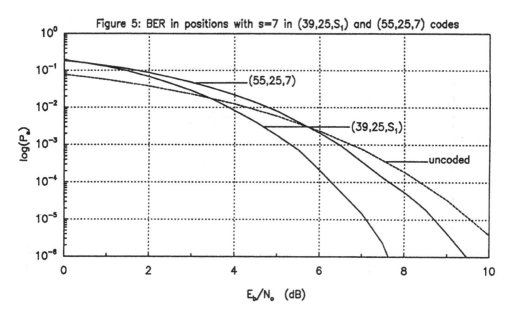

Figure 5: BER in positions with s=7 in $(39,25,S_1)$ and $(55,25,7)$ codes

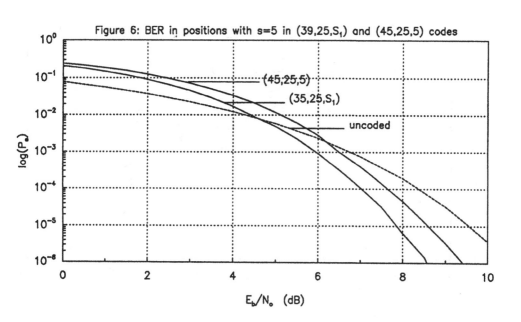

Figure 6: BER in positions with s=5 in $(39,25,S_1)$ and $(45,25,5)$ codes

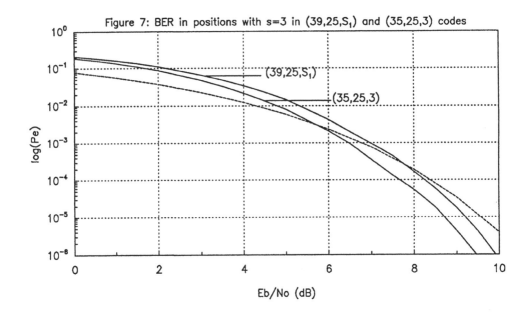

Figure 7: BER in positions with s=3 in $(39,25,S_1)$ and $(35,25,3)$ codes

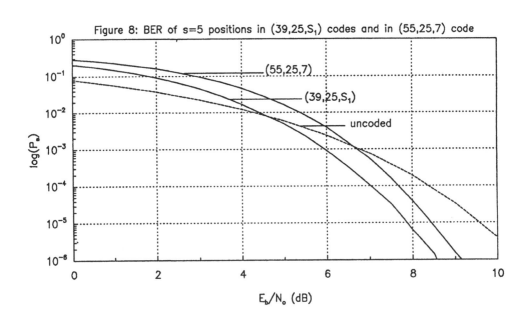

Figure 8: BER of s=5 positions in $(39,25,S_1)$ codes and in $(55,25,7)$ code

Many uep array codes can be formed in this way. As an example of the uep codes that can be achieved, the focus here will be mainly on the array codes introduced above, i.e. the (35,25), (45,25) and (55,25) codes and the $(39,25,S_1)$ code, where

$$S_1 = 2x^7 + 2x^5 + 8x^4 + 13x^3$$

The parity symbols in this code are formed as illustrated in table 3, and the BER performance is shown in figure 4. The first row of table 3 means the parity symbol in position 1 checks the information in positions 1, 2, 3, 4 and 5 of the information word. Column 1 tells us that the information in position 1 is checked by parity symbols in parity positions 1, 5, 6 and 7.

Table 3

position	1	2	3	4	5	6	7	8	9	10	11	12	13	14	15	16	17	18	19	20	21	22	23	24	25
parity no. 1	*	*	*	*	*																				
2						*	*	*	*	*	*														
3											*	*	*	*	*	*									
4																*	*	*	*	*					
5	*					*					*					*					*				
6	*						*					*					*								*
7	*								*				*					*					*		
8		*							*				*				*						*		
9		*								*		*					*								*
10			*					*					*				*					*			
11			*								*				*	*					*				
12				*		*							*						*		*				
13				*						*				*				*					*		
14					*					*			*				*				*				

Concluding Remarks

A comparison of the error rates in positions with the same separation in the different codes confirms the theory that the error rate of a code depends on both its distance structure and its rate. Figure 5, 6 and 7 shows the BER in those information positions with a separation of 7, 5 and 3 respectively. The advantage of uep can be seen in figures 5 and 6, which shows that the protection received by the information in the $s=5$ and $s=7$ positions of the $(39,25,S_1)$ code is greater than that received by information with an identical separation in the (45,25,5) and (55,25,7) codes respectively. This improvement in performance is easily explained by the higher rate of the $(39,25,S_1)$ code. A similar explanation can be given for the lower protection received by the $s=3$ positions of the $(39,25,S_1)$ code when compared to positions with identical separations in the (35,25,3) code as shown in figure 7. Finally, a situation where the advantages of a high rate out-weighs that of a slightly increased separation is illustrated in figure 8, where it is shown that below 10dB, the BER in the s=7 positions of the (55,25) code is higher than that in the s=5 positions of the (39,25) code.

References

1. B. Masnick and J. Wolf, " On Linear Unequal Error Protection Codes," IEEE *Trans. Inform. Theory*, vol. IT-3 no. 4, pp 600 - 607, Oct. 1967.

2. L. A. Dunning and W. E. Robbins, "Optimal encodings of linear block codes for unequal error protection" *Information and control*, vol 37, pp 150 - 177, 1978.

3. F. Pingzhi, C. Zhi and J. Fan, "One Step Completely Orthogonalisable UEP Codes and Their Soft Decision Decoding," *Electronic Letters*, vol. 24, no. 17, pp 1095 - 1096, 18 Aug. 1988.

Unidirectional Error Detecting Codes *

Gerard D. Cohen
Ecole Nationale Supérieure des Telecomunications
75634 Paris Cedex 13, France

Luisa Gargano and Ugo Vaccaro
Dipartimento di Informatica ed Applicazioni
Università di Salerno
84081 Baronissi (SA), Italy

Abstract

A fixed-length binary code is called t–unidirectional error detecting if no codeword can be transformed into another codeword by at most t unidirectional errors. In this paper we consider the problem of mapping information sequences of length k into codewords of a t–unidirectional error detecting code of length $k + p$. In case of systematic codes we show that the parameters p and t must satisfy the relation $t \leq 2^p - 2^{p/2+1} + p$. Moreover, we give a simple systematic encoding to map information sequences into codewords of a t–unidirectional error detecting code. In case of non–systematic codes, we give a method to design t–unidirectional error detecting codes in which the number p of check bits must satisfy the inequality $t \leq 2^p - p - 1$. The encoding and decoding algorithms require time linear in the number k of information bits.

1   Introduction

In many situations it is possible to change a bit 0 into a bit 1 but the contrary is not possible. Examples are given by write-once memories, like digital optical disks, where a 0 can be changed into a 1 but the vice versa is not possible. Moreover, it has been shown that the most likely faults in VLSI chips are unidirectional, that is, both $0 \rightarrow 1$ and $1 \rightarrow 0$ errors are possible but not in the same word [10].

A fixed length binary code is called *t–unidirectional error detecting* (t-UED) if no set of t unidirectional errors can transform a codeword into another codeword.

*The work of the last two authors was partially supported by the National Council of Research (CNR) under grant 90.0071.3.PF69 and 90.01552.CT12.

The problem of the construction of unidirectional error detecting codes has been widely studied [1, 2, 5, 6, 7, 8, 9, 11, 12, 15, 16, 17, 18].

In this paper we consider the problem of the construction and analysis of t-unidirectional error detecting codes. The problem was previously considered by Bose and Lin [6], who gave algorithms to construct systematic codes. In their method the number t of detectable unidirectional errors is related to number p of check bits by the relation $t \leq 5 \cdot 2^{p-4} + p - 4$. Here we consider both systematic and non–systematic codes. We show that for any systematic t-UED code with p check bits, it must hold $t \leq 2^p - 2^{p/2+1} + p$. Moreover, we propose a systematic encoding method which, even though is less efficient of that in [6], is proposed for its extreme simplicity and its good performance for small values of p.

In case of non–systematic codes, we propose an efficient encoding/decoding algorithm to map k–bit information sequences into codeword of a t-UED code having p check bits and able to detect up to $2^p - p - 1$ unidirectional errors. The required time to perform the encoding/decoding operations is linear in the number k of information bits.

Define a partial order relation $\preceq$ on the binary sequences of length n defined by $a_1 \ldots a_n = \mathbf{a} \preceq \mathbf{b} = b_1 \ldots b_n$ if and only if $a_i \leq b_i$ for each i; if the relation $\preceq$ does not hold we write $\mathbf{a} \npreceq \mathbf{b}$. Given the sequence $\mathbf{a} = a_1 \ldots a_n \in \{0,1\}^n$, define the *weight* of $\mathbf{a}$ as the integer $\mu(\mathbf{a}) = \sum_{i=1}^{n} a_i$. It follows that a code C is t-UED if and only if no two codewords $\mathbf{v}$ and $\mathbf{w}$ exist such that $\mathbf{v} \preceq \mathbf{w}$ and $\mu(\mathbf{w}) < \mu(\mathbf{v}) + t$.

Borden [4] shows that a maximum size t-UED code of length n can be obtained by taking all elements in $\{0,1\}^n$ of weight $\lceil n/2 \rceil \pm i(t+1)$, for $i = 0, \ldots, \lceil n/(2(t+1)) \rceil$; moreover a recent result by Griggs [13] implies that this code is unique. We denote such a code by $C(n, t)$.

2 Systematic unidirectional error detecting codes

In this section we consider the problem of encoding k–bits information sequences into code-words of a t–UED systematic code of length $n = k + p$.

We first establish an upper bound on t, once p is fixed. To avoid trivialities, we suppose that $p < \lceil \log k \rceil$, otherwise one has enough check bits to construct an All–Unidirectional Error–Detecting code [3].

Theorem 1 *It is not possible to detect t–unidirectional errors if*

$$t > 2^p - 2^{\frac{p}{2}+1} + p. \tag{1}$$

Proof: Let us consider a chain of information sequences $\mathcal{C} = \{\mathbf{w}_0, \mathbf{w}_1, \ldots, \mathbf{w}_k\}$ with $\mu(\mathbf{w}_i) = i$, for $i = 0, \ldots k$. Let the encoding of any information sequence $\mathbf{w}$ be denoted by $\mathbf{w}p(\mathbf{w})$, where the check sequence $p(\mathbf{w})$ belongs to $\{0,1\}^p$. Consider a sequence $\mathbf{w}_i \in \mathcal{C}$ such that there exists $\mathbf{w}_\ell \in \mathcal{C}$ with $\ell > i$ and $p(\mathbf{w}_i) \preceq p(\mathbf{w}_\ell)$ (such a $\mathbf{w}_i$ exists since $p < \lceil \log k \rceil$; in fact the hypothesis that $p < \lceil \log k \rceil$ implies the even stronger condition that there exist $\mathbf{w}_i$ and $\mathbf{w}_\ell$ such that $p(\mathbf{w}_i) = p(\mathbf{w}_\ell)$). Let j be the smallest integer such that

$$j > i \quad \text{and} \quad p(\mathbf{w}_i) \preceq p(\mathbf{w}_j). \tag{2}$$

We have

$$t + 1 \le \mu(\mathbf{w}_j p(\mathbf{w}_j)) - \mu(\mathbf{w}_i p(\mathbf{w}_i)) = [\mu(\mathbf{w}_j) - \mu(\mathbf{w}_i)] + [\mu(p(\mathbf{w}_j)) - \mu(p(\mathbf{w}_i))] \tag{3}$$

where $\mu(\mathbf{w}_j) - \mu(\mathbf{w}_i) = j - i$ and $\mu(p(\mathbf{w}_j)) - \mu(p(\mathbf{w}_i)) \le p$. Inequality (3) can be then written as

$$t + 1 \le j - i + p. \tag{4}$$

Notice now that if we suppose the existence of two indices r and s, with $i \le r < s \le j$ and $p(\mathbf{w}_r) = p(\mathbf{w}_s)$ the bound on t can only decrease as we get $t + 1 \le s - r \le j - i$. Analogously, if we suppose the existence of r, with $i < r < j$ and $p(\mathbf{w}_r) \preceq p(\mathbf{w}_i)$ the bound on t can only decrease as we get $t + 1 \le j - r + p \le j - i + p$. We can then bound the right hand side of (4) supposing that $p(\mathbf{w}_r) \ne p(\mathbf{w}_s)$, for all $i \le r < s \le j$ and that $p(\mathbf{w}_r) \not\preceq p(\mathbf{w}_i)$, for all $i < r < j$. Therefore, the term $j - i$ is upper bounded by the number of different sequences $\mathbf{x} \in \{0,1\}^p$ such that $\mathbf{x} \not\preceq p(\mathbf{w}_i)$ and, by (2), $p(\mathbf{w}_i) \not\preceq \mathbf{x}$. Hence

$$j - i \le 2^p - 2^{\mu(p(\mathbf{w}_i))} - 2^{p - \mu(p(\mathbf{w}_i))} + 1$$

and by (4)

$$t \le 2^p - 2^{\mu(p(\mathbf{w}_i))} - 2^{p - \mu(p(\mathbf{w}_i))} + p.$$

We can then write

$$t \le \max_{1 \le \mu \le p} 2^p - 2^\mu - 2^{p-\mu} + p.$$

and the result follows. □

We give now a very simple scheme to encode information sequences into a systematic t–UED code. The resulting codes are no better than those in [6], but they allow very simple encoding and decoding and are comparable to them, for small values of p. The encoding scheme is as follows:

1. let $S = \{\mathbf{x}_0, \mathbf{x}_1, \ldots, \mathbf{x}_t\}$ be the set of all the p–bit sequences of weight $\lfloor p/2 \rfloor$, $t = \binom{p}{\lfloor p/2 \rfloor}$;

2. each $\mathbf{w} \in \{0,1\}^k$ is encoded by $\mathbf{w}\mathbf{x}_i$, where $i = \mu(\mathbf{w}) \bmod (t+1)$.

It is immediate to see that the resulting systematic code is t–UED.

3   Encoding information into $C(n,t)$: non–systematic codes

In this section we shall study the problem of encoding k-bit information sequences into codewords of $C(n,t)$, where $n = k + p$ and p denotes the number of check bits. We recall that $C(n,t)$ is the set of all n–bit sequences of weight $\lceil n/2 \rceil \pm i(t+1)$, for $i = 0, \ldots, \lceil n/(2(t+1)) \rceil$.

Since we need at least as many codewords as the number of information sequences, the number of check symbols p should satisfy the obvious inequality

$$2^k \leq \sum_{i \geq 0} \binom{n}{\lceil n/2 \rceil \pm i(t+1)} = \frac{2^{k+p}}{t+1} \sum_{k=0}^{t} \cos^n\left(\frac{k\pi}{t+1}\right). \tag{5}$$

The derivation of the last equality is obtained in [14] (see also [4] and [13]).

The lower bound on p implied by the above relation could be easily reached if one had a table where the correspondence between information sequences and codewords is stored. However, this method becomes rapidly inefficient as the size of the code increases. Thus, it is useful to have encoding algorithms that do not require the storage of large tables. In the following section we provide easily computable encoding rules to map k-tuples into codewords of $C(k+p,t)$ for all $t \leq 2^p - p - 1$. Notice that bound (5) shows that our construction is very close to the optimal and, from the bound (1) on the number t of detectable errors in systematic codes, more efficient than the best possible construction for systematic codes.

3.1 The encoding scheme

We first report an algorithm given in [9] to partition $\{0,1\}^k$ into disjoint chains of length not greater than 2^p, p being the number of check bits that are to be used. A similar algorithm is also presented in [13]. We recall that $C \subseteq \{0,1\}^k$ is a chain if for each $x, y \in C$ either $x \preceq y$ or $y \preceq x$.

Algorithm 1

1. If $k = 1$ the set $\{0,1\}$ consists of the only chain $C_0^{(1)} = Q = \{0,1\}$;

2. If $k > 1$, let $C_0^{(k-1)}$, $C_1^{(k-1)}$, $\ldots$, $C_{r-1}^{(k-1)}$ be the partition of $\{0,1\}^{k-1}$, for some integer r.
   For each chain $C_i^{(k-1)} = \{c_{i,0}, \ldots, c_{i,l_i-1}\}$:

   - If $l_i = 1$ construct the chain $C_{i,1}^{(k)} = \{0c_{i,0}, 1c_{i,0}\}$,

   - If $1 < l_i < 2^p$ construct the chains

   $$C_{i,0}^{(k)} = \{0c_{i,0}, \ldots, 0c_{i,l_i-1}, 1c_{i,l_i-1}\}, \qquad C_{i,1}^{(k)} = \{1c_{i,0}, \ldots, 1c_{i,l_i-2}\},$$

   - If $l_i = 2^p$ construct the chains

   $$C_{i,0}^{(k)} = \{0c_{i,0}, \ldots, 0c_{i,l_i-1}\}, \qquad C_{i,1}^{(k)} = \{1c_{i,0}, \ldots, 1c_{i,l_i-1}\}.$$

Example 1 *Applying Algorithm 1 for $p = 2$ one gets, $C_0^{(1)} = \{0,1\}$, from which one gets $C_0^{(2)} = C_{0,0}^{(2)} = \{00, 01, 11\}$ and $C_1^{(2)} = C_{0,1}^{(2)} = \{10\}$.*
From $C_0^{(2)}$ one has $C_0^{(3)} = C_{0,0}^{(3)} = \{000, 001, 011, 111\}$ and $C_1^{(3)} = C_{0,1}^{(3)} = \{100, 101\}$.
From $C_0^{(3)}$ one has $C_{0,0}^{(4)} = \{0000, 0001, 0011, 0111\}$ and $C_{0,1}^{(4)} = \{1000, 1001, 1011, 1111\}$.

Lemma 1 [9] *Algorithm 1 produces a partition of $\{0,1\}^k$ into chains of size at most 2^p. For each chain $C_i^{(k)} = \{c_{i,0}, \ldots, c_{i,l_i-1}\}$ it holds*

$$\mu(c_{i,j}) = \mu(c_{i,j-1}) + 1 \qquad j = 1, \ldots, l_i - 1. \tag{6}$$

Moreover, for each chain $C_i^{(k)}$ of size $l_i < 2^p$ it holds

$$\mu(c_{i,l_i-1}) = k - \mu(c_{i,0}). \tag{7}$$

Let $C_0^{(k)}, \ldots, C_{r-1}^{(k)}$ be the partition of the set of information sequences $\{0,1\}^k$ obtained applying Algorithm 1. We will determine an encoding function such that the codeword $E(\mathbf{w})$ associated to a sequence $\mathbf{w} \in C_i^{(k)}$ is a sequence of length $k+p$ belonging to $C(k+p,t)$ given by

$$E(\mathbf{w}) = c(\mathbf{w})g(\mathbf{w}) \quad \text{with } c(\mathbf{w}) \in C_i^{(k)}, \quad g(\mathbf{w}) \in \{0,1\}^p. \tag{8}$$

Since we want that $E(\mathbf{w}) \in C(k+p,t)$ it must hold that

$$\mu(E(\mathbf{w})) = \mu(c(\mathbf{w})) + \mu(g(\mathbf{w})) \in \left\{ \left\lceil \frac{k+p}{2} \right\rceil \pm i(t+1) \; : \; i = 0, \ldots, \left\lfloor \frac{k+p}{2(t+1)} \right\rfloor \right\}. \tag{9}$$

Moreover, we need that for each $\mathbf{w}, \mathbf{w}'$ it holds $E(\mathbf{w}) \neq E(\mathbf{w}')$. This is obviously true for $\mathbf{w}$ and $\mathbf{w}'$ belonging to different chains, since (8) tells us that also $c(\mathbf{w})$ and $c(\mathbf{w}')$ belong to different chains and, from Lemma 1, $c(\mathbf{w}) \neq c(\mathbf{w}')$. Therefore, we only need that

$$g(\mathbf{w}) \neq g(\mathbf{w}') \quad \text{for each } \mathbf{w}, \mathbf{w}' \in C_i^{(k)}, \quad i = 0, \ldots, r-1. \tag{10}$$

In order to define functions c and g satisfying (9) and (10), let us distinguish two cases in the size of $C_i^{(k)}$.

Case 1: Encoding of $\mathbf{w} \in C_i^{(k)}$, $|C_i^{(k)}| < 2^p$.

We encode the sequences of such a chain in codewords of weight $\lceil (k+p)/2 \rceil$, satisfying (9), in the following way. Let $u_0, u_1, \ldots, u_{2^p-1}$ be all the sequences in $\{0,1\}^p$, ordered in such a way that $\mu(u_j) \geq \mu(u_{j+1})$, define

$$h(i) = u_i, \quad \text{for } i = 0, \ldots, 2^p - 1. \tag{11}$$

Letting $g(\mathbf{w}) = h(\mu(\mathbf{w}) - \lfloor k/2 \rfloor + 2^{p-1} - 1)$, condition (10) immediately follows from (6). Finally define $c(\mathbf{w})$ as the sequence in $C_i^{(k)}$ of weight $\lceil (k+p)/2 \rceil - \mu(g(\mathbf{w}))$; the existence of such a sequence is proved in [9].

Case 2: Encoding of $\mathbf{w} \in C_i^{(k)}$, $|C_i^{(k)}| = 2^p$.

Let $C_i^{(k)} = \{c_{i,0}, \ldots, c_{i,2^p-1}\}$. We first notice that, from (10), all sequences in $\{0,1\}^p$ must appear as check sequence of some information sequence of $C_i^{(k)}$. Therefore we can define $g(c_{i,j}) = h(j), j = 0, \ldots, 2^p - 1$, where h is defined in (11). We will encode all the sequences of $C_i^{(k)}$ with codewords of the same weight. In order to define the desired encoding we need the following result.

Lemma 2 *It is possible to encode all sequences of the chain $C_i^{(k)} = \{c_{i,0}, \ldots, c_{i,2^p-1}\}$, with $\mu(c_{i,0}) < \ldots < \mu(c_{i,2^p-1})$, into codewords of weight μ, if and only if*

$$\mu - 2^p + 1 \leq \mu(c_{i,0}) \leq \mu - p. \tag{12}$$

Proof. Since we want $g(w) \neq g(w')$, for each $w, w' \in C_i^{(k)}$, we need to use all the different elements of $\{0,1\}^p$ as check sequences. Therefore, the chain must contain an element of weight $\mu - j$ for each $j = 0, \ldots, p$. From this, (6), we get

$$\mu(c_{i,0}) + p \leq \mu \leq \mu(c_{i,2^p-1}) = \mu(c_{i,0}) + 2^p - 1,$$

and (12) holds. □

Suppose now that the encodings of the sequences in the chain $C_j^{(k)} = \{c_{j,0}, \ldots, c_{j,2^p-1}\}$ have weight μ satisfying (9) and (12), that is $\mu - 2^p + 1 \leq \mu(c_{j,0}) \leq \mu - p$, and consider a chain $C_i^{(k)} = \{c_{i,0}, \ldots, c_{i,2^p-1}\}$, with $\mu(c_{i,0}) = \mu - p + 1$ [resp. $\mu(c_{i,0}) = \mu - 2^p$]. Since we want that (9) holds, we will encode the sequences of $C_i^{(k)}$ with codewords of weight $\mu + t + 1$ [resp. $\mu - t - 1$]. By (12) this is possible if $\mu + t + 1 - 2^p + 1 \leq \mu(c_{i,0}) = \mu - p + 1 \leq \mu + t + 1 - p$ [resp. $\mu - t - 2^p \leq \mu(c_{i,0}) = \mu - 2^p \leq \mu - t - 1 - p$], that is, if

$$t \leq 2^p - p - 1 \tag{13}$$

From the above reasoning, we get that, if (13) holds and we define for a sequence $w \in C_i^{(k)} = \{c_{i,0}, \ldots, c_{i,2^p-1}\}$ the weight of the encoding $\mu(E(w)) = \mu(E(c_{i,0}))$ as

$$
\mu(E(w)) =
\begin{cases}
\lceil \frac{(k+p)}{2} \rceil \\
\quad \text{if } \lceil \frac{(k+p)}{2} \rceil - 2^p + 1 \leq \mu(c_{i,0}) \leq \lceil \frac{(k+p)}{2} \rceil - p \\[2mm]
\lceil \frac{(k+p)}{2} \rceil - j(t+1) \\
\quad \text{if } \lceil \frac{(k+p)}{2} \rceil - j(t+1) - 2^p + 1 \leq \mu(c_{i,0}) \leq \lceil \frac{(k+p)}{2} \rceil - (j-1)(t+1) - 2^p \\[2mm]
\lceil \frac{(k+p)}{2} \rceil + j(t+1) \\
\quad \text{if } \lceil \frac{(k+p)}{2} \rceil + (j-1)(t+1) - p + 1 \leq \mu(c_{i,0}) \leq \lceil \frac{(k+p)}{2} \rceil + j(t+1) - p
\end{cases}
\tag{14}
$$

then (12) holds. From this we get that for each $w \in C_i^{(k)}$ there exsists a sequence $c(w) \in C_i^{(k)}$ with $\mu(c(w)) = \mu(E(w)) - \mu(g(w))$ and the encoding is correct. □

Therefore we have the following result.

Theorem 2 *It is possible to encode information sequences of length k into codewords of $C(k + p, t)$ whenever $t \leq 2^p - p - 1$.*

Example 2 *Consider $p = 2$, $t = 1$ and $k = 6$. Applying Algorithm 1 we have, among the others, the chains*

$$
\begin{aligned}
C_1 &= \{001100, 011100, 111100\}, \\
C_2 &= \{010000, 010001, 010011, 010111\}, \\
C_3 &= \{111000, 111001, 111011, 111111\}.
\end{aligned}
$$

We can define the sequences u_i as $u_0 = 11$, $u_1 = 10$, $u_2 = 01$, and $u_3 = 00$. Consider first the encoding of sequences in C_1. Since the size of the chain is $3 < 2^p$ we encode these sequences with codewords of weight $(k + p)/2 = 4$. We have

$$
\begin{aligned}
E(001100) &= c(001100)g(001100) = c(001100)h(2 - 2) = c(001100)u_0 = (001100)(11) \\
E(011100) &= c(011100)g(011100) = c(011100)h(3 - 2) = c(011100)u_1 = (011100)(10) \\
E(111100) &= c(111100)g(111100) = c(111100)h(4 - 2) = c(111100)u_2 = (111100)(01).
\end{aligned}
$$

Consider now the encoding of sequences in C_2 and C_3 of size $4 = 2^p$. From (14), we have

$$
\mu(E(\mu(\mathbf{w}))) = \begin{cases} 4 & \text{if } w \in C_2 \\ 6 & \text{if } w \in C_3 \end{cases} \quad \text{and } g(\mathbf{w}) = \begin{cases} h(\mu(\mathbf{w})) - 1 & \text{if } w \in C_2 \\ h(\mu(\mathbf{w})) - 3 & \text{if } w \in C_3 \end{cases} = \begin{cases} u_{\mu(\mathbf{w})-1} & \text{if } w \in C_2 \\ u_{\mu(\mathbf{w})-3} & \text{if } w \in C_3. \end{cases}
$$

Therefore, we get

$$
\begin{aligned}
E(010000) &= (010001)(11) & E(111000) &= (111001)(11) \\
E(010001) &= (010011)(10) & E(111001) &= (111011)(10) \\
E(010011) &= (010011)(01) & E(111011) &= (111011)(01) \\
E(010111) &= (010111)(00) & E(111111) &= (111111)(00).
\end{aligned}
$$

3.2 Encoding and decoding computation

In this section we consider the problem of determining the complexity of the encoding and decoding operations and show that such operations can be computed in time linear in the length of the information sequences.

The encoding of $\mathbf{w}$ described in Section 2 can be summarized as follows:

1. Find the first element c_0 of the chain $C = \{c_0, \ldots, c_{l-1}\}$ such that $w = c_i \in C$;

2. Let $g(w) = u_{\mu(w)-\lfloor k/2 \rfloor + 2^{p-1} - 1}$ if $l < 2^p$, $g(w) = u_i$ if $l = 2^p$,

3. Find c_j such that $\mu(c_j) = \mu(c_0) + j = \mu(E(w)) - \mu(g(w))$;

(where $\mu(E(w)) = \lceil (k+p)/2 \rceil$ if $l = 2^p$; $\mu(w) = \mu(E(c_0))$ is defined in (14) if $l = 2^p$)

4. Output $c_j g(w)$.

The decoding of xy, with $x \in \{0,1\}^k$ and $y \in \{0,1\}^p$, can be specified as follows:

1. if $\mu(xy) \neq \lceil (k+p)(q-1)/2 \rceil \pm s(t+1)$ for all integers s then Stop [an error has occurred], else

   - Find the first element c_0 of the chain $C = \{c_0, \ldots, c_{l-1}\}$ such that $x \in C$;
   - If $|C| < 2^p$ then let i be the integer such that $x = u_i$. Find $c_j \in C$ such that $\mu(c_j) = \mu(c_0) + j = i + \lfloor k/2 \rfloor - 2^{p-1} + 1$; output c_j;
   - If $|C| = 2^p$ then let i be the integer such that $y = u_i$. Find $c_i \in C$ (i.e., find the $c \in C$ of weight $\mu(c_0) + i$), output c_i.

In both processes, the main points are, given $x \in Q^k$, to find the lightest element c_0 of the chain containing x and to find the element having weight μ and belonging to the chain containing c_0.

Given a sequence $x = x_k \ldots x_1$, we recursively compute the triple (l_i, p_i, a_i), $i = 1, \ldots, k$, where l_i indicates the length of the chain C containing $x_i \ldots x_1$, the integer p_i represents the relative position of $x_i \ldots x_1$ within C, and the first element of C is the sequence $a_i a_{i-1} \cdots a_1$. The triple of index 1 is $(l_1, p_1, a_1) = (2, x_1, 0)$. The elements l_{i+1}, p_{i+1}, and a_{i+1} are computable in one step from the elements l_i, p_i, and a_i, respectively as follows

$$l_{i+1} = \begin{cases} l_i & \text{if } l_i = 2^p \\ l_i - 1 & \text{if } l_i < 2^p, \ x_{i+1} = 1 \text{ and } p_i \leq l_i - 2 \\ l_i + 1 & \text{otherwise.} \end{cases}$$

$$p_{i+1} = \begin{cases} p_i + 1 & \text{if } l_i < 2^p, \ x_{i+1} = 1 \text{ and } p_i = l_i - 1 \\ p_i & \text{otherwise.} \end{cases}$$

$$a_{i+1} = \begin{cases} 0 & \text{if } l_i < 2^p, \ x_{i+1} = 1 \text{ and } p_i = l_i - 1 \\ x_{i+1} & \text{otherwise.} \end{cases}$$

Define now $S^\Delta(a_i \ldots a_1)$ as the element of weight $\mu(a_i \ldots a_1) + \Delta$ belonging to the chain whose first element is $a_i \ldots a_1$ (for all integers Δ for which such a sequence exists). Formally

$$S^1(0) = 1$$

$$S^\Delta(a_i \cdots a_1) = \begin{cases} 1S^{\Delta-1}(a_{i-1} \cdots a_1) & \text{if } a_i = 0, \, l_i = l_{i-1} + 1, \, \Delta = l_i - 1, \text{ and } \Delta > 0 \\ a_i S^\Delta(a_{i-1} \cdots a_1) & \text{otherwise.} \end{cases}$$

The correctness of the function S follows immediately from Algorithm 1.

Define now $\gamma = \mu - \mu(a_k \cdots a_1)$. The desired element belonging to the same chain as $\mathbf{x}$, i.e., the chain whose lightest element is $a_k \cdots a_1$, of weight $\mu(a_k \cdots a_1) + \gamma$ is $S^\gamma(a_k \cdots a_1)$.

Example 3 *Consider the code of Example 2 and the information sequence* $\mathbf{w} = w_6 \cdots w_1 = $ 010011. *We have*

$$(l_1, p_1, a_1) = (2, 1, 0), (l_2, p_2, a_2) = (3, 2, 0), (l_3, p_3, a_3) = (4, 2, 0)$$

$$(l_4, p_4, a_4) = (4, 2, 0), (l_5, p_5, a_5) = (4, 2, 1), (l_6, p_6, a_6) = (4, 2, 0)$$

therefore, $\mathbf{w} \in \{c_0, c_1, c_2, c_3\}$, *with* $c_0 = 010000$ *and, as* $\mu(\mathbf{w}) = \mu(c_0) + 2$, $\mathbf{w} = c_2$. *As in Example 2, we have* $g(\mathbf{w}) = u_2 = 01$ *and* $\mu(c(\mathbf{w})) = 3$, *which implies* $c(\mathbf{w}) = c_2$ *where*

$$c_2 = S^2(010000) = 0S^2(10000) = \cdots = 0100S^2(00) = 01001S^1(0) = 010011.$$

Therefore $E(010011) = 01001101.$

Let us consider now the decoding of the sequence $\mathbf{xy}$ *with* $\mathbf{x} = 011100$ *and* $\mathbf{y} = 01$. *For* $\mathbf{x}$ *we have*

$$(l_1, p_1, a_1) = (2, 0, 0), (l_2, p_2, a_2) = (3, 0, 0), (l_3, p_3, a_3) = (2, 0, 1)$$

$$(l_4, p_4, a_4) = (1, 0, 1), (l_5, p_5, a_5) = (2, 1, 0), (l_6, p_6, a_6) = (3, 1, 0)$$

therefore, $\mathbf{x} \in \{c_0, c_1, c_2\}$, *with* $c_0 = 001100$. *Since* $\mathbf{y} = 01 = u_2$ *we have that the decoding of* $\mathbf{xy}$ *is the sequence* $c_2 = S^2(c_0)$ *with*

$$S^2(c_0) = S^2(001100) = 1S^1(01100) = 11S^0(1100) = 111100.$$

The time required by encoding and decoding processes is dominated by the time necessary to compute the triple (l_i, p_i, a_i) plus a time linear in the number of recursive applications of the function S. Therefore the encoding and decoding can be done in time linear in the length k of the information sequences.

References

[1] S. Al-Bassan and B. Bose, "On Balanced Codes", *IEEE Int. Symp. Inform. Theory,* Japan, June 1988.

[2] S. Al-Bassan and B. Bose "Design of Efficient Balanced Codes", 19th *Int. Symp. on Fault-Tolerant Computing,* Chicago, Ill., 1989.

[3] J. M. Berger, "A Note on Error Detecting Codes for Asymmetric Channel", *Information and Control,* vol. **4**, pp. 68–73, 1961.

[4] J. M. Borden, "Optimal Asymmetric Error Detecting Codes", *Information and Control,* vol. **53**, pp. 66–73, 1982.

[5] B. Bose, " On Unordered Codes", 17th *Int. Symp. on Fault-Tolerant Computing,* Pittsburgh, Penn., pp. 102–107, 1987.

[6] B. Bose and D. J. Lin, "Systematic Unidirectional Error-Detecting Codes", *IEEE Trans. Comp.,* vol. **C-34**, pp. 1026–1032, 1985.

[7] R. M. Capocelli, L. Gargano, G. Landi and U. Vaccaro, "Improved Balanced Encodings", *IEEE International Symposium on Information Theory,* San Diego, USA, 1990.

[8] R. M. Capocelli, L. Gargano and U. Vaccaro, "An Efficient Algorithm to Test Immutability of Variable Length Codes", *IEEE Trans. Inform. Theory,* vol. **IT-35**, pp. 1310–1314, 1989.

[9] R. M. Capocelli, L. Gargano and U. Vaccaro, "Efficient q-ary Immutable Codes", *Discrete Applied Math.,* to appear.

[10] R. W. Cook, W. H. Sisson, T. G. Stoney, and W. N. Toy, "Design of Self-Checking Microprogram Control", *IEEE Trans. Comput.,* vol. **C-22**, pp. 255–262, 1973.

[11] C. V. Freiman, "Optimal Error Detecting Codes for Completely Asymmetric Binary Channels", *Information and Control,* vol.**5**, pp. 64–71, 1962.

[12] P. Godlewski and G. D. Cohen "Some Cryptographic Aspects of Womcodes", *Proceedings of CRYPTO '85,* Lecture Notes in Computer Science, Springer Verlag, 1985.

[13] J. R. Griggs, "Saturated Chains of Subsets and a Random Walk", *Journal of Combinatorial Theory,* Series A, vol. **47**, pp. 262–283, 1988.

[14] D. E. Knuth, *The Art of Computer Programming*, Vol. 1, pp. 70 and 486, Addison-Wesley, Reading, Mass., 1968.

[15] D. E. Knuth, "Efficient Balanced Codes", *IEEE Trans. Inform. Theory*, vol. **IT-32**, pp. 51–53, 1986.

[16] E. L. Leiss, "Data Integrity on Digital Optical Discs", *IEEE Trans. Computers*, vol. **C-33**, pp. 818–827, 1984.

[17] E. L. Leiss, "On Codes which are Unchangeable under Given Subversions", *J. Combin. Inform. & Syst. Sci.*, vol. **10**, pp. 91–109, 1985.

[18] E. L. Leiss, "On Testing for Immutability of Codes", *IEEE Trans. Inform. Theory*, vol. **IT-33**, pp. 934–938, 1987.

Coherent Partitions and Codes

André Montpetit*

Abstract

In this paper we present some results in graphs which extends combinatorial results in coding theory. The basic concept is the structure of coherent partition.

The aim of this paper is to extend the notion of r-partition design introduced by Camion *et al.* [4] for the linear case and in [3] for the nonlinear case. We will replace the Hamming space by a regular connected graph. Subsets of vertices take place of codes.

First we show that the eigenvalues of matrices associated with some partitions of the vertex set, called here coherent partitions, can be viewed as a sort of dual distances of a code (see [5]) .

In a second part, we are interested by the number of cells of coherent partitions associated with a fixed subset of vertices. The minimum number of cells in a coherent partition can be viewed as a parameter of the subset of vertices.

In the last part, we restrict ourselves to Cayley graphs and find a kind of generalization of Theorem 6.11 of [5] giving conditions for the cosets of a normal subgroup to induce a symmetric association scheme.

This kind of structure playing an important role in the study of "regular" code, we recall that a *symmetric association scheme* (see [1]) on X is a set of symmetric relations $\equiv_i$ on X ($0 \leq i \leq n$) such that $\{\equiv_i \mid 0 \leq i \leq n\}$ is a partition of $X \times X$, $\equiv_0 = \mathrm{id}_X$ and there exists p_{ij}^k such that, for all $x \equiv_k y$,

$$p_{ij}^k = \mathrm{card}\{z \in X \mid x \equiv_i z, y \equiv_j z\}. \tag{1}$$

The reader will find at the end a list of the most important notations used in this paper.

This is a part of the Ph.D. thesis of the author. The reader can refer to [6], [7] and [8] for the proofs.

1 Graphs and codes

Let $\Gamma = (X, E)$ be a regular connected graph without loops or multiple edges. Let A be the *adjacency matrix* of Γ i.e. $A(x, y) = 1$ when $\{x, y\}$ is an edge of Γ and 0 otherwise. Since A is symmetric, it is diagonalizable. So the algebra generated by A is semisimple and it has a basis of minimal orthogonal idempotents which are projection matrices onto the eigenspaces of A.

---
*Dép. de mathématiques et informatique, Université de Sherbrooke, Sherbrooke (Québec) Canada J1K 2R1

As an illustration, we give the following example which can be seen as the frame of the combinatorial coding theory. In this work, we extend some results already known in this particular case ([5, 3]).

EXAMPLE 1.1 Let F be an alphabet of q elements ($q \geq 2$). The Hamming distance between x and $y \in F^n$ ($n \geq 1$) is denoted by $d_H(x, y)$. The Hamming graph $H(n, q)$ is the graph whose vertex set is F^n and the 2-subset $\{x, y\}$ is an edge if the Hamming distance between x and y is 1. The eigenvalues of the adjacency matrix A are $n(q-1) - qi$, $i = 0$, $1, \ldots, n$ (see [1]).

The next case is related to the previous one and will be considered at the last part of this paper.

EXAMPLE 1.2 Let F be an abelian group (or a field) of q elements ($q \geq 2$). Let C be a subgroup (or a subspace) of F^n, $n \geq 1$. The coset graph of C $\Gamma(C)$ is the graph whose vertices are the cosets of C and $x + C$, $y + C$ are joined if they have representatives at Hamming distance 1.

Let N be the set of the eigenvalues of the adjacency matrix A and J_α the projection matrix onto the α-eigenspace ($\alpha \in N$) of A. We will call *code* of Γ any nonvoid subset of vertices. The *dual distribution* of the code Y is the vector b (indexed by N) where

$$b_\alpha = \frac{1}{(\text{card } Y)^2} \sum_{x, y \in Y} J_\alpha(x, y) \tag{2}$$

The set $S = \{\alpha \mid b_\alpha \neq 0\}$ will be the *annihilator set* of the code Y.

In the Hamming graph $H(n, q)$, the annihilator set of Y is the set $\{K_1(d_i) \mid 0 \leq i \leq s\}$ where the d_i's are the dual distances of Y (see [5]) and $K_1(x) = n(q-1) - qx$ is the 1st Krawtchouk polynomial.

If $s + 1$ is the size of S, s is the *external distance* of Y. The *annihilator polynomial* of Y is the polynomial

$$q(z) = \prod_{\alpha \in S} (z - \alpha) = \sum_{i=0}^{s+1} \beta_i z^i. \tag{3}$$

The *path matrix* of Y is the card $X \times \infty$ matrix P such that $P(x, i)$ is the number of paths of length i joining x to the code Y. The matrix P has the following properties.

THEOREM 1.1 ([6]) *Let P be the path matrix of Y.*

1. *The jth column of P is $P_j = A^j \Phi$ where Φ is the characteristic vector of Y.*

2. *The columns of P satisfy the recursive formula*

$$P_{s+1+k} = -\sum_{j=0}^{s} \beta_j P_{j+k}, \qquad k \geq 0, \tag{4}$$

   where the β_j's are defined in (3).

3. *The rank of P is $s + 1$.*

2   Coherent partitions

A partition $\pi = \{X_0, X_1, \ldots, X_r\}$ of the vertices of Γ is a *coherent partition* of Γ if there exists σ_{ij} $(i, j = 0, \ldots, r)$ such that, for all $x \in X_j$, σ_{ij} is the number of $y \in X_i$ such that $\{x, y\}$. The matrix $\sigma = (\sigma_{ij})$ will be called the *associate matrix* of π.

If Γ is the Hamming graph $H(n, q)$, the definition of coherent partition is exactly the same as the definition of r-partition design in [3]. The associate matrix is the transpose of the one considered by Camion *et al.*

A lot of coherent partitions are provided by the action of groups.

EXAMPLE 2.1 Let G be a group of automorphisms of Γ. The orbits of G are the cells of a coherent partition of Γ.

EXAMPLE 2.2 Let F be an alphabet of q elements and $\Gamma = H(n, q)$ the Hamming graph. Consider $x \in F^n$. Then $\{E_0, E_1, \ldots, E_n\}$ where $E_i = \{y \mid d_H(x, y) = i\}$ is a coherent partition of $H(n, q)$. It is easy to see that

$$
\sigma_{ij} = \begin{cases}
j & \text{if } i = j - 1; \\
j(q - 2) & \text{if } i = j; \\
(n - j)(q - 1) & \text{if } i = j + 1; \\
0 & \text{otherwise.}
\end{cases}
$$

We will consider farther the following example in conjunction with Example 1.2.

EXAMPLE 2.3 Let F be an abelian group (or a field) of q elements and C be a subgroup (or a subspace) of F^n. The cosets of C are the cells of a coherent partition of the Hamming graph $H(n, q)$.

Let $\mathrm{Coh}(\Gamma)$ be the set of coherent partition of Γ. This set, ordered by refinement ($\pi \le \pi'$ if the cells of π' are unions of cells of π), is a lattice. The supremum in this lattice is the same as in the lattice of partitions of the vertex set X. The discrete partition $\xi = \{\{x\} \mid x \in X\}$ is the least element and $\{X\}$ is the greatest element of this lattice. The lattice structure induces some relations between the associate matrices of comparable coherent partitions.

THEOREM 2.1 *Let* $\pi = \{E_0, \ldots, E_r\}$ *and* $\pi' = \{E'_0, \ldots, E'_s\}$ *be coherent partitions of* Γ *and* σ, σ' *their associate matrices. If* $\pi \le \pi'$, *consider the matrix* M *defined by* $M_{ij} = 1$ *if* $E_i \subseteq E'_j$ *and 0 otherwise. Then*

1. $M\sigma' = \sigma M$;

2. $\sigma' u = \lambda u \iff \sigma M u = \lambda M u$;

3. *the characteristic polynomial of* σ' *is a factor of the characteristic polynomial of* σ.

As a consequence of this theorem, since the discrete partition ξ is a coherent partition of associate matrix A, the next result is immediate.

THEOREM 2.2 *Let π be a coherent partition and σ its associate matrix. The eigenvalues of σ are contained in the set N of the eigenvalues of the adjacency matrix A. Futhermore σ is diagonalizable.*

We will say that the code Y *admits* the coherent partition π if Y is an union of cells of π. The next result shows that we can use the associate matrix of a coherent partition admitted by the code Y to construct the path matrix of Y.

THEOREM 2.3 *Let P be the path matrix of the code Y. Let $\pi = \{X_0, X_1, \ldots, X_r\}$ be a coherent partition admitted by Y and σ its associate matrix. Then,*

1. if x and y are in the same cell, $P(x, j) = P(y, j)$ for all j;

2. the numbers $P_{ij}^* = P(x, j)$ for $x \in X_i$, satisfy the recursive formula

$$P_{ij}^* = \sum_{k=0}^{r} \sigma_{ki} P_{k,j-1}^*; \tag{5}$$

3. the number of cells of π is greater than or equal to the number of distinct rows of P.

Using Theorems 1.1 and 2.3, we can relate the eigenvalues of the associate matrix of a coherent partition admitted by Y and the annihilator set of Y.

THEOREM 2.4 *Suppose that the code Y admits the coherent partition π. Let σ be the associate matrix of π. The elements of S the annihilator set of Y are eigenvalues of the matrix σ.*

If we consider the cells of a coherent partition as codes, we can see the next result.

THEOREM 2.5 ([2]) *Let π a coherent partition and σ its associate matrix. The set of eigenvalues of σ is the union of the annihilator sets of the cells of π.*

COROLLARY 2.5.1 *Let Y be a code of external distance s. Let π be a coherent partition admitted by Y. Suppose that π contains $s + 1$ cells. Then the annihilator set of all codes admitting π is contained in the annihilator set of Y.*

3   Coherence number

All codes of Γ admit the discrete partition $\xi = \{\{x\} \mid x \in X\}$. So the set of coherent partitions admitted by a code Y is nonvoid and there exists a unique coarsest coherent partition π_Y admitted by Y. We will call it the *associate coherent partition* of Y. If $r + 1$ is the number of cells in π_Y, r is the *coherence number* of Y.

We will define some other interesting partitions of the vertex set and relate them to coherent partitions.

Let $d(x, y)$ be the length of the shortest path joining x and y. For any code Y, set $d(x, Y) = \min_{y \in Y} d(x, y)$ and $\rho = \max_{x \in X} d(x, Y)$. The number ρ is the *covering radius* of Y. The *distance partition* relative to Y is the partition defined by

$$x \equiv y \iff d(x, Y) = d(y, Y).$$

This partition has $\rho + 1$ classes.

If P is the path matrix of Y, consider the partition of the vertex set of Γ defined by

$$x \equiv y \iff P(x,j) = P(y,j) \text{ for all } j.$$

This partition is the *path partition* relative to Y. It is a refinement of the distance partition relative to Y. If $\gamma + 1$ is the number of cells of the path partition relative to Y, γ is the *combinatorial number* associated with Y. By Theorem 1.1 the combinatorial number associated with Y is greater than or equal to the external distance of Y.

By Theorem 2.3, all the coherent partitions admitted by Y are finer than the path partition relative to Y. So the combinatorial number associated with Y is less than or equal to the coherence number of Y. Hence we have the sequence of inequalities

$$\rho \leq s \leq \gamma \leq r. \tag{6}$$

The following result was proved in Hamming graph by Camion *et al.* [3].

THEOREM 3.1 *Let Y be a code of Γ. Let s be its external distance and γ the combinatorial number associated with Y. If $s = \gamma$, then the path partition relative to Y is a coherent partition and s is the coherence number of Y.*

We call Y a *coherent code* if the distance partition relative to Y is the path partition relative to Y. Using Theorem 3.1 we have that the path partition is the associate coherent partition of Y. Hence, the inequalities of (6) become equalities. The coherent codes of the Hamming graph are the completely regular codes ([5]).

4 Cayley graphs

Let G be a group and Ω be a generating subset of G such that $1 \notin \Omega$ and $g \in \Omega$ implies $g^{-1} \in \Omega$. Let Γ be the Cayley graph $\Gamma(G, \Omega)$ *i.e.* the graph whose vertex set is G and $\{g, h\}$ is an edge if $gh^{-1} \in \Omega$. This code is regular of valency card Ω.

Consider the subset $H = \{1\}$. We have that its annihilator set is N the set of eigenvalues of the adjacency matrix A since $b_\alpha = J_\alpha(1,1)$ is the multiplicity of α as eigenvalue of A. So its external distance is $(\operatorname{card} N) - 1$. By Theorems 2.2 and 2.4, N is also the set of eigenvalues of the associate matrix of any coherent partition admitted by $\{1\}$ and the coherence number of $\{1\}$ is at least $(\operatorname{card} N) - 1$.

THEOREM 4.1 *Let $\pi = \{X_0, X_1, \ldots, X_r\}$ be a coherent partition admitted by $H = \{1\}$ and σ its associate matrix. These statements are equivalent:*

1. $r = (\operatorname{card} N) - 1$;

2. the multiplicities of the eigenvalues of σ are all 1;

3. the path matrix of $\{1\}$ has card N distinct rows.

If these assertions are satisfied, then

a) π is the path partition relative to $\{1\}$ and so is its associate coherent partition.

b) *the group G with the relations*

$$x \equiv_i y \iff xy^{-1} \in X_i, \qquad 0 \le i \le \text{card } N$$

is a symmetric association scheme.

Now, let the code H be a normal subgroup of G. As in Exemple 2.3, the cosets of H are the cells of a coherent partition. Suppose that H is such that each 2-subset of elements of H is at distance at least 3 in Γ. Hence all the entry of the associate matrix of the coset partition are 0 or 1. In fact this matrix is also the adjacency matrix of the Cayley graph $\Gamma(G/H, \Omega_H)$ where $\Omega_H = \{gH \mid g \in \Omega \setminus H\}$. We are in a situation similar to Examples 1.2 and 2.3. The dual distribution of H has a combinatorial signifiance.

THEOREM 4.2 *Let H be a normal subgroup of G and b its dual distribution. Then b_α is the multiplicity of α as eigenvalue of the associate matrix of the coset partition.*

All the coherent partitions admitted by H coarser than $\{gH \mid g \in G\}$ can be viewed as a coherent partition of $\Gamma(G/H, \Omega_H)$. In particular, the associate coherent partition π_H does. Hence, the coherence number of H in Γ is the same as the coherence number of $\{H\}$ in $\Gamma(G/H, \Omega_H)$.

By Theorem 4.2, the set of eigenvalues of the adjacency matrix of $\Gamma(G/H, \Omega_H)$ is the annihilator set of H (in Γ). The path matrix of H (in Γ) is the same as the path matrix of $\{H\}$ (in $\Gamma(G/H, \Omega_H)$). We can use these facts and Theorem 4.1 to obtain the next result.

THEOREM 4.3 *Let $\pi = \{X_0, X_1, \ldots, X_r\}$ be a coherent partition admitted by the normal subgroup H coarser than the coset partition. Let σ be the associate matrix of π. These statements are equivalent:*

1. *r is the external distance of H;*

2. *the multiplicities of the eigenvalues of σ are all 1;*

3. *the path matrix of H has $r + 1$ distinct rows.*

If these assertions are satisfied, then

a) *π is the path partition relative to H and so the associate coherent partition of H.*

b) *the group G/H with the relations*

$$xH \equiv_i yH \iff xy^{-1}H \subseteq X_i, \qquad 0 \le i \le r$$

is a symmetric association scheme.

If the hypothesis of Theorem 3.1 are satisfied, we have $s = \gamma = r$ where s is the external distance, γ is the combinatorial number and r is the coherence number. Furthermore the path partition is coherent. By Theorem 4.3 we have the next result which can be related to Theorem 6.11 of [5].

COROLLARY 4.3.1 *Let H be a normal subgroup of G. Let s be its external distance and γ the combinatorial number associated with Y in $\Gamma(G, \Omega)$. If $s = \gamma$, the group G/H with the relations*

$$xH \equiv_i yH \iff P(x,j) = P(y,j) \text{ for all } j, \qquad (0 \leq i \leq s)$$

is a symmetric association scheme.

If we have a code which is not a normal subgroup, we can't use the association scheme to investigate "regular" codes. May be the structure of coherent partition is a tool which can help in this case.

Notations

$\Gamma = (X, E)$: a regular connected graph without loops or multiple edges (p. 1)
A: the adjacency matrix of Γ (p. 1)
N: set of eigenvalues of A (p. 2)
J_α: projection matrix on the α-eigenspace of A (p. 2)
Y: code of Γ (p. 2)
b_α: (eq. 2)
S: annihilator set of Y (p. 2)
s: external distance (p. 2)
P: path matrix (p. 2)
π: coherent partition (p. 3)
σ: associate matrix of π (p. 3)
π_Y: associate coherent partition of Y (p. 4)
r: coherence number of Y (p. 4)
ρ: covering radius of Y (p. 4)
γ: combinatorial number associated with Y (p. 5)
$\Gamma(G, \Omega)$: Cayley graph on group G (p. 5)

References

[1] E. Bannai, and T. Ito. "Algebraic combinatorics I: Association schemes". Benjamin-Cummings, Menlo Park, 1984.

[2] P. Camion. (private communication).

[3] P. Camion, B. Courteau, and P. Delsarte. *On r-partition designs in Hamming spaces.* Tech. Rep. 626, INRIA, 1987. (submitted to Discrete Math. in 1986).

[4] P. Camion, B. Courteau, G. Fournier, and S.V. Kanetkar. *Weight distribution of translates of linear codes and generalized Pless identities.* J. Inform. Optim. Sci. **8** (1987), 1–23.

[5] P. Delsarte. *An algebraic approach to the association schemes of coding theory.* Philips Res. Repts Suppl. **10** (1973). (Ph.D. thesis, Université Catholique de Louvain, 1973).

[6] A. Montpetit. "Codes dans les graphes réguliers". Ph.D. thesis, Université de Sherbrooke, 1987.

[7] A. Montpetit. *Codes et partitions cohérentes dans les graphes réguliers.* Ann. sc. math. Québec **14** (1990).

[8] A. Montpetit. *Une généralisation d'un théorème de Delsarte dans les graphes de Cayley.* Ann. sc. math. Québec **15** (1991).

$\frac{1}{\lambda}$-regular and $\frac{1}{\lambda}$-homogeneous tournaments

Annie Astié-Vidal and Vincent Dugat

Laboratoire MLAD, Université Paul Sabatier,

118 route de Narbonne, 31062 TOULOUSE,

FRANCE

Abstract: *We define the notion of $\frac{1}{\lambda}$-regular and $\frac{1}{\lambda}$-homogeneous tournaments that subsumes the notions of homogeneous and near homogeneous tournaments. Then we give some examples of $\frac{1}{\lambda}$-homogeneous tournaments for various values of λ. Finally we give a construction method based on Galois fields.*

The definitions not given here can be found in [Berge 83], [Moon 68] and [Beineke Reid 78].

A *tournament* is a directed graph in which every pair of vertices is joined by exactly one arc. In the following, T =(X,U) will denote a tournament where X is the set of vertices, and U is the set of arcs The *score* s(x) of a vertex x, is the number of vertices dominated by x, and we denote by s(x,A), the number of vertices of A that are dominated by x. A tournament is *regular* if all vertices have equal scores. A tournament is called *rotational* if its vertices can be labelled: 1, 2, ..., n in such a way that, for some subset S of $\{1,..., n-1\}$, vertex i dominates vertex i+j (mod n) if and only if j $\in$ S. In this case, S is said to be the *symbol* of T. If x is any vertex, we let O(x) = {y / (x,y) $\in$ U} and I(x) = {z / (z,x) $\in$ U}. The *cyclone* is the rotational tournament with symbol $\{1, 2, 3, ..., \frac{n-1}{2}\}$. It is unique up to isomorphism for n fixed.

An *automorphism* of a tournament is a permutation of the vertices which preserves the dominance relation. The set of automorphisms of a tournament T forms a group, a(T), called the automorphism group of T. a(T) acts on the set of arcs U, and thus defines a partition of U into arcs orbits. Thus, two arcs (x,y) and (x',y') of U, belong to the same arc orbit iff $\exists \sigma \in$ a(T), $\sigma(x) = x'$ and $\sigma(y) = y'$.

A tournament T is said to be *vertex symmetric* if for every pair of vertices, x, y, there is an automorphism that sends x to y. The term *3-circuit* will denote a circuit of length 3. A tournament with n vertices is *transitive* if its score vector is (1, 2, 3, ..., n).

Anton Kotzig defined in [Kotzig 69] the notion of homogeneous tournaments: a tournament is homogeneous if all its arcs are contained in the same number of 3-circuits. He proved that such a tournament is regular of order n=4α+3, $\alpha \in$ N and that each arc is contained in α+1 3-circuits. Reid and Brown proved in [Reid Brown 72] that a tournament is homogeneous iff the subtournament induced by the successor set of all its vertices is regular. Later Claudette Tabib defined near-homogeneous tournament to be tournament of order n = 4α+1 having two classes of arcs: the class of arcs contained in α 3-circuits, and those of the arcs contained in α+1 3-circuits. The subtournament induced by the successor set of all the vertices of a near-homogeneous tournament is near-regular.

We extend here these notion in defining $\frac{1}{\lambda}$-regular and $\frac{1}{\lambda}$-homogeneous tournaments. We expose the relations between the two notions and establish some results. Furthermore we define a construction method based on the cyclotomic tournaments and give an algorithm.

I $\frac{1}{\lambda}$-regular tournaments

We define the notion of $\frac{1}{\lambda}$-regularity for a tournament:

Definition I.1:
Let λ be a positive integer, a tournament T = (X, U) of order n = λ.α with α ∈ N is said to be

$\frac{1}{\lambda}$-regular if its set of vertices X can be partitioned in λ classes $X_1,....X_\lambda$ of consecutive scores such

as: $|X_i| = \alpha$, i = 1,..., λ.

Then we prove the following results :

Proposition I.2:
The different scores of a $\frac{1}{\lambda}$-regular tournament T = (X, U) of order n = λ.α with α ∈ N are:

$$s_1 = \frac{1}{2}\lambda(\alpha-1),\ s_2 = \frac{1}{2}\lambda(\alpha-1)+1,...,\ s_i = \frac{1}{2}\lambda(\alpha-1)+i-1,\ ...,\ s_\lambda = \frac{1}{2}\lambda(\alpha-1)+\lambda-1 = \frac{1}{2}\lambda(\alpha+1)-1.$$

<u>proof:</u>
Let $s_1, s_2, ..., s_\lambda$ be the different scores of the tournament labelled in non decreasing order. So s_1 is the smallest of them. Since they are all consecutive we have : $s_2 = s_1 + 1, ..., s_\lambda = s_1 +\lambda-1$. Moreover in any tournament T = (X, U) we have [Moon 68] : $\sum_{x\in X} s(x) = \frac{1}{2}n.(n-1)$. Hence here we have:

(1) $\alpha.(s_1+s_1+1+...+s_1+\lambda-1) = \frac{1}{2}\lambda\alpha.(\lambda.\alpha-1)$, since all classes of scores have the same cardinality.

(1) $\Rightarrow \lambda.s_1 + \sum_{i=1}^{\lambda-1} i = \frac{1}{2}\lambda.(\lambda.\alpha-1) \Rightarrow \lambda.s_1 + \frac{1}{2}\lambda.(\lambda-1) = \frac{1}{2}\lambda.(\lambda.\alpha-1) \Rightarrow s_1 = \frac{1}{2}\lambda.(\alpha-1),$

Thus : $s_i = \frac{1}{2}\lambda.(\alpha-1) + i-1 ; 1 \le i \le \lambda.\square$.

Corollary I.3:
If T = (X, U) of order n=λ.α with α ∈ N is a $\frac{1}{\lambda}$-regular tournament then we have that λ(α-1) is even.

Remark I.4:
   A transitive tournament is $\frac{1}{n}$-regular.

proof:
A transitive tournament can be seen as a $\frac{1}{n}$-regular tournament where each class of score has only one element.□.

Remark I.5:
   The notion of $\frac{1}{\lambda}$-regularity subsume those of regularity (with $\lambda=1$) and near-regularity (with $\lambda=2$).

Example I.6:
Let $\lambda=3$, $\alpha=3$ and $n=9$. The following tournaments are $\frac{1}{3}$-regular and non isomorphic :

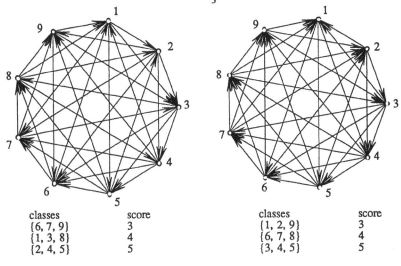

classes	score
{6, 7, 9}	3
{1, 3, 8}	4
{2, 4, 5}	5

classes	score
{1, 2, 9}	3
{6, 7, 8}	4
{3, 4, 5}	5

II $\frac{1}{\lambda}$-homogeneous tournaments :

II.1 Preliminaries:

   In [Dugat 90] is defined a criterion to decompose regular tournaments in the aim to use it in an isomorphism test algorithm. In this method each vertex is labelled with two lists which characterize the tournament. First let us present the unavoidable definitions.

Definition II.1.1:
   The weight $w(x,y)$ of any arc (x,y) of a regular tournament T of order n is the number of 3-circuits containing the arc (x,y).

This gives us a classification of the arcs, but it is better to classify the vertices, so we now create two lists for each vertex x of T:

the In-Weight list of x denoted IW(x): giving the weight of all in going arcs of x,

the Out-Weight list of x denoted OW(x): giving the weight of all out going arcs of x.

These lists are ordered in non decreasing order.

Remark II.1.2:

The regularity of the tournament implies that for any vertex x we have $|IW(x)| = |OW(x)| = \frac{n-1}{2}$

Proposition II.1.3: [Dugat 90]

For any vertex x we have: $\sum_{y \in O(x)} w(x,y) = \frac{1}{8}(n-1)(n+1)$.

Corollary II.1.4: [Dugat 90]

For any arc u of a regular tournament, $1 \leq w(u) \leq \frac{n-1}{2}$.

Definition II.1.5:

Let T = (X, U) be a regular tournament. A W-class C is a subset of the set of vertices such that: x and x' $\in$ C iff IW(x) = IW(x') and OW(x) = OW(x').

Definition II.1.6:

The W-classification of a regular tournament T = (X, U) is the partition of X in W-classes.

Proposition II.1.7:

If T is a vertex-symmetric tournament, all the vertices have the same list as IW-list and as OW-list, so T has only one W-class.

The converse is not true. There exists non vertex-symmetric tournaments that have only one W-class.

By using this criterion we can rewrite the definitions of homogeneous and near-homogeneous tournaments:

Definition II.1.8:

A tournament T = (X, U) is homogeneous iff $\forall$ x $\in$ X, IW(x) = OW(x) = (w, ..., w) where $w = \frac{n+1}{4}$ and $|IW(x)| = |OW(x)| = \frac{n-1}{2}$.

Definition II.1.9:

A tournament T = (X, U) is near-homogeneous iff $\forall$ x $\in$ X, IW(x)=OW(x)=(w, ...,w, w+1, ...,w+1) where $w = \frac{n-1}{4}$, $|IW(x)| = |OW(x)| = \frac{n-1}{2}$ and there is $\frac{n-1}{4}$ terms of value w and $\frac{n-1}{4}$ terms of value w+1.

Our aim now is to generalise these notions.

II.2 New definitions and results

Definition II.2.1:

A tournament $T = (X, U)$ is $\frac{1}{\lambda}$-homogeneous iff :

$\forall\, x \in X$, $IW(x) = OW(x) = (w, ..., w, w+1,..., w+1, ..., w+\lambda-1, ..., w+\lambda-1)$, $|IW(x)| = |OW(x)|$ $= \frac{n-1}{2}$ and there is $\frac{n-1}{2\lambda}$ terms of value w, $\frac{n-1}{2\lambda}$ terms of value w+1, ..., $\frac{n-1}{2\lambda}$ terms of value

$w+\lambda-1$.

Remark II.2.2:

A $\frac{1}{\lambda}$-homogeneous tournament is regular since each IW-list (or OW-list) has a constant length of $\frac{n-1}{2}$.

Proposition II.2.3

With the notations of Definition II.2.1 we have : $w = \dfrac{n+3-2\lambda}{4}$

proof:

Let us calculate the total number C_3 of 3-circuits of a $\frac{1}{\lambda}$-homogeneous tournament $T = (X, U)$. We have :

$$C_3 = \frac{1}{3}n.\frac{n-1}{2\lambda}\,(w+(w+1)+...+(w+(\lambda-1))) = \frac{1}{3}n.\frac{n-1}{2\lambda}\,(\lambda.w+\frac{\lambda(\lambda-1)}{2}) = \frac{1}{12}n.(n-1).(2w+\lambda-1)$$

On the other hand, we know ([Moon 68], [Beineke Reid 78]) that the total number of 3-circuits of a regular tournament is : $C_3 = \frac{1}{24}n.(n-1).(n+1)$. So by identification of the two results we have :

$2w+\lambda-1 = \dfrac{n+1}{2} \Rightarrow w = \dfrac{n+3-2\lambda}{4}$ $\square$.

Proposition II.2.4:

If $T = (X, U)$ is $\frac{1}{\lambda}$-homogeneous, U can be partitioned in λ classes of cardinality $\dfrac{n(n-1)}{2\lambda}$

$U_1, U_2, ..., U_\lambda$ such that $w(u) = \dfrac{n-2\lambda-1}{4}+i$, $\forall u \in U_i$; $i = 1,..., \lambda$.

Proposition II.2.5:

If $T = (X, U)$ is a $\frac{1}{\lambda}$-homogeneous tournament of order n then we have:

(i) $n \equiv 1+2\lambda$ (mod. 4) and $\exists\, k \in N$ such that $w(u) = k+i$, $\forall u \in U_i$; $1 \leq i \leq \lambda$,

(ii) $n=2\lambda\alpha+1$ with $\alpha \in N$, and $w(u)=\frac{1}{2}\lambda(\alpha-1)+i$, $\forall u \in U_i$; $1 \leq i \leq \lambda$.

proof:

(i) Let us call k the quantity $\dfrac{n-2\lambda-1}{4}$. We have $n = 4k+2\lambda+1$, so $n \equiv 2\lambda+1$ (mod. 4). Moreover if $u \in U_i$ we have $w(u) = k+i$ according to definition II.2.4,

(ii) We saw in Definition II.2.1 that each weight was present $\dfrac{n-1}{2\lambda}$ times in the OW-lists (or IW-lists) of a

$\frac{1}{\lambda}$-homogeneous tournament. This quantity must of course be an integer, so we have: $\frac{n-1}{2\lambda} = \alpha$, with

$\alpha \in \mathbb{N}$. Hence $n = 2\lambda\alpha+1$. On the other hand, $n = 4k+2\lambda+1$, $k \in \mathbb{N}$ implies that $2\lambda\alpha+1 = 4k+2\lambda+1$, that is to say: $k = \frac{1}{2}\lambda(\alpha-1)$, the result follows.$\square$

Corollary II.2.6:

λ even implies $n \equiv 1 \pmod{4}$

λ odd implies $n \equiv 3 \pmod{4}$

Theorem II.2.7:

Let $T=(X,U)$ be a tournament of order $n=2\lambda\alpha+1$. T is $\frac{1}{\lambda}$-homogeneous iff it is regular and $\forall\, x \in X$ we have $O(x)$ is $\frac{1}{\lambda}$-regular.

proof:

a)$\Rightarrow$

Let $T = (X, U)$ be a $\frac{1}{\lambda}$-homogeneous tournament, let x be any vertex and y be a successor of x, we have :

$$s(y, O(x)) = s(y, X) - w(x,y),$$

since T is regular $\forall\, x \in X$, $s(x,X) = \frac{n-1}{2} = \lambda\alpha$, so :

$$s(y, O(x)) = \lambda\alpha - w(x,y), \text{ with } w(x,y) = \frac{1}{2}\lambda(\alpha-1)+j \; ; \; 1 \le j \le \lambda.$$

$$s(y, O(x)) = \lambda\alpha - (\frac{1}{2}\lambda(\alpha-1)+ j) \; ; j = 1, ..., \lambda$$

$$s(y, O(x)) = \frac{1}{2}\lambda(\alpha+1) - j \; ; j = 1, ..., \lambda$$

Finally: $s(y,O(x)) = \frac{1}{2}\lambda(\alpha-1), ..., \frac{1}{2}\lambda(\alpha-1)+\lambda-1.$

This corresponds to the scores of a $\frac{1}{\lambda}$-regular tournament.

b)$\Leftarrow$

Let $T = (X, U)$ be a regular tournament such that : $\forall\, x \in X$, $O(x)$ is $\frac{1}{\lambda}$-regular. Now let (x, y) be an arc of U. We have : $w(x,y) = s(y, X) - s(y, O(x)) = \frac{n-1}{2} - s(y, O(x))$. $s(y, O(x))$ can take λ different and consecutive values since $O(x)$ is $\frac{1}{\lambda}$-regular, thus $w(x,y)$ can take λ consecutive values and T is $\frac{1}{\lambda}$-homogeneous.$\square$

Definition II.2.8:

The complementary $\overline{T}$ of a tournament T is the tournament obtained by inverting all the arcs of T.

Remark II.2.9:

If T is $\frac{1}{\lambda}$-homogeneous then $\overline{T}$ is $\frac{1}{\lambda}$-homogeneous.

proof:

When we inverse the arcs of T, a 3-circuit is transformed in a 3-circuit. Thus the weight w(u) of any arc of T is unchanged. The IW and OW lists of $\overline{T}$ are the same than those of T, so $\overline{T}$ is $\frac{1}{\lambda}$-homogeneous.$\square$

Corollary II.2.10:

Let T= (X, U) be a $\frac{1}{\lambda}$-homogeneous tournament with n=2λα+1 vertices the following points are

equivalent :   (1) O(x) is $\frac{1}{\lambda}$-regular, $\forall$ x ∈ X,

(2) I(x) is $\frac{1}{\lambda}$-regular, $\forall$ x ∈ X.

Remark II.2.11:

The notion of $\frac{1}{\lambda}$-homogeneity subsumes those of homogeneity (with λ=1), and near-homogeneity

(with λ=2).

These definitions and results rise a question: do $\frac{1}{\lambda}$-homogeneous tournaments exist and can we construct

some of them ?

II.3 Construction of $\frac{1}{\lambda}$-homogeneous tournaments:

We can first precise some aspects of the structure of these graphs:

Proposition II.3.1:

If T = (X, U) is a $\frac{1}{\lambda}$-homogeneous tournament with n vertices then λ divides $\frac{n-1}{2}$.

proof: see Prop. II.2.5.

The existence of 1-homogeneous (or simply homogeneous) tournament is studied in [Reid Brown 72],
[Tabib 84].

Corollary II.3.2:

With the same notations, if $\frac{n-1}{2}$ is prime then the only possible values of λ are 1 and $\frac{n-1}{2}$.

Remark II.3.4:

There always exists a $\frac{1}{\frac{n-1}{2}}$-homogeneous tournament of order n, and this tournament is the cyclone.

Proposition II.3.5:

There is exactly two non isomorphic $\frac{1}{2}$-homogeneous (near-homogeneous) tournaments of order 9.

proof:

The figure 2 shows these two graphs. The tournament nh1 is rotational and is presented in [Tabib 80] (see
also [Astié Dugat 90a] and [Astié Dugat 90b]). The tournament nh2 is not rotational neither vertex-
symmetric. All the non isomorphic regular tournaments with 9 vertices have been studied in [Astié Dugat
90b]. Among the 15 such tournaments, only 2 are near-homogeneous.□

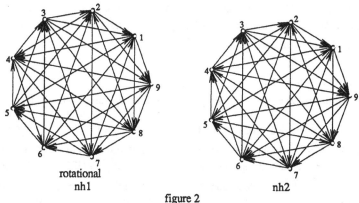

rotational
nh1   nh2

figure 2

We look for $\frac{1}{\lambda}$-homogeneous tournaments of order 17, 19 and 25. Our investigation method is based on the

inversion of the direction of hamiltonian circuits in the cyclone of order n. It is clear that this method limits our research of $\frac{1}{\lambda}$-homogeneous tournaments to the rotational (thus vertex-symmetric) tournaments.

$\underline{n = 17:}$ we have $\frac{n-1}{2} = 8$, so the possible values of λ are: 1, 2, 4, 8.

$\lambda = 1$: there is no homogeneous rotational tournament of order 17.

$\lambda = 2$: C. Tabib in [Tabib 80] says that there does not exist any near-homogeneous (ie $\frac{1}{2}$-homogeneous)

rotational tournament of order 17.

$\lambda = 4$: we have found one $\frac{1}{4}$-homogeneous tournament of order 17. The IW and OW lists of each vertex

are (3,3, 4,4, 5,5, 6,6)

$\lambda = 8$: the cyclone of order 17 is $\frac{1}{8}$-homogeneous.

$\underline{n = 19:}$ we have $\frac{n-1}{2} = 9$, so the possible values of λ are: 1, 3, 9.

$\lambda = 1$: there is a rotational homogeneous tournament with 19 vertices. It is the quadratic residue tournament with 19 vertices.

$\lambda = 3$: we have found a $\frac{1}{3}$-homogeneous tournament with 19 vertices:

$\lambda = 9$: corresponds to the cyclone.

$\underline{n = 25:}$ we have $\frac{n-1}{2} = 12$, so the possible values of λ are: 1, 2, 3, 4, 6, 12. According Corollary II.2.6

since $25 \equiv 1 \pmod{4}$ λ must be even, so there is no homogeneous neither $\frac{1}{3}$-homogeneous rotational

tournaments with 25 vertices.

$\lambda = 2$: we ignore if there are near-homogeneous rotational tournaments,

$\lambda = 4, 6, 12$: we have found $\frac{1}{\lambda}$-homogeneous tournaments for these values of λ.

We now present a method that allows the construction of $\frac{1}{\lambda}$-homogeneous tournaments by using Galois fields and results of [Astié 75] and [Astié Dugat 90a]. These tournaments are rotational. In the following, we call the *dyadic valuation* of an integer α, denoted $v(\alpha)$, the greatest power of 2 that divides α. That is to say: $\alpha = 2^{v(\alpha)}.\beta$ where β is an odd integer.

Definition II.3.6: cyclotomic tournaments [Astié 75].

Let p be a prime ($p \neq 2$) and H be the subgroup of order h such that $p^m-1 = 2^{v(p^m-1)}.h$ (where h is odd), of the multiplicative group of the field $GF^*(p^m)$: $H = \{a^{2^{v(p^m-1)}}$; $a \in GF^*(p^m)\}$. The decomposition of the multiplicative group of the field with respect to its subgroup H has $2^{v(p^m-1)}$ cosets. Moreover, the oddness of h implies that $x.H \neq -x.H$, $\forall x \in GF^*(p^m)$. Now let us define k as $2.k = 2^{v(p^m-1)}$, the decomposition of $GF^*(p^m)$ as the union of its cosets is: $GF^*(p^m) = x_1.H \cup (-x_1).H \cup ... \ x_k.H \cup (-x_k).H$. The cyclotomic tournaments are then defined by $T = (X,U)$ with $X=GF(p^m)$ and $(x,y) \in U \Leftrightarrow y-x \in S := \bigcup_{i=1}^{k} \varepsilon_i.x_i.H$, $\varepsilon_i = \pm 1$, $i=1,...,k$. There are 2^k choices for S, and they do not necessarily give isomorphic tournaments.

The following Remark contains some results proved in [Astié 75]

Remark II.3.7:

The cyclotomic tournaments are vertex-symmetric tournaments, and if $m \neq 2$ or $p \neq 2^s-1$, $s \in N$ and $s>1$, a cyclotomic tournament of order p^m has $k(T) = 2^{v(p^m-1)-1}$ arc orbits.

As we have done in [Astié Dugat 90a] with near-homogeneous tournaments, we can use cyclotomic tournaments to construct $\frac{1}{\lambda}$-homogeneous tournaments.

Remark II.3.8:

Let $T=(X, U)$ be a $\frac{1}{\lambda}$-homogeneous tournament with k arc-orbits, we have: $k \geq \lambda$. Clearly each arc-orbit is included in a class of U_i (defined in proposition II.2.5).

Cyclotomic $\frac{1}{\lambda}$-homogeneous tournaments:

We are now going to search $\frac{1}{\lambda}$-homogeneous cyclotomic tournaments with exactly λ arc-orbits (that is $k=\lambda$). In this case we have $n = p^m = 2k.h +1$; $h \in N$ and h is odd, so we can find a $\beta \in N$ such that $h=2\beta+1$, and we can write $n = p^m = 2\lambda(2\beta+1)+1$, $\beta \in N$.

Theorem II.3.9:

A rotational tournament $T_n = (X, U)$ of order n is $\frac{1}{\lambda}$-homogeneous iff the subtournament induced by S is $\frac{1}{\lambda}$-regular.

proof:
From the theorem II.2.7 and since T_n is vertex-symmetric it suffices to verify on a unique vertex $x \in X$ that

$O(x)$ is $\frac{1}{\lambda}$-regular. We take $x = n$, that is $O(n) = S.\square$

For instance:

if $\underline{k = \lambda = 2}$ we have $p^m = 4(2\beta+1)+1 = 8\beta+5$ ie $p^m \equiv 5$ (mod. 8) $\Rightarrow p^m \equiv 1$ (mod. 4) and the tournament is near-homogeneous (see [Astié Dugat 90a] where we have found 6 new near-homogeneous cyclotomic tounaments for orders $p = 29, 53, 173, 229, 293, 733$).

If $\underline{k = \lambda = 3}$ we have $p^m = 6(2\beta+1)+1 = 12\beta+7$, so if $\beta = 0$ then $p^m = 7$ and we have the cyclone with 7 vertices. If $\beta = 1$, $p^m = 19$ and we must verify if this cyclotomic tournament with 19 vertices and 3 arc-orbits is $\frac{1}{3}$-homogeneous (see below).

If $\underline{k = \lambda = 4}$ we have $p^m = 8(2\beta+1)+1 = 16\beta+9$. If $\beta = 0$ we have $p^m = 9$, and the tournament is the cyclone of order 9.

A detailed construction:

Let's show the construction of the cyclotomic tournament with 19 vertices and 3 arc-orbits. We have $H = \{x^6 ; x \in GF^*(19)\} = \{x^6 \pmod{19}\}$ since 19 is prime and $GF^*(19) = \{1, 2, ..., 19\}$. Thus $H = \{1, 7, 11\}$. $GF^*(19) = x_1.H \cup (-x_1).H \cup x_2.H \cup (-x_2).H \cup x_3.H \cup (-x_3).H$. We have $2^3 = 8$ possible choices for S. Let's choice $S = H \cup x_2.H \cup x_3.H$. The values of x_2 and x_3 remain to determine: $x_2 = 2$ and $x_3 = 4$ are convenient, so $S = \{1,2,3,4,6,7,9,11,14\}$. The cyclotomic tournament is $T = (X, U)$ with $X = GF(19)$ and $(x,y) \in U$ iff $y-x \in S$. The question is now: is this tournament $\frac{1}{3}$-homogeneous ? The subtournament induced by $S = O(19)$ is $\frac{1}{3}$-regular as we can see in figure 3, so T is $\frac{1}{3}$-homogeneous. This tournament is not isomorphic to the one previously found, so there are at least two non isomorphic $\frac{1}{3}$-homogeneous tournaments with 19 vertices.

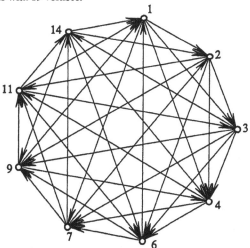

The vertices 4, 6, 9 have a score of 3 ; 1, 7, 11 a score of 4 ; and 2, 3, 14 a score of 5.

figure 3

References :

[Astié 75]   A. ASTIE-VIDAL, "Vertex-Symmetric Tournaments of Order n with the Minimum Number of Arc Orbits", *Recent Advances in Graph Theory* (ed Fielder) Academia, Prague (1975), p17-30.

[Astié Dugat 90a] A. ASTIE-VIDAL and V. DUGAT, "Near-homogeneous tournaments and permutation group", to appear in *Discrete Mathematics*.

[Astié Dugat 90b] A. ASTIE-VIDAL and V. DUGAT, "Automatic construction of non vertex-symmetric regular tournaments", to appear in the proceedings of the conference *Combinatorics'90*, may 1990, Gaeta (Italy).

[Beineke Reid 78] L.W. BEINEKE and K. B. REID, *Tournaments*, in "Selected Topics in Graph Theory", (L.W. Beineke and R.J. Wilson Eds) chap 7, p385-415 Academic Press, London (1978).

[Berge 83]   C. BERGE, "Graphes", 3rd Ed., (Bordas, Paris, 1983).

[Dugat 90]   V. DUGAT, "A polynomial decomposition of regular tournaments and application to isomorphism test", submitted to *Journal of Algorithms*.

[Kotzig 69]   Anton KOTZIG, "Sur les tournois avec des 3-cycles régulièrement placés", *Mat. Casopis* 19, (1969), p126-134.

[Moon 68]   John W. MOON, "Topics on tournaments", Holt, Rinehart and Wilson, New York, 1968.

[Reid Brown 72]   K.B. REID and E. BROWN, "Doubly regular tournaments are equivalent to skew Hadamard matrices", *Journal of Combinatorial Theory* 12, (1972), p332-338.

[Tabib 80]   C. TABIB, "Caractérisation des tournois presqu'homogènes", *Annals of Discrete Mathematics* 8, (1980), p77-82.

[Tabib 84]   C. TABIB, "Pluralité des tournois homogènes de même ordre", *Annales Scientifiques du Québec* 8, (1984), n°1, p81-95.

SECTION 3

GEOMETRIC CODES

DECODING OF CODES ON HYPERELLIPTIC CURVES

Dominique LE BRIGAND

Université Pierre et Marie Curie ; 45-55, 5-ième étage

4 place Jussieu 75250 PARIS Cedex 05

ABSTRACT: In 1989, R. PELLIKAAN gave an algorithm which decodes geometric codes up to $\lfloor(\frac{d^*-1}{2})\rfloor$-errors, where d^* is the designed distance of the code. Unfortunately this algorithm is not completely effective. I present facts about the jacobian of a hyperelliptic curve which permits in some cases to perform the algorithm.

RESUME: En 1989, R. PELLIKAAN a donné un algorithme qui permet de décoder des codes géométriques en corrigeant jusqu'à $\lfloor(\frac{d^*-1}{2})\rfloor$-erreurs, si d^* désigne la distance prévue du code. Malheureusement cet algorithme n'est pas totalement effectif. Je présente des résultats sur la jacobienne des courbes hyperelliptiques qui permettent dans certains cas de décoder effectivement.

KEY WORDS: Geometric code, Decoding, Hyperelliptic Curve, Jacobian.

INTRODUCTION:

Pellikaan's algorithm [P] applies a number of times in parallel Skorobogatov and Vladuts' basic decoding algorithm [S-V]. To perform the algorithm, we have to find some integer s and s-uple of divisors and this not easy in practice except for codes on an elliptic curve. In this paper we answer that question for a particular hyperelliptic curve and point out some properties of the jacobian which will be useful to solve the problem for any other hyperelliptic curve.

1-EXISTENCE OF THE ALGORITHM:

Let X be a smooth projective curve of genus g defined and absolutely irreducible on $K = GF(q)$; let $X(K)$ be the set of K-rational points of X and let p and D be two divisors on X such that:

$$p = P_1 + \ldots + P_n$$

where $P_1, \ldots, P_n$ are n distinct points of X(K)

$$D , \deg D = a, \ 4g - 1 \leq a \leq n + 2g - 2$$

D and p have disjoint supports.

We consider the Code $C = C_\Omega(X, p, D) = \text{Im } res_p$, where

$$res_p : \Omega(p - D) \longrightarrow K^n, \quad res_p(\omega) = \Big(res(P_1, \omega), \ldots, res(P_n, \omega)\Big);$$

the minimum distance of the Code is such that: $d \geq d^* = a - 2g + 2$; d^* is the designed distance; let: $t^* = \lfloor(\frac{d^*-1}{2})\rfloor$. Let $\mathbf{D}_k$ denote the set of effective divisors of degree k; if s is an integer $s > 1$, we consider the map:

$$\Psi_k^s : \mathbf{D}_k^s \longrightarrow \text{Jac}(X)^{s-1}; \quad \Psi_k^s(F_1, \ldots, F_s) = \Big([F_1 - F_2], \ldots, [F_{s-1} - F_s]\Big)$$

where $\text{Jac}(X)$ is the jacobian of the curve X, that is the set of classes of zero degree divisors modulo principal divisors, and $[\ ]$ is the class of a divisor modulo principal divisors. Finally let $m = g - 2$ in case of even a (resp: $m = g - 1$ in case of odd a).

THEOREM [P]: *If $q \geq 3$ and s is an integer, $s > 1$, such that the map Ψ_m^s is not surjective and if $\underline{F} = (F_1, \ldots, F_s)$ is an element of $\mathbf{D}_{g+t^*}^s$ such that $\Psi_{g+t^*}^s(\underline{F})$ is not in the image of Ψ_m^s then there exists an algorithm of complexity $O(n^4)$ in time and $O(n^3)$ in space which decodes up to t^*-errors.*

R.Pellikaan[P] and S.G. Vladuts[V] give conditions on q and on the curve for the existence of an integer s such that Ψ_m^s is not surjective. Then the problem is to find the s-uple $\underline{F}$.

In fact to perform the algorithm we need a s-uple $\underline{G}$ such that for any i, $i = 1, \ldots, s$, the supports of G_i and p are disjoint, but it is not necessary for the G_i's to be effective: the important fact is that $\Big([G_1 - G_2], \ldots, [G_{s-1} - G_s]\Big)$ doesn't belong to the image of Ψ_m^s and then the conclusion of Proposition 5-[P] is true; further, proposition 3 of [P] is true if $a \leq n + 2g - 2$ even if the divisor isn't effective.

2-ELLIPTIC CURVES ($g = 1$):

Let X be an elliptic curve and $C = C_\Omega(X, p, D)$ a geometric code. If $D = aP$, where a is even and P is a rational point of X not in the support of p, then Skorobogatov and Vladuts [S-V] showed that their basic decoding algorithm, using just one divisor $F = bP$ where $b = \frac{a}{2}$, performs a complete decoding.

More generally, if we suppose that the curve X has more than one rational point on K, then the map:

$$\Psi_0^2 : \mathbf{D}_0^2 \longrightarrow \text{Jac}(X)$$

is not surjective as $\text{Jac}(X)$ is isomorphic to $X(K)$ and $\mathbf{D}_0^2 = \{0\}$. So, to perform the algorithm, we choose two divisors F_1 et F_2 of degree $t^* + 1$, the supports of which are disjoint from the support of p and such that $[F_1]$ is not equal to $[F_2]$.

3-HYPERELLIPTIC CURVES:

Let Y be a plane curve with affine equation:

$$v^2 + h(u) \times v = f(u)$$

such that h and f are polynomials of $K[u]$, $deg(h) \leq g$ where g is an integer, $g > 1$, f is a monic polynomial of degree $2g + 1$, finally Y has no singular point in its affine part.

The curve Y has a unique point P_∞ at the infinity which is a singular point of multiplicity $2g - 1$; if we denote by (x, y, z) the projective coordinates, then:

$$P_\infty = (0, 1, 0).$$

Let $P = (x, y, 1)$ be an affine point of the curve the coordinates of which are in any extension of K, then we denote by $\tilde{P} = (x, -y - h(x), 1)$ the other point of Y with the same x-coordinate; if $P = P_\infty$ then $\tilde{P} = P_\infty$.

A smooth projective model X of Y is called a **hyperelliptic curve of genus g**. Let :

$$\Pi : X \longrightarrow Y$$

be the normalization morphism, then the restriction of Π to $\Pi^{-1}(Y - \{P_\infty\})$ is an isomorphism and $\Pi^{-1}(P_\infty)$ consists of a single point we still denote by P_∞. Because of that, we will make no difference between the points of X and those of Y in the rest of this paper.

If $P = (\alpha, \beta, 1)$ is an affine point of X, then the divisor:

$$P + \tilde{P} - 2P_\infty \text{ is equivalent to } 0$$

as it is the divisor of the function $\frac{(x-\alpha z)}{z}$; so P_∞ is a hyperelliptic point (see [F]). Further, this result implies that every zero-degree divisor of X is equivalent to a divisor F:

$$F = \sum_i n_i P_i - \left(\sum_i n_i\right) P_\infty$$

such that the points P_i are affine points of X and if P_i occurs in the support of F, then $\tilde{P}_i$ doesn't occur unless $\tilde{P}_i = P_i$ but then $n_i = 1$. Such a divisor is called **quasi-reduced**. If F is quasi-reduced and if:

$$F = \sum_i n_i P_i - \left(\sum_i n_i\right) P_\infty \text{ where } \sum_i n_i \leq g$$

then F is called **reduced**.

LEMMA 1: *A non zero quasi-reduced divisor isn't principal.*

Proof: Since P_∞ is a hyperelliptic point, we have:

$$\dim L(2lP_\infty) = \dim L((2l+1)P_\infty) = l+1 \text{ for } 0 \le 2l \le (2g-2).$$

This implies that a rational divisor $F = D_0 - kP_\infty$, $k \le (2g-1)$, isn't principal if k is odd since there is no function in $K(X)$ which has exactly a pole of odd order $k \le 2g-1$ in P_∞. Further, by Riemann-Roch theorem we have:

$$\dim L(kP_\infty) = k - g + 1 \text{ for } k \ge (2g-1).$$

Suppose F is a non zero quasi-reduced divisor:

$$F = \sum_{i=1}^{k} P_i - kP_\infty$$

where the points $P_i = (x_i, y_i, 1)$ are distinct or not and their coordinates belong to a same finite extension L of K. The preceding assertions concerning the dimension of $L(kP_\infty)$ is true on any extension L of the base field. We set:

$$\dim L(kP_\infty) = r + 1$$

a L-basis of $L(kP_\infty)$ is:

$$1, f_1, \ldots, f_r \text{ where } f_i = \prod_{j=1}^{i} \frac{(x - x_j z)}{z}$$

and we have seen that:

$$(f_i) = \sum_{j=1}^{i} (P_j + \widetilde{P_j}) - 2iP_\infty.$$

So, if F was the divisor of a function f, then f would be equal to:

$$f = \lambda_0 1 + \lambda_1 f_1 + \ldots + \lambda_r f_r.$$

Since P_1 is a zero of f, $\lambda_0 = 0$, then $\widetilde{P_1}$ is a zero of f contradicting the fact that F is quasi-reduced.

Cantor [C] gives the following lemma concerning the jacobian of a hyperelliptic curve X:

LEMMA 2: *In each class of the jacobian of a hyperelliptic curve there is a unique representative F in a reduced form.*

In fact Cantor assumes that the characteristic of K is not equal to 2 but it can easily be proved that the result is available on any finite field (see [K]).

Proof of Lemma 2: let $[F]$ be an element of the Jacobian of X, $\deg F = 0$; we show first that we can find a representative of $[F]$ in the form $D_0 - gP_\infty$, where D_0 is an effective divisor. Using Riemann-Roch theorem we have that:

$$\dim_K L(F + gP_\infty) \geq 1$$

so let f be a non zero element of $L(F + gP_\infty)$, then:

$$D_0 = (f) + F + gP_\infty$$

is an effective divisor and $D_0 - gP_\infty$ is equivalent to F so it answers the question (this result is true for any curve). Then we can find a reduced divisor equivalent to $D_0 - gP_\infty$.

It remains to prove that every reduced rational divisor is a representative of a class of the jacobian and that such a representative is unique. This is a consequence of Lemma 1.

We use the preceding description of the jacobian of a hyperelliptic curve in the following example.

4-EXAMPLE:

Let $q = 16$, $K = GF(q)$, X the hyperelliptic curve with affine equation:

$$v^2 + v = u^5$$

the curve X is maximal according to the Weil-upper-bound: its genus g is equal to 2 and it has 33 rational points on K so:

$$\text{card} D_1 = 33 \text{ and card } \text{Jac}(X) = (1 + \sqrt{q})^{2g} = 5^4$$

Each class of $\text{Jac}(X)$ has a unique representative in a reduced form:

$$F = \sum_i n_i P_i - \left(\sum_i n_i\right) P_\infty \text{ where } \sum_i n_i \leq 2$$

Let $C = C_\Omega(X, p, D)$ be a geometric code and $k = t^* + 2$:

4-1: if a is even the map:

$$\Psi_0^2 : D_0^2 \longrightarrow \text{Jac}(X)$$

is not surjective, since $D_0 = \{0\}$; so if there is two k-degree divisors G_1 and G_2 the supports of which are disjoints from the support of p and such that $[G_1]$ is not equal to $[G_2]$ then the algorithm will correct up to t^*-errors. Let λ be an element in $GF(q^k)$ not in K and such that:

$$tr_k(\lambda^5) = 0$$

where tr_k denote the map trace of $GF(q^k)$ on $GF(2)$; let Q_1 be one of the two points of the curve X (the other is $\widetilde{Q_1}$) such that: $Q_1 = (\lambda, y, 1)$; the point Q_1 has k conjugates

$Q_1, \ldots, Q_k$ by the Frobenius map. We set:

$$G_1 = \sum_{i=1}^{k} Q_i$$

Then let β be an element in $GF(q^2)$ not in K and such that:

$$tr_2(\beta^5) = 0 :$$

let R_1 be one of the two points of the curve X such that: $R_1 = (\beta, y', 1)$; the point R_1 has exactly two conjugates R_1 et R_2 by the Frobenius map. We set:

$$G_2 = G_1 - \sum_{i=1}^{2} R_i + 2P_\infty$$

It is clear that $\underline{G} = (G_1, G_2)$ answers the question.

4-2: if a is **odd** we look for an integer s, $s > 1$, such that the map:

$$\Psi_1^s : D_1^s \longrightarrow \operatorname{Jac}(X)^{s-1}$$

is not surjective. We can have $s = 3$ since:

$$\operatorname{card} \mathbf{D_1} = 33 \quad \operatorname{card} \operatorname{Jac} X = 5^4 \text{ and } 33^3 < 5^8.$$

But we are going to prove that $s = 2$ is suitable too. Since $\mathbf{D_1}$ is the set of K-rational points of the curve, then:

$$im\Psi_1^2 = \left\{ [P_1 - P_2] \text{ where } P_i \in X(K) \right\}$$

and:

$$[P_i - P_j] = 0 \quad \text{if } P_j = P_i$$
$$= [P_i - P_\infty] \quad \text{if } P_j = P_\infty$$
$$= [2P_i - 2P_\infty] \quad \text{if } P_j = \tilde{P}_i \text{ and } \tilde{P}_i \neq P_i \cdot$$
$$= [P_i + \tilde{P}_j - 2P_\infty] \quad \text{if } P_j \neq P_i, \tilde{P}_i.$$

Let G_1 and G_2 be the same divisors as previously:

$$G_1 = \sum_{i=1}^{k} Q_i \text{ and } G_2 = G_1 - \sum_{i=1}^{2} R_i + 2P_\infty.$$

Then:

$$[G_1 - G_2] = \left[\sum_{i=1}^{2} R_i - 2P_\infty \right] = [R_1 - \tilde{R}_2]$$

and since the points R_1 and R_2 aren't rational on K, $[G_1 - G_2]$ isn't in the image of Ψ_1^2. Then $\underline{G} = (G_1, G_2)$ answers the question.

Remark 1: Let α be a root of the polynomial:

$$x^8 + x^6 + x^5 + x^4 + 1 = 0;$$

then α is a generator of $GF(q^2)$; since $tr_2(\alpha^{45}) = 0$ and $tr_2(\alpha^{55}) = 0$, we can choose: $\beta = \alpha^9$ or α^{11}.

In the next section we try to extend the result of section 4-2 to an arbitrary hyper-elliptic curve.

5-GENERAL CASE:

PROPOSITION: Let X be a hyperelliptic curve of genus g de fined on $K = GF(q)$. We suppose the existence of a rational and reduced divisor:

$$F = \sum_{i=1}^{g} R_i - gP_\infty$$

where the coordinates of the points R_i are in $GF(q^g)$ and don't belong to any smaller extension of K. Then the map Ψ^2_{g-1} isn't surjective.

Remark 2: Let N_r be the number of points of the curve X which are rational on $GF(q_r)$; if g is prime, then a divisor F will exist iff $N_1 < N_g$. Using Weil's bound, we can see that a sufficient condition for the existence of an F is:

$$g < \frac{\sqrt{q^g} - \sqrt{q}}{2}$$

which is certainly fulfilled for q large. Now, how can we construct such a divisor F?

We have found one in section 4-2:

$$F = R_1 + R_2 - 2P_\infty.$$

More generally, suppose $char K = 2$ and X has the affine equation:

$$v^2 + v = f(u).$$

We can make a similar construction if β is an element of $GF(q^g)$ which doesn't belong to any intermediate field between K and $GF(q^g)$ and such that:

$$tr_g\big(f(\beta)\big) = 0.$$

Proof of the proposition: First F is a representative of a non zero class of $Jac(X)$. We are going to prove that $[F]$ isn't in the image of Ψ^2_{g-1}. Let $\underline{D} = (D_1, D_2)$

be any element of $\mathbf{D}_{g-1}^2$. We set:

$$D_1 = \sum_{i=1}^{g-1} P_i \text{ and } D_2 = \sum_{i=1}^{g-1} Q_i.$$

Since D_1 and D_2 are rational divisors on K, the points P_i and the points Q_i have coordinates in K or in an extension $GF(q^r)$ of K where $r < g$. We have:

$$D_2 - D_1 = \sum_{i=1}^{g-1} Q_i - \sum_{i=1}^{g-1} P_i = \sum_{i=1}^{g-1} Q_i + \sum_{i=1}^{g-1} \tilde{P}_i - (2g-2)P_\infty.$$

The divisor $D_2 - D_1$ is equivalent to a quasi-reduced divisor:

$$D = \sum_{j=1}^{2l} T_j - (2l)P_\infty$$

where $0 \le 2l \le (2g-2)$ and the points T_j are some of the points Q_i or $\tilde{P}_i$. Then, because of the hypothesis on F, the divisor $F + D$ is quasi-reduced, so it isn't principal by Lemma 1 and $[F]$ can't be equal to $[D_1 - D_2]$; then the map Ψ_{g-1}^2 isn't surjective.

If X is an hyperelliptic curve and F a zero degree divisor satisfying the propertie of the proposition then we can decode a geometric code $C_\Omega(X, p, D)$ and correct up to t^* errors: we construct an effective divisor G_1 of degree $g + t^*$ the support of which is disjoint from the support of p; we set:

$$G_2 = G_1 - F$$

then we can perform the algorithm using $\underline{G} = (G_1, G_2)$.

Remark 3: In [S-V], the modified decoding algorithm can correct up to t^* errors if some properties are fulfilled concerning the curve X and the divisor D. Recall that this algorithm applies to a geometric code such that $D = a_1 H$ where H is an effective divisor of degree h and:

$$2g - 2 < a_1 h \le n + g - 1.$$

The algorithm consists in doing the basic one using successively:

$$H, 2H, \ldots, b_1 H$$

where b_1 is the less integer such that the equation which gives the coefficients of the error evaluator function has a non trivial solution.

This algorithm corrects up to t^* errors in the following case:

1) X is an elliptic curve and $h = 1$ or 2.

2) X is a hyperelliptic curve and H is hyperelliptic point or a Weierstrass divisor.

3) X is an elliptic curve and a_1 is even such that:

$$a_1 > \frac{2g-2}{h} + 1$$

4) X is a curve such that $g \leq 4$ and X has a non hyperelliptic rational point P such that:

$$\max_{j \in \mathbb{Z}} \left(j + 1 - \dim L((2j+1)P) \right) = 0$$

moreover $D = aP$ where a is even.

CONCLUSION:

We proved that if X is an hyperelliptic curve of genus g for which we can find a reduced divisor $F = R - gP_\infty$ where R is an effective divisor of degree g which can't be obtain as sum of effective divisors of degree less than g, then we can decode any geometric code on X using twice in parallel Skorobogatov and Vladuts' basic decoding algorithm and correct up to $\lfloor (\frac{d^*-1}{2}) \rfloor$-errors.

BIBLIOGRAPHY:

[C] D.G. CANTOR: *Computing in the Jacobian of a Hyperelliptic Curve*, Math. of Comp., vol. 48,177 (1987), 95-101.

[F] W. FULTON: *Algebraic Curves*, Benjamin, New-York (1969).

[K] N. KOBLITZ: *Hyperelliptic Cryptosystems*, J. Cryptology, 1 (1989), 139-150.

[P] R. PELLIKAAN: *On a decoding algorithm for codes on maximal curves*, IEEE Trans. Info. Theory, vol. 35, 6 (1989),1228-1232.

[S-V] A.N. SKORÓBOGATOV, S.G. VLADUTS: *On the decoding of Algebraic-Geome tric Codes*, IEEE Trans. Info. Theory, vol. 36, 5 (Sept. 1990), 1051-1060.

[V] S.G. VLADUTS: *On the decoding of Algebraic-Geometric Codes over F_q for $q \geq 16$*, IEEE Trans. Info. Theory, vol. 36, 6 (Nov. 1990), 1461-1463.

DECODING OF CODES ON THE KLEIN QUARTIC

- D. ROTILLON
   University Paul Sabatier
   Toulouse - France

- J.A. THIONG LY
   University Toulouse le Mirail
   Toulouse - France

Equipe CNRS 035932 Arithmétique et Théorie de l'Information
CIRM Marseille Luminy - France

Abstract

Following R. Pellikaan who gave, in 1989, an algorithm which decodes geometric codes up to $t^* = \left\lceil \dfrac{d^*-1}{2} \right\rceil$ errors where d^* is the designed distance of the code, we describe an effective decoding procedure for some geometric codes on the Klein quartic.

Key-words : Geometric codes, decoding, Jacobian, Klein quartic, bitangent, zeta function.

0. INTRODUCTION

Let X be a non singular, absolutely irreducible, projective curve defined over the finite field $\mathcal{F}_q$. Let g denote the genus of X. Let D and G be two divisors with disjoint supports. More precisely :

$$D = P_1 + \ldots + P_n$$

where the P_i are n rational points of X over $\mathcal{F}_q$. Following Goppa ([1]), the geometric codes $C_L(D,G)$ and $C_\Omega(D,G)$ are linear codes over $\mathcal{F}_q$ defined by :

$$C_L(D,G) = \{(f(P_1),\ldots,f(P_n)) \in \mathcal{F}_q^n \mid f \in L(G)\}$$

$$C_\Omega(D,G) = \{((\text{Resp}_{p_1}(\omega),...,\text{Resp}_{p_n}(\omega)) \in \mathcal{F}_q^n \mid \omega \in \Omega(G-D)\}$$

$C_\Omega(D,G)$ is the dual code of $C_L(D,G)$.

If $2g-2 < \deg G < n$, then $C_\Omega(D,G)$ is a (n,k,d) code such that :

$k = n-\deg G+g-1$

$d \geq d^* = \deg G-2g+2$

d^* is called the designed distance of $C_\Omega(D,G)$.

Then a parity check matrix of $C_\Omega(D,G)$ is :

$$(g_i(P_j)) \quad 1 \leq i \leq n-k, \ 1 \leq j \leq n$$

where $\{g_1,...,g_{n-k}\}$ is a basis of $L(G)$.

We denote by $x = (x_1...,x_n)$ or $x = (x_{p_1},...,x_{p_n})$ every vector in $\mathcal{F}_q^n$.

Let : $a = (a_1,...,a_n)$ be a transmitted codeword belonging to $C_\Omega(D,G)$.

$W = (w_1,...,w_n)$ be the received word

$e = (e_{p_1},...,e_{p_n})$ be the error vector with t errors in the positions

$Q_1,...,Q_t$, which means

$$e_{Q_1} \neq 0,..., e_{Q_t} \neq 0, \ e_{p_i} = 0 \text{ for } P_i \neq Q_j$$

The main problem in the decoding procedure is to determine a function $\tilde{g}$ in $L(G)$ such that :

$$\tilde{g}(Q_1) = 0... ..... \tilde{g}(Q_t) = 0$$

We note that in general $\tilde{g}$ have more than zeroes among $P_1,...P_n$.

Let F be a divisor with support disjoint from the support of D.

We denote by $\quad f_1,...,f_v$ a basis of $L(F)$

and by $\quad h_1,...,h_u$ a basis of $L(G-F)$

Consider the following matrix of syndromes :

$$S_t = \begin{pmatrix} S_{11}........S_{1v} \\ \vdots \\ \vdots \\ \vdots \\ S_{u1}........S_{uv} \end{pmatrix}$$

where $S_{ij} = \sum_{k=1}^{n} h_i(P_k)f_j(P_k)W_k$. We will write : $S_{ij} = \langle h_if_j, w \rangle$

Since h_if_j belongs to $L(G)$ $(1 \leq i \leq u, 1 \leq j \leq v)$, we have :

$$S_{ij} = <h_i f_j, e>$$

We identify a function $f = \lambda_1 f_1 + ... + \lambda_v f_v$ in $L(F)$ with the vector $(\lambda_1,...,\lambda_v)$, and we consider :

$$\text{Ker } S_t = \left\{ \tilde{g} = \lambda_1 f_1 + ... + \lambda_v f_v \in L(F) \mid S_t \begin{pmatrix} \lambda_1 \\ \cdot \\ \cdot \\ \lambda_v \end{pmatrix} = \begin{pmatrix} 0 \\ \cdot \\ \cdot \\ 0 \end{pmatrix} \right\}$$

Proposition 1 :

1) Consider the map $\varphi(F,Q)$ defined by :

$$L(G-F) \rightarrow \mathcal{F}_q^t : h \rightarrow (h(Q_1),...,h(Q_t)).$$

If the map $\varphi(F,Q)$ is surjective then :

$$\text{Ker } S_t = L \left(F - \sum_{i=1}^{t} Q_i \right)$$

2) If dim $L(F) > t$ then :

$$L \left(F - \sum_{i=1}^{t} Q_i \right) \neq \{0\}.$$

Proof : see [3]

In conclusion :

1) To find a function $\tilde{g} \neq 0$ whose set of zeroes contains the set $Q_1,...,Q_t$ of the errors positions, it is enough to determine a divisor F such that dim $L(F) > t$.

2) Any element of the kernel of the syndrome matrix S_t gives such a $\tilde{g}$, if the map $\varphi(F,Q)$ is surjective.

A condition for $\varphi(F,Q)$ to be surjective is given by S.N. Skorobogatov and S.G. Vladut ([3]) by assuming :

$$\deg(G-F) > t + 2g - 2.$$

This leads to an algorithm which decodes up to $\left[\dfrac{d^*-1-g}{2}\right]$ errors with complexity $O(n^3)$. Another way to find a map $\varphi(F,G)$ surjective is given by Pellikaan in [2].

1. OUTLINES OF PELLIKAAN'S ALGORITHM

Notations :
- D_ℓ is the set of positive divisors on X with degree ℓ
- a_ℓ is the cardinal of D_ℓ
- $J(X)$ is the Jacobian of X over $\mathcal{F}_q$.
- $h = \#J(X)$
- [Y] is the class of Y in $J(X)$.

Consider for any integer $s \geq 2$, the map

$$\psi_\ell^s : D_\ell^s = D_\ell \times \dots \times D_\ell \longrightarrow J(X)^{s-1}$$

$$(F_1,\dots,F_s) \longrightarrow ([F_1-F_2],\dots,[F_{s-1}-F_s])$$

Proposition 2 :

Let $t^* = \left[\dfrac{d^*-1}{2}\right]$, and let G be such that deg $G \geq 4g-2$. For an integer $s \geq 2$, suppose that ψ_{g-1}^s or ψ_{g-2}^s is not surjective in case deg G is odd or even respectively. Then there exists an s-uple $(F_1,\dots,F_s)$ in $D_{g+t^*}^s$ such that for every t^*-uple $Q = (Q_1,\dots,Q_{t^*})$ with $\{Q_1,.,Q_{t^*}\} \subset \{P_1,\dots,P_n\}$ the map $\varphi(F_{i_0},Q)$ is surjective for at least one i_0, $1 \leq i_0 \leq s$.

Proof : see [2]

By proposition 1 and 2, for every set of t errors positions $Q_1,\dots,Q_t$ with $t \leq t^*$, one has :

$$\text{Ker } S_t = L\left(F_{i_0} - \sum_{j=1}^{t} Q_j\right).$$

So one can determine a function $\tilde{g}$ belonging to Ker S_t such that the set of errors positions is included in the set of zeroes of $\tilde{g}$.

Now it remains to find the errors values $e_{Q_1}, ..., e_{Q_t}$. Let $\tilde{g} \in \mathrm{Ker}\ S_t$ and let $Q_1, ..., Q_{t'}$ be the set of zeroes of $\tilde{g}$ included in the set $\{P_1, ..., P_n\}$.

Consider the following system :

$$
\begin{pmatrix}
g_1(Q_1) & \cdots\cdots\cdots & g_1(Q_{t'}) \\
& & \\
& & \\
& & \\
g_{n-k}(Q_1) & \cdots\cdots\cdots & g_{n-k}(Q_{t'})
\end{pmatrix}
\begin{pmatrix}
x_1 \\
\vdots \\
x_{t'}
\end{pmatrix}
=
\begin{pmatrix}
<g_1,w> \\
\vdots \\
<g_{n-k},w>
\end{pmatrix}
\tag{1}
$$

The vector $(e'_1, ..., e'_{t'})$ whose non zero components are the t values $e_{Q_1}, ..., e_{Q_t}$ is a solution of (1) since :

$$
\sum_{k=1}^{n} g_i(Q_k)\ e_{Q_k} = <g_i,e> = <g_i,w> = \sum_{k=1}^{t'} g_i(Q_k) e'_k.
$$

In fact one has the following proposition :

<u>Proposition 3</u>

Let $\tilde{g} \in \mathrm{Ker}\ S_t = L\left(F - \sum_{j=1}^{t} Q_j\right)$ for some divisor F.

If $\deg(G\text{-}F) > t^* + g\text{-}2$ and if $t^* \geq g$, then the system(1) has the unique above solution.

Proof : see [2]

<u>Remark 1</u> : The conditions of the proposition 3 are fulfilled with F in D_{g+t^*}, and with G such that $\deg G \geq 4g$ in case $\deg G$ is even, or $\deg G \geq 4g\text{-}1$ in case $\deg G$ is odd.

Now, if $a_{g-1} < ch$ for some constant c, then there exists an integer s such that ψ^s_{g-1} and ψ^s_{g-2} are not surjective. This is the case when the curve is maximal (see Pellikaan [2]) and more generally if either $q \geq 37$, or $q \geq 16$ and the genus is large enough (see Vladut [5]).

Two problems in Pellikaan's algorithm remain unsolved in practice :

1) Finding the smallest s such that ψ^s_{g-1} and ψ^s_{g-2} are not surjective.

2) Finding an s-uple $(F_1, ..., F_s)$ of divisors in D_{g+t^*} according to proposition 2.

In the following, we show that for some family of codes over the Klein quartic :

1) The smallest suitable s is s = 2 when deg G is even, and when deg G is odd, s = 4 is suitable.

2) When deg G is even, two suitable divisors F_1, F_2 in D_6 can be constructed in order to decode up to $t^* = 3$ errors.

II. THE ZETA FUNCTION OF A CURVE X OF GENUS g = 3

Let $Z(X, \mathcal{F}_q) = \dfrac{P(T)}{(1-T)(1-qT)}$

where

$$P(T) = 1 + P_1 T + \dots + P_5 T^5 + q^3 T^6$$

$$= \prod_{i=1}^{3} (1 - \alpha_i T)(1 - \bar{\alpha}_i T)$$

with

$$|\alpha_i| = |\bar{\alpha}_i| = q^{\frac{1}{2}}.$$

$N_r = q^r + 1 - S_r$ is the number of points of X over $\mathcal{F}_{q^r}$, $S_r = \displaystyle\sum_{i=1}^{3} (\alpha_i^r + \bar{\alpha}_i^r)$ (2)

$$\beta_i = \alpha_i + \bar{\alpha}_i \qquad (i = 1, 2, 3).$$

From the equality :

$$\sum_{i=0}^{6} P_i T^i = \prod_{i=1}^{3} (1 - \beta_i T + qT^2)$$

one finds :

$$P_1 = - (\beta_1 + \beta_2 + \beta_3)$$

$$P_2 = \beta_1 \beta_2 + \beta_1 \beta_3 + \beta_2 \beta_3 + 3q \qquad (3)$$

$$P_3 = -[2q(\beta_1 + \beta_2 + \beta_3) + \beta_1 \beta_2 \beta_3]$$

It is known that $P_4 = qP_2$, $P_5 = q^2 P_1$.

From (2), one has obviously : $-P_1 = q + 1 - N_1$ (4).

It is easy to verify that :

$$\sum_{i=1}^{3} \beta_i^2 = q^2 + 1 - N_2 + 6q$$

$$\sum_{i=1}^{3} \beta_i^3 = (q+1)^3 - N_3 - 3qN_1. \tag{5}$$

By expanding $(\beta_1 + \beta_2 + \beta_3)^2$ and $(\beta_1 + \beta_2 + \beta_3)^3$, and using (3), (4), (5), we obtain :

$$P_2 = \frac{N_2}{2} + \frac{P_1^2}{2} - \frac{q^2+1}{2}$$

$$P_3 = \frac{N_3}{3} - \frac{P_1^3}{3} + P_1 P_2 - \frac{q^3+1}{3}. \tag{6}$$

It is known that
$$a_2 = q^2 + q + 1 + (q+1)P_1 + P_2$$
and
$$h = P(1) = q^3 + 1 + (q^2+1)P_1 + (q+1)P_2 + P_3.$$

By using the expressions (4) and (6) of P_1, P_2, P_3, we obtain :

$$\begin{cases} a_2 = \dfrac{N_1^2 + N_2}{2} \\[4mm] h = \dfrac{N_1^3 + 3N_1 N_2 + 2N_3 - 6qN_1}{6} \end{cases}$$

More generally, for a curve X of genus 3 over any extension $\mathcal{F}_{q^r}$ of $\mathcal{F}_q$, let $a_{g-1} = a_2 = \# D_2$, $a_1 = a_{g-2} = \# D_1$, $h_r = \# J(X, \mathcal{F}_{q^r})$, then we have the following lemma.

<u>Lemma 1</u> : For any $r \geq 1$:

$$\begin{cases} a_2 = \dfrac{N_r^2 + N_{2r}}{2} \\[4mm] h_r = \dfrac{N_r^3 + 3N_r N_{2r} + 2N_{3r} - 6q^r N_r}{6} \end{cases}$$

<u>Corollary 1</u> :

For a curve X of genus g = 3 over $\mathcal{F}_{q^r}$ (r ≥ 1), one has : $a_{g-2}^2 < h_r$.

Hence ψ_{g-2}^2 is not surjective.

Proof : Trivial since $a_{g-2} = a_1 = N_r$.

<u>Remark 2</u> : There is an easy computational method for determining the values of $S_r = \sum_{i=1}^{g} \left(\alpha_i^r + \bar{\alpha}_i^r \right)$ of any curve of genus g if we know the zeta function over $\mathcal{F}_q$. Indeed :

Let $Z(X,\mathcal{F}_q) = \dfrac{P(T)}{(1-T)(1-qT)}$

where $P(T) = \prod_{i=1}^{g} (1-\alpha_i T)(-\bar{\alpha}_i T)$.

Then by long division, we have :

$- \dfrac{P'(T)}{P(T)} = S_1 + S_2 T + S_3 T^2 + ...$

III. CLASS NUMBER OF THE KLEIN QUARTIC OVER $\mathcal{F}_{q^r}$

Here the curve X is the Klein quartic over $\mathcal{F}_{q^r}$ (q = 2) :

$$X^3 Y + Y^3 Z + Z^3 X = 0$$

In the following we show that the cardinal of $J(X,\mathcal{F}_{2^r})$ and the cardinal of $J(Y,\mathcal{F}_{2^{3r}})$ are closely related, where Y is the elliptic curve over $\mathcal{F}_{2^{3r}}$ whose equation is :

$$Y^2 Z + XYZ = X^3 + X^2 Z + Z^3.$$

The zeta function of X over $\mathcal{F}_2$ is :

$$Z(X,\mathcal{F}_2) = \dfrac{1+5T^3+8T^6}{(1-T)(1-2T)}$$

The zeta function of Y over $\mathcal{F}_{23}$ is :

$$Z(Y,\mathcal{F}_{23}) = \frac{1+5T+8T^2}{(1-T)(1-8T)}$$

$$1+5T^3+8T^6 = \prod_{i=1}^{3} (1-\alpha_i T)(1-\overline{\alpha}_i T)$$

Let

$$S_r = \sum_{i=1}^{3} (\alpha_i^r + \overline{\alpha}_i^r)$$

recall that $N_r = q^r+1 - S_r$

Let :
$$1 + 5T + 8T^2 = (1 - \lambda T)(1 - \overline{\lambda} T)$$
$$s_r = \lambda^r + \overline{\lambda}^r.$$

Clearly :
$$\#J(Y,\mathcal{F}_{q^{3r}}) = q^{3r} + 1 - s_r.$$

This is the number of rational points of Y as well.

Lemma 2

$$\begin{cases} S_r = 0 & \text{when } r \not\equiv 0 \mod 3 \\ S_{3r} = 3s_r \end{cases}$$

Proof : see remark 2.

Proposition 4

For any $r \geq 1$,

1) $h_r = q^{3r} + 1 - s_r$ if $r \not\equiv 0 \mod 3$
2) $h_{3r} = (q^{3r} + 1 - s_r)^3$

Proof

From lemma 1 :

$$6h_r = (q^r+1-S_r)^3 + 3(q^r+1-S_r)(q^{2r}+1-S_{2r}) + 2(q^{3r}+1-S_{3r}) - 6q^r(q^r+1-S_r)$$

1)   If $r \not\equiv 0 \mod 3$ then $S_r = 0$ and $S_{2r} = 0$, so

$$6h_r = (q^r+1)^3 + 3(q^r+1)(q^{2r}+1) + 2(q^{3r}+1-S_{3r}) - 6q^r(q^r+1)$$

After simplifications, and using $S_{3r} = 3s_r$, we have :

$$h_r = q^{3r}+1 - s_r.$$

2) By lemma 1 and 2 :

$$6h_{3r} = (q^{3r}+1-3s_r)^3 + 3(q^{3r}+1-3s_r)(q^{6r}+1-3s_{2r})+2(q^{9r}+1-3s_{3r}) - 6q^{3r}(q^{3r}+1-3s_r)$$

It is easy to check that :

$$s_{2r} = s_r^2 - 2q^{3r}$$

$$s_{3r} = s_r^3 - 3q^{3r} s_r.$$

We substitute these values in the above expression of $6h_{3r}$.
After simplifications, we obtain : $h_{3r} = (q^{3r}+1-s_r)^3$.

In other words : $\# J(X, F_{2^{3r}}) = (\#J(Y, F_{2^{3r}}))^3$.

Corollary 2

The smallest integer s for which $a_{g-1}^s < h_r^{s-1}$ is $s = 4$. Hence, ψ_{g-1}^4 is not surjective.

Proof

1) For $r \not\equiv 0 \bmod 3$, $a_2 = \dfrac{(q^r+1)^2+(q^{2r}+1)}{2} = q^{2r}+q^r+1$.

Since $s_r = \lambda^r + \bar\lambda^r$ with $|\lambda| = |\bar\lambda| = q^{\frac{3r}{2}}$, we have
$$|s_r| \le 2q^{\frac{3r}{2}} \text{ and } |s_r^2| \le 4q^{3r}.$$

So

$$h_r \le q^{3r} + 1 + 2q^{\frac{3}{2}}.$$

One has :

$$a_2^2 = q^{4r} + 2q^{3r} + ... > q^{3r} + 2q^{\frac{3r}{2}} + 1 \ge h_r.$$

$$a_2^3 = q^{6r} + 3q^{5r} + 6q^{4r} + ...$$

$$h_r^2 = (q^{3r} + 1)^2 - 2(q^{3r} + 1)\, s_r + s_r^2$$

$$h_r^2 \geq q^{6r} + 4q^{\frac{3r}{2}} + 6q^{3r} + ...$$

Hence : $a_2^3 > h_r^2$

Now the leading terms of a_2^4 and of h_r^3 are q^{8r} and q^{9r} respectively.

We conclude that $a_2^4 < h_r^3$.

2) $a_2 = \dfrac{(q^{3r}+1-3s_r)^2+(q^{6r}+1-3s_{2r})}{2}$

$$\begin{cases} a_2 = q^{6r}+4q^{3r}+3s_r^2-3(q^{3r}+1)s_r+1 \\ h_{3r} = (q^{3r}+1-s_r)^3 \end{cases}$$

one checks easily as in 1) that :

$$a_2^i > h_{3r}^{i-1} \quad \text{for } i = 2,3$$

and

$$a_2^4 < h_{3r}^3 \qquad\qquad \square$$

__Remark 3__ : Pellikaan ([2]) sets the question of whether the map ψ_{g-1}^{g+1} is always not surjective. This is true in our case.

IV. DECODING A FAMILY OF CODES OVER THE KLEIN QUARTIC

We apply what we have obtained to the following code $C_\Omega(D,G)$ defined on the Klein quartic over F_{2^r}, $r \equiv 0 \pmod 3$, with :

D = P_1 + ... + P_{21}, P_i being the rational points over F_8, not over F_2.

G = $4Q_0 + 4Q_1 + 4Q_2$, Q_i being the rational points over F_2.

Q_0 = (1, 0, 0), Q_1 = (0, 1, 0), Q_2 = (0, 0, 1)

The parameters of the code C_Ω (D,G) are :

$$\begin{cases} n = 21 \\ k = 11 \\ d \geq d^* = 8 \end{cases}$$

We know by corollary 1 that there exists a couple (F_1, F_2) in $D_6 \times D_6$ (since $d^* = 8$, $t^* = 3$, $g+t^* = 6$) such that $[F_1-F_2]$ in $J(X)$ is not in the image $\psi_1^2(D_1^2)$, that is to say, for any couple (P, Q) of distinct points of X, the divisor F_1-F_2 of degree 0, is not equivalent to the divisor P-Q.

In the following, K denotes the canonical linear system g_4^2 cut out on the curve by the lines of the plane. This means that the projective dimension of this linear system is two, and that all divisors in the class K are of degree four. By a slight abuse we will use the same notation to denote any divisor of this class.

Proposition 5

Let A, B be two rational points over F_{2^r} such that the tangents at A and B have no common point on the curve X.
If $F_1 = 2(K-A)$, $F_2 = 2(K-B)$, then $[F_1-F_2] \neq [P-Q]$ for any couple (P,Q) of distinct points of X.

Proof

Suppose there exists P, Q, $P \neq Q$ such that $[F_1-F_2] = [P-Q]$.
This implies that :

$$2(K-A) - 2(K-B) \equiv P-Q$$

or

$$2A + P \equiv 2B + Q.$$

Then it means that the projective dimension $|2A+P|$ is ≥ 1. Since there is no g_3^2 on the trigonal curve because every g_3^1 on X is a complete linear system (if not it would infer the existence of a g_2^1 on the curve by removing a base point, which is impossible since X is not hyperelliptic), thereby 2A + P and 2B + Q belong to a g_3^1, and we can find a point p such that :

$$2A + P + p \propto K$$

[In fact we know that every g_3^1 on X can be obtained through a projection π_p of center p uniquely determined, and 2A + P $\propto$ 2B + Q $\propto$ K-p].

Hence, there exist two adjoint curves passing through the points A, P and B, Q respectively, which means that the tangents at A and at B intersect at p : contradiction   □

Example

We give an explicit construction of F_1 and F_2 by using the bitangents of the Klein quartic.

Firstly we get the general condition for a line with the equation $aX+bY+cZ=0$ to be a bitangent.

A description of the bitangents can be obtained through the use of the Hasse-Witt matrix of the curve (see Stohr-Voloch [4]), but in the following we give a direct and less sophisticated way for the sake of simplicity. We know the tangents at the F_2 rational points ; so we may assume that $XYZ \neq 0$, and we write down the general equation of the tangent at such a general point $(x,y,1)$ by using the equation $x^3y+x+y^3 = 0$ and get $y^4X+x^2Y+x^4y^2Z = 0$.

Moreover we notice that this line passes through the points (y^2, z^2, x^2) and (x^8, y^8, z^8) which both belong to the curve. So we have a new set of conditions to be fulfilled in order to get a bitangent line, which are :

* $(x,y,z) \neq (x^8, y^8, z^8)$ which means that there cannot be a bitangent which is tangent in points rational over F_8.

* $(y^2, z^2, x^2) = (x^8, y^8, z^8)$, hence setting $z = 1$, this implies $x^6y^6 = 1$, and so we have : $abc = x^6y^6 = 1$.

On the other hand it is easy to establish the following relations :

$$a^3 = cb^2, b^3 = ac^2, c^3 = ba^2.$$

Hence : $a^4 = b, c^4 = a, b^4 = c$; which proves that a, b, c are necessarily in F_{64}.

Trying to find solutions in F_8 to make calculations simpler, we have the conditions :

$$a^2 = c, b^2 = a, c^2 = b \quad \text{with a, b, c in } F_8.$$

From these relations, we have the following proposition which gives a complete description of the seven bitangents :

Proposition 6

The seven bitangents are :

$$\beta^i X + \beta^{4i} Y + \beta^{2i} Z = 0 \qquad (i = 0,...,6)$$

where β is a primitive root of F_8.

Remark 4 : The points of contact of the bitangents constitute 14 new points of X rational over F_{64} paired off, and they are the only ones since we know that :

$$N_6 = 38 = 24 + 14 = N_3 + 14$$

The following lemma gives explicitly the points of contact.

Lemma 3

For $i = 0,...,6$ the two points of contact of the bitangent

$$\beta^i X + \beta^{4i} Y + \beta^{2i} Z = 0$$

are :

$$\begin{pmatrix} \alpha^2 \ \beta^i \\ \alpha \ \ \beta^{5i} \\ 1 \end{pmatrix} \quad \text{and} \quad \begin{pmatrix} \alpha \ \ \beta^i \\ \alpha^2 \ \beta^{5i} \\ 1 \end{pmatrix}$$

where α is a primitive root of $\mathcal{F}_4$.

Proof

From $\quad \beta^i X + \beta^{4i} Y + \beta^{2i} Z = 0$, we have

$X = \beta^{3i} Y + \beta^i Z.$

By substituting this value of X in $X^3 Y + Y^3 Z + Z^3 X = 0$, we obtain :

$$(\beta^i Y^2 + \beta^{6i} YZ + \beta^{4i} Z^2)^2 = 0$$

Setting $Z = 1$, we have to solve the following quadratic equation :

$y^2 + \beta^{5i} y + \beta^{3i} = 0.$

Let : $\quad\quad y = \beta^{5i} w$; then :

$w^2 + w + 1 = 0$

The two roots of this equation are $w_1 = \alpha$; $w_2 = \alpha^2$.

So : $\quad y_1 = \beta^{5i} \alpha \quad\quad$ and $\quad\quad y_2 = \beta^{5i} \alpha^2$

and : $\quad x_1 = \beta^{3i} y_1 + \beta^i = \beta^i(\alpha+1) = \beta^i \alpha^2$

$x_2 = \beta^i(\alpha^2+1) = \beta^i \alpha \quad\quad\quad \square$

The two points $R_1 = \begin{pmatrix} \alpha^2 \ \beta^i \\ \alpha \ \ \beta^{5i} \\ 1 \end{pmatrix}$ and $R_2 = \begin{pmatrix} \alpha \ \ \beta^i \\ \alpha^2 \ \beta^{5i} \\ 1 \end{pmatrix}$ are conjugated under the

exponentiation by 8. So $R = R_1 + R_2$ can be considered as a point of degree 2 over $\mathcal{F}_8$. We have $K \equiv 2R$.

Proposition 7

Let $\quad F_1 = 2K - 2Q_0 \quad\quad$ and $\quad F_2 = 2K - R$.

Then $\quad F_1 - F_2 \neq P - Q$ for every pair P, Q of distinct points of X.

Proof

Suppose there exists P, Q, P $\neq$ Q such that :

$$F_1 - F_2 = R - 2Q_0 \equiv P - Q, \text{ or } R + Q \equiv 2Q_0 + P.$$

By the same argument as in proposition 6, we deduce that there exists a point p over $\mathcal{F}_8$ such that :

$$R + Q + p \equiv 2Q_0 + P + p \equiv K.$$

The line which cuts out the divisor $2Q_0 + P + p$ is necessarily the tangent at Q_0. But it is an inflexion tangent at the flex Q_0. So either $p = Q_0$ or $P = Q_0$.

* In the first case, it would imply that Q_0 is on the bitangent which is impossible because there are not bitangents which are tangent at points rational over $\mathcal{F}_8$.

* In the second case, if $P = Q_0$, then $p=Q_2$ because $\text{div}(Y)=3 \ Q_0+Q_2$, and the bitangent defined by R would pass through Q_2, which is impossible, as well. $\square$

References :

[1] V.D. Goppa. "Geometry and Codes". Mathematics and its applications, soviet Series 24. Kluwer Ac Publ. Dordrecht, The Netherlands, 1989.

[2] R. Pellikaan : "On a decoding algorithm for codes on maximal curves". IEEE Trans Inf Theory, Vol. 35, N° 6, 1989.

[3] A.N. Skorobogatov, S.G. Vladut : "On the decoding of algebraic - Geometric Codes", IEEE Trans Inf Theory, Vol 36, 5 (Sept. 1990), 1051-1060.

[4] K.O. Stöhr, J.F. Voloch. "A formula for the Cartier operator on plane algebraic curves". J. Reine Angew Math, 377, 49-64, 1987.

[5] S.G. Vladut : "on the decoding of algebraic geometric codes over $\mathcal{F}_q$ for $q \geq 16$". IEEE Trans Inf Theory, Vol. 36, 6 (Nov. 1990), 1461-1463.

Asymptotically good families of geometric Goppa codes and the Gilbert-Varshamov bound

Conny Voß

Universität GHS Essen, Fachbereich 6-Mathematik

D-4300 Essen 1, Germany

Abstract: This note presents a generalization of the fact that most of the classical Goppa codes lie arbitrarily close to the Gilbert-Varshamov bound (cf.[2, p. 229]).

0. Introduction

One interesting application of geometric Goppa codes is the construction of asymptotically good families of codes. In section 1 we show that almost all long linear codes are good ones in the sense that they get close to the Gilbert-Varshamov bound. In section 2 we present the main result of [4], which gives a generalization of the well-known fact that classical Goppa codes meet the Gilbert-Varshamov bound (cf. [3, p. 111]). Finally in section 3 we show the main result of this note, namely that the sequences of codes considered in [4] are often asymptotically good ones.

1. Behaviour of long linear codes

In this section our aim is to show that, when the transmission rate remains unchanged, almost all long linear codes over a fixed finite field lie arbitrarily close to the Gilbert-Varshamov bound. This gives a generalization of a similar statement for the field GF(2) (cf.[1]).

We need some notations:

GF(q) is the field with q elements;

$$V_q(n,d) = \sum_{j=0}^{d} \binom{n}{j}(q-1)^j;$$

$c_{n,k}$ = {C| C is an [n,k]-code over GF(q)};

$c_{n,k}^d$ = {C $\in$ $c_{n,k}$| d(C) $\leq$ d} where d(C) denotes the minimum distance of C;

$H_q(\lambda) = \lambda \log_q(q-1) - \lambda \log_q \lambda - (1-\lambda)\log_q(1-\lambda);$

$\lfloor x \rfloor = \max\{n \in \mathbb{N}|\ n \leq x\};$

$\lceil x \rceil = \min\{n \in \mathbb{N}|\ n \geq x\}.$

Counting subspaces in GF(q)^n of fixed dimension it is not difficult to get the following lemma (cf. [5]):

LEMMA 1.1.
$$\frac{\# c_{n,k}^d}{\# c_{n,k}} \leq V_q(n,d)\, \frac{q^k - 1}{q^n - 1}\,.$$

Now we can state our first result:

THEOREM 1.1. Fix R $\in$ [0,1] and $\epsilon > 0$, and let $0 \leq \delta \leq 1-q^{-1}$ be such that $R = 1 - H_q(\delta)$. Then

$$\lim_{n\to\infty} \frac{\# c_{n,\lceil nR \rceil}^{\lfloor (\delta-\epsilon)n \rfloor}}{\# c_{n,\lceil nR \rceil}} = 0.$$

REMARK. The theorem means that for n sufficiently large almost all long linear codes with transmission rate R $\approx$ 1-$H_q(\delta)$ have relative minimum distance $\geq \delta-\epsilon$ and, therefore, they get close to the Gilbert-Varshamov bound.

PROOF OF THEOREM 1.1. According to Lemma 1.1, it suffices to show that

$$\lim_{n \to \infty} V_q(n, \lfloor (\delta-\epsilon)n \rfloor) \cdot q^{-n+\lceil nR \rceil} = 0.$$

We have that $\lim_{n \to \infty} n^{-1} \log_q V_q(n, \lfloor (\delta-\epsilon)n \rfloor) = H_q(\delta-\epsilon)$ (cf. [3, p. 55]) and, by the choice of δ, that $H_q(\delta-\epsilon) - (1-R) < 0$. Consequently

$$\lim_{n \to \infty} V_q(n, \lfloor (\delta-\epsilon)n \rfloor) \cdot q^{-n+\lceil nR \rceil} \leq \lim_{n \to \infty} q^{n \cdot (H_q(\delta-\epsilon)-(1-R))+1} = 0.$$

Hence the assertion follows.   □

2. Asymptotically good families of subfield subcodes

Now we want to formulate the main result of [4]. It gives a generalization of the well-known fact that the class of classical Goppa codes meet the Gilbert-Varshamov bound (cf. [3, p. 111]), upon considering the class of subfield subcodes of algebraic geometric Goppa codes. These codes are natural generalizations of classical Goppa codes. We first introduce some notations, cf.[4]:

$F/GF(q^m)$ is an algebraic function field of one variable such that $GF(q^m)$ is algebraically closed in F;

$D = P_1+\ldots+P_n$ and G are divisors of F where the pairwise distinct places P_i are of degree one and where

suppG ∩ suppD = ∅;

$L(G) = \{x \in F | (x) \geq -G\}$ where (x) denotes the principal
divisor of x;

$C(G,D) = \{(x(P_1),\ldots,x(P_n)) | x \in L(G)\} \subset (GF(q^m))^n$, called
geometric Goppa code over $GF(q^m)$ associated with
G and D;

$C^{\perp}$ = the dual code of the code C;

$\Gamma(G,D) = C(G,D)^{\perp} \cap GF(q)^n$, the subfield subcode of
$C(G,D)^{\perp}$.

Now we can state the main theorem of [4]:

THEOREM 2.1. Let $(m_i)_{i \geq 1}$ be a sequence of natural numbers such
that $m_i \to \infty$. For any i let F_i be a function field over $GF(q^{m_i})$
of genus g_i, and let A_i be a divisor of F_i with $A_i \geq 0$ and
$\deg A_i = a_i$. Assume that

$$\lim_{i \to \infty} g_i q^{-m_i/2} = \lim_{i \to \infty} a_i q^{-m_i/2} = 0.$$

Let $0 \leq \delta \leq 1-q^{-1}$. Then there are divisors Q_i and D_i of F_i such
that the following holds:

(i) $D_i = \sum_{j=1}^{n_i} P_{ij}$ with pairwise distinct places P_{ij}
of degree 1.

(ii) supp $D_i \cap$ supp A_i = supp $D_i \cap$ supp $Q_i = \emptyset$.

(iii) $\Gamma(Q_i-A_i,D_i)$ is an $[n_i,k_i,d_i]$-code over $GF(q)$ with
$\lim_{i \to \infty} n_i q^{-m_i} = 1$, $\lim \inf d_i/n_i \geq \delta$, and
$\lim \inf k_i/n_i \geq 1 - H_q(\delta)$.

REMARKS. (1) The theorem means that the codes $\Gamma(Q_i-A_i,D_i)$ meet
the Gilbert-Varshamov bound.

(2) The proof of the theorem (cf. [4, p. 115]) shows that the divisors Q_i can be chosen to be prime divisors.

(3) Taking $F_i = GF(q^{m_i})(z)$ and $A_i = P_\infty$ (the pole of z in F_i), the theorem gives the result for the class of classical Goppa codes.

3. Good sequences of subfield subcodes

Our next aim is to show that "almost all" sequences of subfield subcodes as in Theorem 2.1 are good ones. This also gives a generalization of an appropriate statement for classical Goppa codes, cf. [2].

In the situation of Theorem 2.1, for $0 \leq \delta \leq 1-q^{-1}$, $\epsilon > 0$ and $t_i \in \mathbb{N}$, let

$$X(t_i) = \{Q_i | Q_i \text{ prime divisor of } F_i, \deg Q_i = t_i\}$$

and let

$$X(t_i, \epsilon, \delta) = \{Q_i \in X(t_i) | \Gamma(Q_i - A_i, D_i) \text{ is an } [n_i, k_i, d_i]\text{-code}$$
$$\text{over } GF(q) \text{ with } d_i/n_i \geq \delta \text{ and}$$
$$k_i/n_i \geq 1 - H_q(\delta) - \epsilon\}.$$

Now we state our main result:

THEOREM 3.1. Let the situation be as above.

Then for $t_i = \left[\dfrac{(H_q(\delta) + \epsilon) n_i}{m_i} \right] - 1$, we have:

$$\lim_{i \to \infty} \frac{\# \, X(t_i, \epsilon, \delta)}{\# \, X(t_i)} = 1.$$

REMARK. The theorem means that for i sufficiently large almost all the codes $\Gamma(Q_i - A_i, D_i)$ with relative minimum distance $d_i/n_i \geq \delta$ have the transmission rate $k_i/n_i \geq 1 - H_q(\delta) - \epsilon$, that is, they lie arbitrarily close to the Gilbert-Varshamov bound.

For the proof of Theorem 3.1 we need the following lemmas:

LEMMA 3.1. Let $F/GF(q^m)$ be a function field of genus g, $D = P_1 + \ldots + P_n$ a divisor with pairwise distinct places P_i of degree 1, $A \geq 0$ a divisor with deg A = a and with supp D $\cap$ supp A = $\emptyset$. Let t, e be integers with $0 < e < n$ and $t > 0$. Let $u \in GF(q^m)^n$ be a vector of (Hamming-)weight e and denote $X(t) = \{Q |$ Q prime divisor of F, deg Q = t$\}$. Then

$$\# \, \{Q \in X(t) \mid u \in C(Q-A,D)^{\perp}\} \leq t^{-1}(a+e+2g-2)q^{m(a+g)}.$$

LEMMA 3.2. With the notations above, we have:

$$| \, \# \, X(t) - t^{-1}q^{mt} \, | \leq t^{-1}(7g+2)q^{mt/2}.$$

LEMMA 3.3. Same situation as in Lemma 3.1. Then

$$\dim \Gamma(Q-A,D) \geq n - m(t+1).$$

For the proof of the lemmas we refer to [4].

PROOF OF THEOREM 3.1. We can assume that $\delta > 0$. Choose $e_i \in \mathbb{N}$ with $e_i - 1 < \delta n_i \leq e_i$ and let

$$X(t_i)^{e_i} = \{Q_i \in X(t_i) \mid \min \text{ dist } \Gamma(Q_i - A_i, D_i) \leq e_i - 1\}.$$

For $Q_i \in X(t_i) \setminus X(t_i)^{e_i}$ we have $Q_i \in X(t_i, \epsilon, \delta)$, since $d_i/n_i \geq e_i/n_i \geq \delta$ and by Lemma 3.3 and by the choice of t_i,

$$k_i/n_i \geq 1 - \frac{m_i(t_i+1)}{n_i} \geq 1 - H_q(\delta) - \epsilon.$$

Hence it is sufficient to show that

$$\lim_{i \to \infty} \frac{\# X(t_i)^{e_i}}{\# X(t_i)} = 0.$$

From Lemma 3.1 and 3.2, we obtain:

$$\frac{\# X(t_i)^{e_i}}{\# X(t_i)} \leq \frac{(a_i+e_i+2g_i)\, q^{m_i(a_i+g_i)}\, V_q(n_i, e_i-1)}{q^{m_i t_i}(1-(7g_i+2)q^{-m_i t_i/2})}.$$

For i sufficiently large we have that $7g_i+2 < \frac{1}{2}q^{m_i/2}$. Hence

$$q^{m_i t_i}(1-(7g_i+2)q^{-m_i t_i/2}) > \frac{1}{2}q^{m_i t_i}$$

and therefore, by the choice of t_i, we have that

$$\frac{\# X(t_i)^{e_i}}{\# X(t_i)} \leq \frac{2(a_i+e_i+2g_i)\, q^{m_i(a_i+g_i)}\, V_q(n_i, e_i-1)}{q^{(H_q(\delta)+\epsilon)n_i-2m_i}}.$$

Since $\lim_{i \to \infty} n_i^{-1}\log_q V_q(n_i, e_i-1) = H_q(\delta)$ (cf. [3, p. 55]), it suffices to show that

$$\lim_{i \to \infty} 2(a_i+e_i+2g_i)\, q^{(a_i+g_i+2)m_i-\epsilon n_i} = 0.$$

Now we observe that

$$\lim_{i\to\infty} g_i q^{-m_i/2} = \lim_{i\to\infty} a_i q^{-m_i/2} = \lim_{i\to\infty} m_i q^{-m_i/2} = \lim_{i\to\infty} n_i q^{-n_i/2} = 0.$$

and therefore that $\quad \lim_{i\to\infty} 2(a_i+e_i+2g_i)\, q^{-(m_i+\epsilon n_i)/2} = 0.$

The theorem now follows from

$$\lim_{i\to\infty} n_i q^{-m_i} = 1 \quad \text{and} \quad \lim_{i\to\infty} q^{m_i}(\tfrac{\epsilon}{2} - (a_i+g_i+\tfrac{5}{2})m_i q^{-m_i}) = \infty. \qquad \square$$

References

[1] Gallager, R. G., Low density parity-check codes, M.I.T. Press Cambridge, Massachusetts (1963), 9-12.

[2] Goppa, V. D., A rational representation of codes and (L, g)-codes, Problems of information transmission 7 (1973), 223-229.

[3] Lint, J. H. van, Introduction to coding theory, Graduate Texts in Mathematics 86, Springer, Berlin, 1982.

[4] Stichtenoth, H., Voß, C., Asymptotically good families of subfield subcodes of geometric Goppa codes, Geometriae Dedicata 33 (1990), 111-116.

[5] Voß, C., Asymptotisch gute Codes im Zusammenhang mit der Gilbert-Varshamov Schranke, Diplomarbeit, Essen 1989.

Multiplicative character sums
and
non linear geometric codes

Marc Perret[1]

Abstract. Let q be a power of a prime number, $\mathbf{F}_q$ the finite field with q elements, n an integer dividing q - 1, n ≥ 2, and χ a character of order n of the multiplicative group $\mathbf{F}_q^*$. If X is an algebraic curve defined over $\mathbf{F}_q$ and if G is a divisor on X, we define a non linear code $\Gamma(q, X, G, n, \chi)$ on an alphabet with n + 1 letters. We compute the parameters of this code, through the consideration of some character sums.

Introduction. If f is a rational function on the curve X, define

$$W(f) = \sum_{P \in X_*(\mathbf{F}_q)} \chi(f(P)),$$

where $X_*(\mathbf{F}_q)$ is the set of rational points on the curve X which are neither a zero or a pole of f. The study of such a character sum, or more precisely of a slightly different sum, enables us to derive some estimations for the number of points of a Kummer covering [3].

In an other way, the extra data of a divisor G on X, prime to the $\mathbf{F}_q$-rational points of X, allows to define, both following Goppa [1] and Tietäväinen [6], some non linear codes, whose parameters can be bounded via some estimates for the above mentioned modified character sums.

In the first part, we study the modified character sums and their related Kummer coverings. The genus of a Kummer covering is given in § I. 1, Theorem 1. In § I. 2, we define the modified character sums, and obtain a bound for them (Theorem 2), from which we deduce a generalized Weil's inequality for Kummer coverings (§ I. 3, Theorem 3).

We construct in the second part the non linear geometric codes (§ II. 1), compute their parameters in § II. 2 (Theorem 4), and give some more explicit examples in § II. 3.

---

[1] Equipe CNRS "Arithmetique & Théorie de l'Information"

C.I.R.M. Luminy Case 916

13288 Marseille Cedex 9

France.

No proofs are given here. The proofs of the first part can be found in [3], and those of the second part in a forthcoming paper [4].

I. Multiplicative character sums

1. Kummer coverings. Let X be a smooth irreducible algebraic projective curve defined over F_q, K its rational function field and n an integer dividing $q - 1$. If P is a point of X, we denote by v_P the normalised valuation of K defined by P. Namely, for $f \in K$

$$v_P(f) = m \text{ if P is a zero of order m of f,}$$
$$- m \text{ if P is a pole of order m of f,}$$
$$0 \text{ if P is not a zero nor a pole.}$$

We then define the reduced order of f at P by

$$v_P'(f) = \underset{g \in K^*}{\text{Min}} \, (|v_P(fg^n)|),$$

(where $| x |$ is the absolute value of $x \in R$), in such a way that $0 \le v_P'(f) < n$. This number $v_P'(f)$ is the remainder of the euclidean division of $v_P(f)$ by n.

Let H_0 be a subgroup of K^* containing K^{*n} as a subgroup of finite index, let Y be the smooth model of $L = K(H_0^{1/n})$, and assume that L and K have the same constant field : we then say that H_0 is regular. The extension L/K is a Kummer extension [3], so that the corresponding covering $\pi : Y \to X$ is a Kummer covering. The following theorem gives the genus g_Y of Y, where

$$U'(f) = \{P \in X(\overline{F_q}), v_P'(f) \ne 0\},$$

and where $\overline{F_q}$ is an algebraic closure of F_q.

Theorem 1. *Let $r = (H_0 : K^{*n})$, g_X and g_Y the genus of X and Y. Then*

$$2g_Y - 2 = r(2g_X - 2) + \sum_f \sum_{u \in U'(f)} \deg u,$$

where the first sum runs over a system of representatives of H_0/K^{*n}.

2. Character sums. Let χ be a character of F_q^* of order n, and $k = F_q$. For $f \in K^*$, we define the sum

$$W(f) = \sum_{P \in X_*(F_q)} \chi(f(P)) \ ;$$

here, $X_*(F_q) = \{ P \in X(F_q), f(P) \neq 0, \infty \}$. A modified sum is more closely related to the Kummer covering $Y \to X$ defined in I.1. For $f \in K$ and $P \in X(F_q)$, let

$$\chi'(f(P)) = \chi(h(P)) \quad \text{if } f = g^n.h, \text{ h invertible at P,}$$
$$\chi'(f(P)) = 0 \quad \text{if } v_P'(f) = 0 \ ;$$

then define

$$W'(f) = \sum_{P \in X(F_q)} \chi'(f(P)).$$

Theorem 2. *Let H_0 be a regular subgroup of K^*, containing K^{*n} as a subgroup of finite index, and H' a system of representatives of the non vanishing classes of $H_0 \pmod{K^{*n}}$. Then*

$$| \sum_{f \in H'} W'(f) | \leq \frac{B(H_0)}{2} [2\sqrt{q}],$$

with

$$B(H_0) = (r - 1)(2g_X - 2) + \sum_{f \in H'} \sum_{u \in U'(f)} \deg u \ .$$

The proof of this theorem involves the theory of abelian L-function and the Riemann hypothesis for curves (the Weil theorem), by considering the L function related to f :

$$L'(T,f) = \exp(\sum_{s=1}^{\infty} \frac{T^s}{s} W_s'(f)),$$

with

$$W_s'(f) = \sum_{P \in X(F_{q^s})} \chi'(N_{F_{q^s}/F_q}(f(P))).$$

In the particular case where H_0 is the subgroup of K^* generated by $\phi \in K^*$ and K^{*n}, it is easyly seen that H_0 is regular if ϕ is not a constant function, so we obtain

Corollary 1. *Let ϕ be a non constant rational function on X. Then*

$$| \sum_{i=1}^{n-1} W'(\phi^i) | \leq (n-1)(g_X - 1 + \sum_{P \in U'(\phi)} \deg P)[2\sqrt{q}].$$

3. Number of points of a Kummer covering.

As a consequence of theorem 2, we obtain the following

Theorem 3. *Let X as above, K its rational function field, and $\pi : Y \to X$ be the Kummer covering defined by a subgroup H_0 of K^* containing K^{*n} as a subgroup of finite index. Then*

$$|\#Y(F_q) - \#X(F_q)| \leq \frac{B(H_0)}{2} [2\sqrt{q}].$$

Because of theorem 1, this estimate can also be written

$$|\#Y(F_q) - \#X(F_q)| \leq (g_Y - g_X)[2\sqrt{q}].$$

This is an improvement, in this case, of Weil's inequality $|\#Y(F_q) - \#P_1(F_q)| \leq g_Y[2\sqrt{q}]$. The same inequality has been proved by Lachaud in the case of a general abelian covering $\pi : Y \to X$ (see [2]).

II. Non linear geometric codes

1. The codes. Let X be as above and set $N = \#X(F_q)$. Let G be a divisor on X prime to $X(F_q)$, and χ a character of order n of F_q^* with value in the group $\mu_n(C)$ of n-th roots of unity in C. Consider the map

$$c : L(G) \to (\mu_n(C) \cup \{0\})^N$$
$$f \to (c_P(f))_{P \in X(F_q)},$$

where

$$c_P(f) = \chi'(f(P)),$$

and χ' is as defined in I.2. We define the code $\Gamma = \Gamma(q, X, G, n, \chi)$ as the image of L(G) under c.

2. Parameters of Γ. The following is a lower bound for the Hamming distance between two codewords in terms of the above character sum :

Lemma 1. *For $f, g \in L(G)$, letting $\phi = f.g^{n-1}$; then*

$$d(c(f), c(g)) \geq \frac{n-1}{n} N - \frac{1}{n} \sum_{i=1}^{n-1} W'(\phi^i) - \deg G.$$

Remark. It is possible to give an upper bound for $d(c(f), c(g))$.

Corollary 1 and Lemma 1 enable to give an estimate for the parameters of Γ :

Theorem 4. *If*

$$N > (g - 1 + 2 \deg G)[2\sqrt{q}]) + \frac{n}{n-1} \deg G,$$

then $\Gamma(X, G, \chi)$ is a non linear code of length $N = \#X(F_q)$ on an alphabet with $(n + 1)$ elements, of minimum distance

$$d_{min}(\Gamma) \geq \frac{n-1}{n} (N - (g - 1 + 2 \deg G)[2\sqrt{q}]) - \deg G,$$

and of cardinality

$$M = \#\Gamma \geq q^{\deg G + 1 - g}.$$

3. Examples. Let $N_g(q)$ (resp $n_g(q)$) be the maximum (resp. minimum) number of F_q-rational points of an algebraic smooth projective irreducible curve of genus g defined over F_q. Such a curve will be called maximal (resp. minimal) if it reaches this bound. Moreover, let $k = \frac{\log M}{\log (n + 1)}$ be the "dimension" of the $(N, M, d)_{n+1}$ code.

a. Codes from the projective line. The projective line $P_{F_q}^1$ has genus 0 and $q + 1$ rational points over F_q. If we choose a divisor G of degree m and a quadratic character χ of F_q^*, (that is for $n = 2$ and q odd), we obtain the following

Proposition 1. *For all powers q of an odd prime number and all integers $m < \frac{q + 1 + [2\sqrt{q}]}{2([2\sqrt{q}] + 1)}$, there exists a F_3 non linear code with parameters*

$$\left(q + 1, k = (m + 1) \frac{\log q}{\log 3}, d \geq \frac{q + 1 + [2\sqrt{q}]}{2} - m([2\sqrt{q}] + 1) \right)_3.$$

b. **Codes on elliptic curves.** The above construction on an extremal elliptic curve X over F_q, $n = 2$, a divisor G of degree m and a quadratic character χ of F_q^*, gives

Proposition 2. *For all powers q of an odd prime number and all integers* $m < \dfrac{N_1(q)}{2([2\sqrt{q}] + 1)}$ *, there exists a* F_3 *non linear code with parameters*

$$\left(N_1(q), \ k \geq m \frac{\log q}{\log 3}, \ d \geq \frac{N_1(q)}{2} - m([2\sqrt{q}] + 1) \right)_3,$$

and a similar result holds if we replace $N_1(q)$ *by* $n_1(q)$.

Note that the exact values of $N_1(q)$ and $n_1(q)$ are known (cf for example [5]).

Examples.
1) For q = 127, $N_1(127) = 150$ and $n_1(127) = 106$.

$m = 2$ gives a $(150, k \geq 2\dfrac{\log 127}{\log 3}, d \geq 29)_3$ and a $(106, k \geq 2 \dfrac{\log 127}{\log 3}, d \geq 7)_3$

code.

2) q = 1033 $N_1(1033) = 1098$ $n_1(1033) = 970$

$m = 6$ $(1098, k \geq 6\dfrac{\log 1033}{\log 3}, d \geq 159)_3$ $(970, k \geq 6\dfrac{\log 1033}{\log 3}, d \geq 95)_3$

c. **Codes over F_4 from curves of genus 2.** If we consider extremal curves of genus 2 over F_q for $q \equiv 1 \pmod 3$, and multiplicative characters of F_q^* of order 3, we obtain

Proposition 3. *For all powers q of a prime number,* $q \equiv 1 \pmod 3$*, all integers* $g \in N$*, and all integer m such that*

$$m < \frac{N_2(q) - [2\sqrt{q}]}{2[2\sqrt{q}] + 3/2},$$

there exists a non linear F_4*-code with parameters*

$$\left(N_2(q), \ k \geq (m-1)\frac{\log q}{\log 4}, \ d \geq \frac{2}{3}(N_2(q) - (1+2m)[2\sqrt{q}]) - m \right)_4,$$

and a similar result holds if we replace $N_2(q)$ by $n_2(q)$.

A general formula for $N_2(q)$ and $n_2(q)$ can be found in [5].

Examples.
1) $q = 511$ $N_2(511) = 602$ $n_2(511) = 422$

$m = 2$ $(602, k \geq \frac{1}{2} \frac{\log 511}{\log 2}, d \geq 210)_4$ $(422, k \geq \frac{1}{2} \frac{\log 511}{\log 2}, d \geq 100)_4$

$m = 5$ $(602, k \geq 2 \frac{\log 511}{\log 2}, d \geq 37)_4$

2) $q = 1033$ $N_2(1033) = 1162$ $n_2(511) = 906$

$m = 5$ $(1162, k \geq 2 \frac{\log 1033}{\log 2}, d \geq 263)_4$ $(906, k \geq 2 \frac{\log 1033}{\log 2}, d \geq 87)_4$

$m = 6$ $(1162, k \geq \frac{5}{2} \frac{\log 1033}{\log 2}, d \geq 178)_4$

4. Conclusion. The above construction gives non linear codes on any alphabet with $n + 1 \geq 3$ elements. Lemma 1 links the Hamming distance and some character sums, and estimations on these character sums enable us to give a lower bound for the minimum distance, and to compute the cardinality of these codes under a technical assumption (theorem 4). Because of the generality of the estimations we used, one can expect the true minimum distance to be much greater than the given lower bound in many cases. Numerical computations could give information, for example, in the case of the code constructed from the space of polynomials of given bounded degree on the projective line, and from the quadratic character.

Bibliography

[1]   Goppa, V. D.*Algebraico-geometric codes*, Math. USSR Izv., **21** (1983), pp.75-91.

[2]   Lachaud, G. *Artin-Schreier curves, exponential sums and the Carlitz-Uchiyama bound for geometric codes*, accepted in Jour. of Number Theory.

[3]   Perret, M. *Multiplicative Character Sums and Kummer Coverings*, accepted in Acta Arithmetica.

[4]   Perret, M. *Non linear geometric codes*, in preparation.

[5]   Serre, J. P. *Nombre de points des courbes algébriques sur F_q*, Sém de Theorie des Nombres de Bordeaux 1982/83, n° **22** = Œuvres, n° 129, vol. III, pp. 664-668, Springer, New york 1986.

[6]   Tietäväinen, A. *Character sums and applications of coding theory*, Ann. Univ. Turkuensis, Ser. A I, **186** (1984), pp. 110-117.

SECTION 4

PROTECTION OF INFORMATION

A survey of identification schemes

Marc Girault

Service d'Etudes communes des Postes et Télécommunications (SEPT)
42 rue des Coutures, BP 6243, 14066 CAEN, France.

Abstract. *The goal of this lecture is to present and compare a large variety of recent cryptographic schemes used to corroborate the identity claimed by an entity. The protocols used in these schemes are minimum (or even zero)-knowledge protocols, in that they provably leak no substantial knowledge on the entity's secret key, provided the mathematical problems on which they lie are effectively hard. Are considered schemes based on factorization problem (Fiat-Shamir, Guillou-Quisquater, ...), on discrete logarithm problem (Chaum et al., Beth, Schnorr, ...), or both (Brickell-Mc Curley, Girault). We also show how less traditional problems -at least in cryptography- are also used to build such schemes, as general linear decoding problem (Stern) or permuted kernels problem (Shamir). In this summary, we introduce the reader to the topic, say a little word about each of the schemes mentioned above, describe some of them in detail and provide a bibliography.*

1. INTRODUCTION

Alice and Bob are having a nice telematic discussion but before obeying an order coming from Alice, Bob would like to make sure that he communicates with her (and not with the impostor Charlie). Bob could request a password from Alice but an eavesdropper could catch this password and later impersonate Alice (Note that this "bad" solution is still the most popular one at the present time).

A much better idea consists to share a secret K used as a parameter of a cryptographic algorithm f, such as DES [DES]. When Bob wants to authenticate Alice, he sends her a random number X (the "challenge") and she replies with the number $f_K(X)$ (the "response"). Now, Bob compares this response with the one he has himself calculated. If the two responses match, he is reasonably convinced that he communicates with somebody who knows K -that is to say : Alice.

The latter solution is often satisfactory but nonetheless has some drawbacks. One of these drawbacks is the necessity for each pair of users to share a specific secret key K. In a 1 million-user network, each user should store 1 million (minus one) keys, which leads to a total of one half trillion secret keys -a rather large number.

There are several ways to avoid such undesirable situations. One of them is to use so-called public-key algorithms [DH], such as RSA [RSA], discovered in the later-70s. With these algorithms, Bob needs not share a secret information with Alice. He only needs to know a public information on Alice (her public key) whilst Alice owns the corresponding secret key (that nobody else knows).

The two keys, public and secret, are mathematically strongly connected, but knowing only the first one does not enable anyone to infer the second one from it in reasonable time. This (very surprising) trick is made possible by the existence of mathematical intractable problems -those problems that no efficient algorithm can solve at the present time. The most famous intractable problem used in cryptography is factorization problem (given n, find the prime factors of n), on which RSA is based and for which no polynomial or "almost" polynomial algorithm is known.

The "challenge-response" procedure can also be used with public-key algorithms. Alice calculates the response with her secret key (as she did before) but Bob is no longer able to calculate it (he does not know the secret key). Nonetheless, Bob is able to *check* the response of Alice with the help of her public key. In summary, Bob sends a question to Alice, the answer of which he is unable to calculate, but he is able to verify.

Alas, we have not yet reached the paradise of authentication. There remains a somewhat theoretical but important problem. With "challenge-response" procedures, it could occur that some enemy extract some information about the secret key involved, from the data which are exchanged. In other words, we are not sure that such a procedure does not leak some information about the secret key, especially if the enemy is the challenger himself, choosing "random" numbers in a special biased way.

In 1985, the concept of "zero-knowledge interactive proofs" is introduced by Goldwasser, Micali and Rackoff [GMR]. With such proofs, Alice can convince Bob that she knows the solution of a hard problem without revealing anything about this solution. Does it lead to practical identification schemes ? Yes, answered Fiat and Shamir a year later [FS]. Since that time, a lot of other schemes have been proposed. The goal of our lecture is to present and compare a large variety of them from a theoretical and practical point of view.

2. A BRIEF SURVEY

2.1 The ancestor : the Fischer-Micali-Rackoff protocol

At EUROCRYPT'84 conference, Fischer, Micali and Rackoff [FMR] designed a protocol in order to render secure the so-called oblivious transfer protocol due to Rabin [R1]. In this paper, strangely unpublished in EUROCRYPT'84 proceedings, one can find the basic ideas of both zero-knowledge concept and Fiat-Shamir scheme.

Two years later, Fiat and Shamir have the idea to apply this protocol to identification [FS]. Their protocol is essentially the same as FMR's one, and is proved to be zero-knowledge -in the mean time, Goldwasser, Micali and Rackoff had invented zero-knowledge concept- assuming factorization is a hard problem.

2.2 General set-up for identification schemes

All the identification schemes use more or less a similar general set-up. When Alice wishes to use one of these schemes, she visits a trusted authority (or authentication center) in order to register her public key P. The authority verifies Alice's identity, generates an identification string I and computes an electronic signature of the pair (I,P), called certificate. This certificate will later be used by verifiers to convince themselves that P is effectively the public key of Alice (and not of somebody else). Such schemes are called certificate-based schemes.

It is also possible in some schemes to choose the public key P in first and then calculate the secret key with the help of a secret information held by the center (and only by it). In that case, P can be chosen equal to I, so that the authority needs not any more generate a certificate. These schemes, in which public key and identification strings are the same number, are called identity-based schemes and have been introduced by Shamir [Sh1]. An identity-based scheme can always be converted into a certificate-based scheme, but the converse is not true.

2.3 Schemes based on factorization problem

2.3.1 Set-up of Fiat-Shamir scheme

Fiat-Shamir scheme is based on the intractability of factorization problem. Let n be the product of two (say) 256-bit primes p and q. In the identity-based version, the one we will describe, the authority must hold p and q. In the non identity-based version, p and q can be "destroyed" as soon as n is calculated.

Alice's secret key s is calculated as the "square root" modulo n of her identification string I (in other words : $s^2 = I \mod n$). We do not deal in this summary with some technical points (e.g. only one integer over four is a square and almost each square has four square roots) but emphasize the fact that only those who know p and q can efficiently calculate s. More precisely, factorization and "square root modulo n" problems have been proved to be, in some sense, equivalent ones [R2].

Now, Alice will convince anybody she is Alice by proving she knows s (loosely speaking : the square root of her name) without revealing anything about s. This can be done with a three-move Fischer-Micali-Rackoff-like protocol, which can be briefly described in the following (and somewhat picturesque) way [Gi1] :

2.3.2 A picturesque description of Fiat-Shamir protocol

First, Alice randomly splits her secret key s into two "half-keys" s_0 and s_1 such that $s = s_0 s_1$ (mod n). Then she computes $s'_0 = s^2_0$ (mod n), takes a playing card, writes s_0 on one side and s'_0 on the other one. She does the same with s_1 and another card, then puts down both cards on the table, letting s'_0 and s'_1 visible. Note that $s'_0 s'_1 = I$ (mod n).

At this time, the verifier (Bob) tosses a coin and, depending on the result, requires either the first card or the second one to be turned upside down. Only one card must be turned, because turning both would reveal s.

After execution, the verifier checks the consistency of the two sides of the card which has been turned. Explicitly : is the number on the first side equal to the square modulo n of the other side ? If it is, he is satisfied. Otherwise, he rejects Alice. In both cases, he learnt nothing about s, since knowledge of only s_0 or s_1, which are individually perfectly random, does not reveal even a little information on s.

At this point, the reader may understand that only Alice is able to follow this protocol without rejection when it is repeated a number of times. An impostor, who does not know s, is unable to present cards which are both consistent. Actually, the best he can do is to try to guess the outcome of the coin toss (one chance over two), and to prepare the corresponding card in a consistent way. But as soon as his guess is wrong, he will be detected. Consequently, if the protocol is repeated twenty times, he will succeed with only one chance over 2^{20} (= one million).

In order to avoid too many repetitions, Fiat and Shamir have proposed to use several identification strings instead of only one. The impostor non-detection probability is proved to be equal to 2^{-kL}, where k is the number of strings and L the number of repetitions.

2.3.3 Some variants

Another interesting variant of the basic protocol has been proposed by Guillou and Quisquater [GQ] and, slightly later, by Ohta and Okamoto [OO]. In their modification, 2 is replaced by a larger prime exponent e, and we have $s^e = I$ (mod n). Instead of sending a random bit to A, B sends a random integer c smaller than e. This enables to achieve a very good level of security with only three moves, at the cost of more computations than Fiat-Shamir scheme. Some other variants ([MS], [OS],...) have also be proposed.

2.4 Schemes based on discrete logarithm problem

The discrete logarithm problem is the following one : let n be an integer, g and x be two integers smaller than n, and $y = g^x$ (mod n) ; given n, g and y, find x. This problem seems to be of similar complexity to factorization problem, and is also widely used in cryptography.

After preliminary works of Chaum, Evertse and Graaf [CEG], Beth [Be] and Günter [Gu], Schnorr proposed in 1989 an identification scheme based on discrete logarithm problem [Sc]. In that case, we have $P = g^s$ (mod p), where p, a prime, and g are both public universal numbers, and s is Alice's secret key. This scheme has some advantages on Fiat-Shamir one (e.g. most of Alice's computations can be done in advance) but also some drawbacks (Bob performs much more computations).

Some variants of Schnorr's scheme have already been proposed. The first one is secure if either of factorization and discrete logarithm problems is intractable [BC]. The second one is a non certificate-based version of Schnorr's scheme [Gi2] or [GP] (see also the last paragraph).

2.5 Schemes based on other problems

Considering that public key cryptography had become dangerously dependent on the difficulty of only two problems (factorization and discrete logarithm), some researchers have recently attempted to design new identification schemes, based on other types of problems, which can be generically called "matrix problems". We now present two such schemes.

The first one is due to Stern [St], and is based on general linear decoding problem, a NP-hard one. In this scheme, all the users share a (k-m) randomly built matrix A, with k<m, and a parameter w<<m. Alice's secret key s is a m-bit word s of weight w, and her public key is P = As. If A is seen as the parity-check matrix of a linear error-correcting code, computing s from A and P comes to finding a small-weight word of given syndrome P, a very hard problem. Now, Alice proves to Bob that she knows s by following a protocol in three moves (in the most recent version).

The second one is due to Shamir [Sh2] and is also based on a NP-hard problem, called "permuted kernels problem". All the users share a prime p and a (k-m) matrix A with coefficients in the finite field GF(p). Alice's secret key is a permutation s of the interval [1,m], and her public key is the m-vector P = sU where U is an element of the kernel of A. Computing s from p, A and P comes to finding the permutation of a vector which comes into the kernel of A, a very hard problem. Now, Alice proves to Bob that she knows s by following a protocol in five moves.

2.6 A new type of schemes

In certificate-based schemes, the trusted authority has to certify Alice's public key and Bob has to check this certificate as a preliminary step. This process leads to extra storage and extra computation. In identity-based schemes, there are no longer certificates but Alice's secret key is computed by the authority, which therefore can impersonate her without being detected. Is it possible to design schemes in which both drawbacks simultaneously vanish ? In [Gi3] we answer positively, using the concept of "self-certified" public keys.

3. A DETAILED DESCRIPTION OF SOME SCHEMES

Only brute versions of the protocols are described in this section (not the optimizations). By convention, $x^0 = 1$ (mod n).

3.1 Identification schemes based on factorization problem

All the schemes of this section are identity-based (but, consequently, also have a certificate-based version).

3.1.1 <u>Fiat-Shamir scheme</u>

Actually, Fiat-Shamir scheme is based on the difficulty of extracting square roots modulo a composite number, but this problem is provably equivalent to factorization problem.

Set-up

- The authority computes the product n of two large primes p and q. The typical size of n is 512 bits.
- Let F be a pseudo-random function and I be Alice's identification string. The authority computes $I_j = F(I,j)$ for small values of j, until it obtains k quadratic residues, and computes square roots s_j of I_j^{-1} (mod n) for these values of j. For commodity, we will assume that j=1...k. Therefore, for any such j : $s_j^2 I = 1$ (mod n).
- Alice's memory contains :
  * universal parameters : n,F
  * public parameters : I,k
  * secret parameters : s_j for j=1...k

Basic protocol

When k=1, Alice holds only one secret denoted by s. She proves to Bob she is Alice by convincing him that she knows s, a square root of I^{-1} modulo n, without revealing it. This can be done in the following way :

1. Alice sends I to Bob

   Repeat L times :

2. Alice randomly selects an integer x<n, computes $t = x^2$ (mod n) and sends t to Bob.
3. Bob randomly selects a bit c and sends c to Alice.
4. Alice computes $y = xs^c$ (mod n) and sends y to Bob.
5. Bob checks that $y^2 I^c = t$ (mod n).

Note that Bob will accept Alice with probability 1 (completeness property) and will detect the impostor Charlie with probability $1-2^{-L}$ (soundness property). Moreover, the protocol does not reveal anything on s, since it is "scrambled" with the random number x (zero-knowledgeness property).

General protocol

1. Alice sends (I,k) to Bob, who computes $I_j = F(I,j)$ for $j = 1...k$.

   Repeat L times :

2. Alice randomly selects an integer x<n, computes $t = x^2$ (mod n) and sends t to Bob.
3. Bob randomly selects a binary vector $c = (c1,...,ck)$ and sends c to Alice.
4. Alice computes $y = xs_1^{c1}...s_k^{ck}$ (mod n) and sends y to Bob.
5. Bob checks that :

$$y^2 I_1^{c1}...I_k^{ck} = t \text{ (mod n)}$$

Note that the security level (impostor detection probability) is now $1-2^{-kL}$.

This scheme is much faster than RSA scheme while it is theoretically safer. But there are many bits to store and to exchange.

3.1.2 Guillou-Quisquater scheme

Guillou-Quisquater scheme is based, like RSA, on the difficulty of extracting e-th roots modulo a composite number, which is practically equivalent to factorization problem. Apart from that, it is similar to Fiat-Shamir scheme.

Set-up

- The authority computes the product n of two large primes p and q, and two integers e and d such that ed = 1 (mod [p-1][q-1]). The typical size of n (resp. e) is 512 bits (resp. 30 bits).
- Let I be Alice's identification string. The authority computes the e-th root s of I^{-1} (mod n), that is : $s = I^{-1/e}$ (mod n) = I^{-d} (mod n). Therefore $s^e I = 1$ (mod n).
- Alice's memory contains :
    * universal parameters : n,e
    * public parameter : I
    * secret parameter : s

Protocol

Alice proves to Bob she is Alice by convincing him that she knows s, the e-th root of I^{-1} modulo n, without revealing it. This can be done in the following way :

    1. Alice sends I to Bob.
    2. Alice randomly selects an integer x<n, computes $t = x^e$ (mod n) and sends t to Bob.
    3. Bob randomly selects an integer c<e and sends c to Alice.
    4. Alice computes $y = xs^c$ (mod n) and sends y to Bob.
    5. Bob checks that $y^e I^c = t$ (mod n).

Note that there is only one secret and only one round, but the number of multiplications is larger than in Fiat-Shamir scheme. The security level is equal to $1-2^{-e}$.

3.2 Identification schemes based on discrete logarithm problem

These schemes are all certificate-based but the way certificates are computed by the authority and checked by Bob is out of the description of these schemes.

3.2.1 Chaum-Evertse-van der Graaf scheme

This scheme is based on the difficulty of computing discrete logarithms modulo a prime number.

Set-up

    - The authority computes a large prime p and a generator g of Z_p^*, the multiplicative group of the ring of non-zero integers smaller than p. The typical size of p is 512 bits.
    - Alice randomly selects a secret key s<p-1, then computes her public key $P = g^{-s}$ (mod p), and gives P to the authority.
    - The authority computes a certificate Cert for this public key and gives it to Alice.
    - Alice's memory contains :
            * universal parameters : p,g
            * public parameters : I,P,Cert
            * secret parameter : s

Protocol

Alice proves to Bob she is Alice by convincing him that she knows s, the opposite of the discrete logarithm of P modulo p. This can be done in the following way :

    1. Alice sends I, P and Cert to Bob, who checks the certificate.

Repeat L times:

2. Alice randomly selects an integer x<p-1, computes $t = g^x$ (mod p) and sends t to Bob.
3. Bob randomly selects a bit c and sends c to Alice.
4. Alice computes $y = x + sc$ (mod p-1) and sends y to Bob.
5. Bob checks that $g^y P^c = t$ (mod p).

Properties of this protocol are similar to those of Fiat-Shamir basic protocol.

3.2.2 Schnorr scheme

We only describe Schnorr scheme without its specific optimizations The set-up is the same as in Chaum-Evertse-Van der Graaf scheme, except that there is an additional universal parameter e.

Protocol

1. Alice sends I, P and Cert to Bob, who checks the certificate.
2. Alice randomly selects an integer x<p-1, computes $t = g^x$ (mod p) and sends t to Bob.
3. Bob randomly selects an integer c<e and sends c to Alice.
4. Alice computes $y = x + sc$ (mod p-1) and sends y to Bob.
5. Bob checks that $g^y P^c = t$ (mod p).

Properties of this protocol are similar to those of Guillou-Quisquater protocol.

3.3 Identification schemes based on matrix problems

For lack of room, we only describe Stern scheme, which is based on general linear decoding problem (see section 2.4).

Set-up

- The authority selects a random binary (k-m) matrix A with k<m, an integer w<<m and a pseudo-random function F. The typical size of k (resp. m, resp. w) is 256 bits (resp. 512 bits, resp. 50 bits).
- Alice randomly selects a secret m-bit word s of weight w, computes her public key P = As and gives P to the authority.
- The authority computes a certificate Cert for this public key and gives it to Alice.
- Alice's memory contains :
    * universal parameters : A,w,F
    * public parameters : I,P,Cert
    * secret parameter : s

Protocol

Alice proves to Bob she is Alice by convincing him that she knows a m-bit word of weight w and of syndrome P. This can be done in the following way :

1. Alice sends I, P and Cert to Bob, who checks the certificate.

      Repeat L times:

2. Alice randomly selects a m-bit word and a random permutation U of the interval [1,m]. Then she computes $T_1 = F(U,Ax)$, $T_2 = F(Ux)$ and $T_3 = F(U[x+s])$, where + stands here for bitwise Exclusive-Or operation, and sends T_1, T_2 and T_3 to Bob.
3. Bob randomly selects an element of {0,1,2} and sends c to Alice.
4. Alice sends y to Bob, where :
        * if c=0, y = (U,x)
        * if c=1, y = (U,x+s)
        * if c=2, y = (Ux, U[x+s])
5. Bob checks :
        * if c=0, that $F(U,Ax) = T_1$ and $F(Ux) = T_2$
        * if c=1, that $F(U,A[x+s]+P) = T_1$ and $F(U[x+s]) = T_3$,
        * if c=2, that $F(Ux) = T_2$, $F(U[x+s]) = T_3$, and the weight of the word (Ux + U[x+s]) is equal to w.

Properties of this protocol are similar to those of Fiat-Shamir basic protocol, except that the impostor detection probability is 1/3 instead of 1/2 for each round. As a consequence, a greater number of rounds will be necessary to achieve the same security level.

4. BIBLIOGRAPHY

[Be] T. Beth, "A Fiat-Shamir-like authentication protocol for the ElGamal scheme", Advances in Cryptology, Proc. of EUROCRYPT'88, LNCS 330, Springer-Verlag, 1988, pp.77-86.

[BC] E. F. Brickell and K. S. McCurley, "An interactive identification scheme based on discrete logarithms and factoring", Proc. of EUROCRYPT'90, to appear.

[CEG] D. Chaum, J.H. Evertse and J. van de Graaf, "An improved protocol for demonstrating possession of discrete logarithms and some generalizations", Advances in Cryptology, Proc. of EUROCRYPT'87, LNCS 304, Springer-Verlag, 1988, pp. 127-141.

[DES] Data Encryption Standard, National Bureau of Standards, Federal Information Processing Standards, Publ. 46, 1977.

[DH] W. Diffie and M. Hellman, "New directions in cryptography", IEEE Transactions on Information Theory, Vol.IT-22, Nov.1976, pp.644-654.

[FMR] M.J. Fischer, S. Micali and C. Rackoff, "A secure protocol for the oblivious transfer", presented at EUROCRYPT'84, Paris, 9-11 Apr.1984.

[FS] A. Fiat and A. Shamir, "How to prove yourself : Practical solutions to identification and signature problems", Advances in Cryptology, Proc. of CRYPTO'86, LNCS 263, Springer-Verlag, 1987, pp.186-194.

[Gi1] M. Girault, "Recent solutions to entity authentication problem", Proc. of ARESAD seminar on Analysis and Management of International Crises, to appear.

[Gi2] M. Girault, "An identity-based identification scheme based on discrete logarithms modulo a composite number", Proc. of EUROCRYPT'90, to appear.

[Gi3] M. Girault, "Self-certified public keys", submitted to EUROCRYPT'91.

[Gu] C.G. Günter, "Diffie-Hellman and ElGamal protocols with one single authentication key", Proc. of EUROCRYPT'89, to appear.

[GP] M. Girault and JC. Paillès, "An identity-based identification scheme providing zero-knowledge authentication and authenticated key exchange", Proc. of ESORICS 90, pp. 173-184.

[GQ] L.C. Guillou and J.J. Quisquater, "A practical zero-knowledge protocol fitted to security microprocessors minimizing both transmission and memory", Advances in Cryptology, Proc. of EUROCRYPT'88, LNCS 330, Springer-Verlag, 1988, pp.123-128.

[GMR] S. Goldwasser, S. Micali and C. Rackoff, "The knowledge of interactive proof-systems", 17th ACM Symposium on Theory of Computing, 1985, pp.291-304.

[MS] S. Micali and A. Shamir, "An improvement of the Fiat-Shamir identification and signature scheme", Advances in Cryptology, Proc. of CRYPTO'88, LNCS 403, Springer-Verlag, 1989, pp.244-247.

[OO] K. Ohta and T. Okamoto, "A modification of the Fiat-Shamir scheme", Advances in Cryptology, Proc. of CRYPTO'88, LNCS 403, Springer-Verlag, 1989, pp.232-243.

[OS] H. Ong and C.P. Schnorr, "Fast signature generation with the Fiat-Shamir scheme", Proc. of EUROCRYPT'90, to appear.

[R1] M.O Rabin, "How to exchange secrets by oblivious transfer", manuscript, Harvard Center for Research in Computer Technology, 1981.

[R2] M.O Rabin, "Digitalized signatures and public-key functions as intractable as factorization", MIT, Laboratory for Computer Science, MIT/LCS/TR-212, Jan.1979.

[RSA] R.L. Rivest, A. Shamir and L. Adleman, "A method for obtaining digital signatures and public-key cryptosystems", CACM, Vol.21, n°2, Feb.1978, pp.120-126.

[Sc] C.P. Schnorr, "Efficient identification and signatures for smart cards", Advances in Cryptology, Proc. of CRYPTO'89, LNCS 435, Springer-Verlag, pp.239-252.

[Sh1] A. Shamir, "Identity-based cryptosystems and signature schemes", Advances in Cryptology, Proc. of CRYPTO'84, LNCS 196, Springer-Verlag, 1985, pp.47-53.

[Sh2] A. Shamir, "An efficient identification scheme based on permuted kernels", Advances in Cryptology, Proc. of CRYPTO'89, LNCS 435, Springer-Verlag, pp.606-609.

[St] J. Stern, "An alternative to the Fiat-Shamir protocol", Proc. of EUROCRYPT'90, to appear ; new version presented at Oberwolfach seminar on Cryptography, 24-30 Sept. 89.

A Correlation Cryptographic Scheme

Sami Harari

Groupe d'Etude du Codage de Toulon
Université de Toulon et du Var
B.P. 132
83957 La Garde Cedex France

0. Summary

In [1] McEliece introduced a public key cryptosystem using error correcting codes. The message is coded into codewords by using a public generator matrix G. The key to the system is an error vector of weight smaller than the error capacity which is added to the coded information vector. Polynomial time attacks have been devised against this scheme.

In this work we introduce a secret key coding scheme which relies on a subset S of a particular set of random codes of known minimum distance and for which a decoding algorithm is known. The cryptogram vector is this case too made of the sum of a codeword and a noise vector. However, the vector containing information is the "noise" vector and the masking vector is the "codeword". The code itself is randomly chosen at each message in the corresponding set.

The robustness of the system is studied and, in particular, the attacks on the McEliece scheme [2] do not hold in this scheme, as well as the attacks that succeed when the set of codes is reduced to exactly one code.

This cryptosystem has the property that, if a set of matrices can be put in a tamper proof environment, then a user does not need to know any secret information to operate the system.

A conventional public key version of the cryptosystem, which needs more on line redundancy to achieve the same level of secrecy, will be given in a forthcoming paper.

As it is the case for the McEliece cryptosystem the new system introduces redundancy. These redundancy requirements are also studied.

1. Introduction.

In [1] McEliece introduced a cryptographic scheme using error correcting codes. The message is coded into codewords by using a public generator matrix G. The key to the system is an error vector of weight smaller than the error capacity. The cryptogram is obtained by adding the "key" to the codeword, and applying to it a secret permutation. To decipher such a cryptogram the legitimate user has to first apply the inverse permutation to the cryptogram, then the error vector is added to the resulting vector and the information is then extracted from the codeword. At first glance the cryptanalyst is faced with the decoding of an apparently random code which is a NP complete problem in the general case. However a polynomial time attack has been devised by Brickell [3] and refined by Adams and Meijer [4]. The attack exploits the fact that the "key" vector is of weight less than the error capacity of the code, and therefore statistically most of the bits of the secret codeword are not affected by it, and that the linear code that is used is MDS.

In this work, after recalling the McEliece algorithm and its weakness, we introduce first a secret key cryptosystem which relies on a particular set of random codes of known minimum distance and for which a decoding algorithm, obtained by the study of the weight of Hadamard product of vectors, is possible. The code that is used for enciphering is chosen at random in the set of possible codes.

The cryptogram vector is in this case too made of the sum of a codeword and a noise vector. However, the vector containing information is the "noise" vector and the masking vector is the "codeword". It is shown that breaking the system is an NP complete problem and in particular the attacks on the McEliece scheme [3][4] do not hold against this scheme since no assumption is made on the characteristics of the code, and the weight of the "noise vector" is in no way limited. This cryptosystem introduces redundancy. However deciphering can be done in a very efficient manner, even for codes of length that are much larger then the lengths of the codes that are used in the original McEliece scheme.

This secret key cryptosystem is, in a certain way, a public key cryptosystem since the system can be operated by a user who has no knowledge of any secret information, if a set of matrices can be stored in a tamper proof environment.

For a given level of security, there is a tradeoff between the volume of of secret information to be stored and the on line redundancy.

182

This work begins with the study of some properties of random codes, and the expected values of the weight of correlation vectors between codewords and random vectors for the special set of random codes that are considered in the algorithms.

2. The McEliece algorithm [1].

McEliece's system works as follows: The system user (receiver) secretly constructs a linear t-error correcting Goppa code with $k \times n$ code generator G, a $k \times k$ scrambler matrix S that has an inverse over $GF(2)$, and a $n \times n$ permutation matrix P. Then he computes

$$\overline{G} = S.G.P$$

which is also a linear code but supposedly hard to decode. He publishes $\overline{G}$ as his encryption key. The sender encrypts a k- bit message vector m into an n-bit cryptogram c as

$$c = m.\overline{G} + e$$

where e is a random n-bit error vector of weight less than or equal to t. The receiver computes $c.P^{-1} = (m.S).G + e.P^{-1}$ and uses the decoding algorithm to get rid of $e.P^{-1}$. Finally to get m he descrambles $m.S$ by multiplying it by S^{-1}.

There have been several methods proposed for attacking McEliece's system [4]. Among them, the best attack with least complexity, is to repeatedly select k bits at random from the n-bit cryptogram vector c to form c_k in hope that the selected bits are not in error. If there is no error among them, then $c_k.\overline{G_k}^{-1}$ is equal to m where G_k is the $k \times k$ matrix obtained by choosing k columns of $\overline{G}$ according to the same selection of c_k.

3. Random codes.

3.1 Definition Let $\vec{v} = (v_0, \ldots, v_{n-1})$ be a binary vector of length n. The *support* S of $\vec{v}$ is the subset of indices of $\{0, \ldots, n-1\}$ such that if i is in S then $v_i = 1$.

3.2 Definition Let $\vec{v} = (v_0, \ldots, v_{n-1})$ and $\vec{v}' = (v'_0, \ldots, v'_{n-1})$ be two binary vectors of length n. The *correlation vector* vector $h(\vec{v}, \vec{v}')$ of $\vec{v}$ and $\vec{v}'$, is defined by $h(\vec{v}, \vec{v}') = (v_0.v'_0, \ldots, v_{n-1}.v'_{n-1})$

3.2.1 Lemma Let $\vec{a}$ be a random binary vector of length n of weight n_1, and $\vec{b}$ be a random binary vector of length n of weight n_2 chosen independantly. The expected weight of $\vec{c} = h(\vec{a}, \vec{b})$ is equal to

$$w(\vec{c}) = \frac{n_1.n_2}{n}$$

3.2.2 Proof. In order to evaluate the expected weight of $\vec{c}$, consider the coordinates of the vectors as the outcome of the corresponding probability sources. We have that

$$P[c_i = 1] = P[(a_i = 1) \cap (b_i = 1)] = P[a_i = 1].P[b_i = 1] = (n_1/n).(n_2/n)$$

since the vectors are independant of each other. Therefore the weight of $\vec{c}$ is equal to

$$w(\vec{c}) = n.\frac{n_1.n_2}{n^2} = \frac{n_1.n_2}{n}$$

The algorithm relies on a set of (n, k) codes which have the property that, for each of them, there exists a generator matrix such that each of its rows has a support that is disjoint with the support of all the other rows. We will suppose that, in the rest of the paper, k divides n.

3.3 Generation of random codes.

In order to generate a (n, k) code with a random generator matrix having the property that the support of each of its rows is disjoint from the support of all the others one can operate as follows. Take a random permutation π of the set of indices of codewords. Let $\pi(0), \ldots, \pi(n-1)$ be the corresponding permutation table. Let k divide n and $\mu = n/k$. The i-th row of G for $i = 0, \ldots, k-1$ will have all its coordinates equal to 0 except for the indices $\pi(i.\mu), \pi(i.\mu + 1) \ldots, \pi(i.\mu + (\mu - 1))$.

3.3.1 Lemma. The weight enumerator $A(z)$ of a (n, k) code C having a generator matrix G with each of its rows having constant weight w and such that the support of each of the rows of G is disjoint from the support of all the others is given by the formula:

$$A(z) = \sum_{i=0}^{k} \binom{k}{i}.z^{wi}$$

3.4 Proof

The weight of the sum of two binary vectors $\vec{v}$ and $\vec{v'}$ of length n is given by:

$$weight(\vec{v} + \vec{v'}) = weight(\vec{v}) + weight(\vec{v'}) - 2.weight(h(\vec{v}, \vec{v'}))$$

where $h(\vec{v}, \vec{v'})$ is the correlation vector of $\vec{v}$ and $\vec{v'}$. If the code has a generator matrix G with the property stated in the hypothesis, then the weight of the sum of any j- tuple of the rows of G is equal to $j.w$. The lemma follows from the fact that the vectors of the code are obtained by taking all possible sums of the rows of the generator matrix.

3.5 Decoding of random codes.

3.5.1 Lemma (algebraic correlation) Let C be a code with generator matrix G made of rows $\vec{g}_i$ $i = 0 \ldots k-1$ of minimum weight μ having disjoint support. Let $\vec{r} = \vec{c} + \vec{e}$ be the sum of a codeword with an error vector with $w(\vec{e}) < \mu/2$. The algorithm to determine $\vec{c}$ given $\vec{r}$ is as follows:

a) Compute the correlation vectors $\vec{\epsilon}_i = h(\vec{r}, \vec{g}_i)$ $i = 0 \ldots k-1$ of the received codeword with each of the rows of the generator matrix. If the $w(\epsilon_i) < \mu/2$ then define $\lambda_i = 0$ else $\lambda_i = 1$.

b) With the coefficients λ_i determined as above, the closest codeword to $\vec{r}$ for the Hamming distance is then

$$\vec{c} = \sum_{i=0}^{k-1} \lambda_i \vec{g}_i.$$

3.5.2 Proof. Let $\vec{c} = \sum_{i=0}^{k-1} \lambda_i \vec{g}_i$. If for a certain index i the equality $\lambda_i = 1$ holds then the correlation of $\vec{c}$ and $\vec{g}_i$ is equal to μ. It follows that $w(\epsilon_i) > \mu/2$. On the other hand if $\lambda_i = 0$ then the correlation of $\vec{c}$ and $\vec{g}_i$ is equal to 0 and therefore $w(\epsilon_i) < \mu/2$. This ends the proof.

This algorithm has also the property that it can discriminate between codes having generator matrices with the same property.

3.5.3 Theorem (statistical correlation) Let G_j $j = 1, \ldots, l$ be a set of (n, k) generator matrices, each of them having rows with disjoint support of weight μ. Suppose these matrices are chosen independantely of each other. Let $\vec{r}$ be a vector in the linear space spanned by one of them say G_{j_0}. Let j be any index in the set $1, \ldots, l$. The weight of the correlation vector of $\vec{r}$ with each of the the rows of G_j has the following values:

i) 0 or μ if $j = j_0$.

ii) close to $\mu/2$ if $j \neq j_0$.

3.5.4 Proof. If $j = j_0$ the property follows from the proof of lemma 3.5.1. If $j \neq j_0$ then for the code G_j the vector $\vec{r}$ is a random vector of weight $n/2$. The correlation vector of this vector with a row of G_i, which is a random vector of weight μ, is therefore of weight close to $\mu/2$ by lemma 3.2.1.

4. The new algorithm : secret key case.

The new algorithm relies on the properties stated. Suppose A and B wish to exchange information in a secret manner.

1) A trusted center chooses a set of random permutations, $\sigma_j \quad j = 1, \ldots, s$, of the set $\{0, \ldots, n-1\}$. The center gives these permutations to A and B. These permutations are kept secret by everyone.

For these permutations, A and B compute generator matrices $G_j \quad j = 1, \ldots, s$ of random codes, each matrix having the property that each pair of its rows has disjoint support (cf the construction of 3.3). Let μ be the exact weight of the rows of G_j.

2) To encipher an information vector $\vec{\imath}$, A first associates to it a vector $\vec{m}$ of length n and of weight less than $[\mu/2]$. User A then chooses at random a matrix G_j in the set of matrices.

3) A then chooses $\vec{r}$ to be a random linear combination of the rows $\vec{g}_i$ of the matrix $G = G_j$ of weight $n/2$.

$$\vec{r} = \sum_{i=0}^{k-1} \lambda_i \vec{g}_i.$$

The cryptogram associated to $\vec{m}$ is the vector

$$\vec{c} = \vec{r} + \vec{m}$$

4) At the deciphering end B tries to find $\vec{m}$ by trying to recompose the vector $\vec{r}$.

a) He first determines the matrix G, that was used by A to construct $\vec{r}$, using a statistical correlation algorithm based on theorem 3.5.3. For each of the matrices $G_j \quad j = 1 \ldots, s$ B computes the weight of the the correlation vectors of $\vec{c}$ with each of the rows of G_j.

For exactly one index j_0 at least k of these weights have values either close to 0 or close to μ. If $k > \mu/2$ then $k - \mu/2$ of these products have values exactly 0 or μ. This index is the index of the matrix used by A for computing $\vec{c}$.

b) Then B proceeds to find $\vec{r} = \sum_{i=0}^{k-1} \lambda_i \vec{g}_i$. This means finding $\lambda_i \quad i = 0, \ldots, k-1$. In order to find the set of values that were used by A, B operates as follows using an algebraic correlation algorithm:

For each i in the set $0, \ldots, k-1$, B computes the correlation vector $h(\vec{c}, \vec{g}_i)$ of $\vec{c}$ with each of the vectors $\vec{g}_i$. If the weight of the result is greater than $\mu/2$ then $\lambda_i = 1$ else $\lambda_i = 0.$

At the end of the process B has an estimation of $\vec{r}$ given by $\vec{r} = \sum_0^{k-1} \lambda_i \vec{g}_i$ which he then adds to $\vec{c}$ and deduces $\vec{m}$ hence the information sent.

4.2 Correctness of the scheme.

The correctness of the scheme relies on the following properties:

a) The expected weight of the correlation vector of two random vectors chosen independantely: If a vector of weight $n/2$ is correlated with a vector of weight μ the result is of expected weight $\mu/2$ by theorem 3.5.3. This is the case when B is trying to determine the index j of the matrix that was used by A for enciphering the data.

b) The expected weight of the correlation vector of a vector of the code with one of the rows of the generator matrix: If a codeword of such a code is correlated with a row vector of weight μ of its generator matrix, the result is of expected weight either 0 or μ. If such a codeword is modified in less than $\mu/2$ positions the corresponding the correlation products will have values less than $\mu/2$ or greater than $\mu/2$. If $k > \mu/2$ then $k - \mu/2$ products will not be perturbed by the modification and will have their initial value (0 or μ). In an implementation extensive tests must be done on the codes to ensure that the statistical properties hold for the codes that will be actually be used.

c) The unique decoding property of an error correcting code. A codeword, here $\vec{r}$, is modified by a vector of weight less than half the minimum distance, here $\vec{m}$. Therefore a maximum likelihood decoding algorithm applied to $\vec{c}$ will in all cases yield $\vec{r}$.

The decoding algorithm can lead to very fast implementations in software and hardware since it is based upon the computation of correlation vectors. This computation of correlation can be done in efficient manner using bit AND operations on 32 bit words. It has also the property that it is invariant when permutations are applied to the indices of all the vectors, therefore it is the same decoding algorithm whatever the set of matrices used, as long as the disjoint support property is valid.

4.3 Robustness of the scheme.

The aim of the attacker is to obtain the set of secret matrices G_j for $j = 0, \ldots, s$ in order to be able to decrypt all cryptograms. For each of these matrices a brute force attack is equivalent to finding minimum weight vectors in a random code. This problem is believed to be NP complete [2]. The two attacks that can be thought of are the Hadamard product attack and the correlation attack. The probability of success of these attacks will be determined as a function of the system's parameters.

4.3.1 The correlation product attack.

a) The set of codes is reduced to one code.

Suppose that the set of matrices G_j is reduced to a single matrix G which is used many times to code information. Suppose that the intruder collects all the information interchanged between A and B. Suppose also that the vector $\vec{i}$ is known to the attacker. This is equivalent to choosing $\vec{m} = 0$ since enciphering is a linear operation.

Define recursively the Hadamard product of a set of l vectors to be the correlation of any vectors with the correlation vector of the remaining $l-1$. For example, in the case $l = 3$ the Hadamard product $h(\vec{c}_i, \vec{c}_j, \vec{c}_k)$ of $\vec{c}_i, \vec{c}_j, \vec{c}_k$ is defined by $h(\vec{c}_i, \vec{c}_j, \vec{c}_k) = h(\vec{c}_i, h(\vec{c}_j, \vec{c}_k))$.

Let $\vec{c}_1, \vec{c}_2, \vec{c}_3 \ldots$ be a sequence of cryptograms recorded by the attacker. Consider the following sequence of vectors:

$$h(\vec{c}_1, \vec{c}_2), \quad h(\vec{c}_1, \vec{c}_2, \vec{c}_3), \ldots$$

By 4.2.a this sequence of codewords has decreasing weights and, eventually, has for weights the following sequence of weights:

$$k\mu/2, k\mu/4, k\mu/8, \ldots, \mu, 0$$

The penultimate vector is a row of the matrix G. Therefore, if G has k rows, by taking $log_2 k$ correlation products of transmitted vectors we obtain a row of G and by iterating this procedure k times for different subsets of transmitted vectors one obtain G completely.

b) General case.

Suppose that the set of matrices used is of size s and let $G_1, \ldots, G_s$ be the generator matrices used to generate vectors $\vec{r}$ of the algorithm. Again suppose the intruder collects a sequence $\vec{c}_i$ of cryptograms interchanged between A and B. Let

$$\vec{c}_1, \vec{c}_2, \ldots, \vec{c}_{t+1}$$

be a sequence of $t+1$ collected vectors. The sequence of correlation vectors is easily seen to have decreasing weight and therefore there exists a t such that $h(\vec{c}_1, \vec{c}_2, \ldots, \vec{c}_{t+1})$ is equal to zero and such that

$$\vec{c} = h(\vec{c}_1, \vec{c}_2, \ldots, \vec{c}_t)$$

is not equal to zero. If all the vectors $\vec{c}_i$ in the sequence belong to the same code, then the attack succeeds like in the preceeding case. However if the vectors belong to more than one code it fails, since the sequence converges to a vector that does not belong to any code. Hence the intruder cannot derive any useful information from the given set of vectors. One can notice that, in all cases, we have

$$t = O(log_2 k).$$

The probability that one $\vec{c}_i$ belongs to a specific matrix G_j is seen to be $1/s$, hence the probability of obtaining one of row of a secret matrix, by applying the previous procedure, is s^{-1}. By using this attack repeatedly the probability of obtaining all the rows of one of the $G'_i s$ is therefore

$$P_c = s^{-t.k} = O(2^{-t.k.log_2 s.}).$$

This probability can be made arbitrarily small with reasonable values of s and k which will be studied in the next section.

4.3.2 The correlation attack.

The following attack has been elaborated by J. Stern [5]. Let the attacker collect a sequence of cryptograms $\vec{c}_1, \vec{c}_2, \vec{c}_3 \ldots$ Choose a couple of coordinates u, v. Let $S_{u,v}$ be a subset of the sequence of collected vectors made of cryptograms with the coordinates of index u and v both equal to 1. Then there is a high probability that the vectors of $S_{u,v}$ correspond to the same $\vec{g}_i$ and by studying the other components of these vectors, in the set $S_{u,v}$, one can obtain the support of $\vec{g}_i$ by majority voting decision on the components equal to 1. Applying this result to k well chosen subsets of the indices one can obtain the set of rows of one matrix G_j and then apply the same method to deduce all the G_j $j = 1, \ldots, s$.

It will be shown that there are other vectors in the set $S_{u,v}$. These vectors appear with a much higher probability, if the parameters of the cryptosystem are chosen adequately. In this case no decision can be taken on a majority vote.

Let $\vec{g}$ be a row vector of one of the matrices G_j having the property that $g_u = g_v = 1$. The probability that $\vec{g}$ is chosen, among other vectors, to construct $\vec{r}$ is equal to $1/ks$.

Let P_{cr} be the probability that the cryptogram $\vec{c}$, has both components of index u and v equal to 1, while $\vec{g}$ is chosen to construct $\vec{r}$.

In order that the corresponding cryptogram $\vec{c}$, has both components equal to 1, the message vector $\vec{m}$ must have the coresponding components equal to 0, therefore we have the equality and approximation:

$$P_{cr} = \frac{1}{ks}(1 - \frac{w(\vec{m})}{n})^2 \approx \frac{1}{ks}.$$

This approximation is an equality in the case of a known plaintext attack i.e. when $w(\vec{m}) = 0$, which will be the case in the rest of this section. Consider the case where $\vec{c}$ has both components of index u and v equal to 1, and such that $\vec{g}$ is not chosen as a vector to construct $\vec{r}$. Let P_{nc} be the probability of that event, which can happen in two ways.

a) One row vector of some other generator matrix has both these components equal to 1, and has been chosen to construct $\vec{r}$. Let p_1 be the probability of that event. We derive the bounds, depending on the number of secret matrices for which this is true:

$$\frac{1}{ks}.(\frac{\mu}{n}) \leq p_1 \leq \frac{1}{ks}.(s-1).(\frac{\mu}{n})$$

b) Two rows of some other matrix are necessary to construct $\vec{r}$ with both these components equal to 1. Let p_2 be the probability of this event. We have the following bounds, depending on the number of matrices for which this happens:

$$(\frac{1}{k})^2(1 - \frac{\mu}{n})^2 \leq p_2 \leq (\frac{1}{k})^2(s-1).(1 - \frac{\mu}{n})^2$$

We have that $P_{nc} = p_1 + p_2$ and, considering that $\frac{\mu}{n}$ is negligible, we obtain the approximation:

$$P_{nc} \approx \frac{s}{2}.(\frac{1}{k})^2 = \frac{s}{2k^2}$$

If, by choosing the parameters of the system both events of the examined types a) and b) examined are much more probable then events described by J.Stern, no statistical inference on any other component can be made from considering the set $S_{u,v}$.

5. Bounds on the redundancy.

Suppose each of the matrices G_j $j = 1, \ldots, l$ is made of vectors of length n and has k rows. If the supports of the rows are disjoint and the rows have the same weight μ then it follows that $\mu = n/k$. Set $t = [\mu/2 - 1]$. Each of the coded messages has weight less than t. The size δ of the space of information set is defined by:

δ *is the greatest integer such that*

$$2^\delta \leq 1 + \binom{n}{1} + \binom{n}{2} + \ldots + \binom{n}{t}$$

Choosing in all cases $\mu \geq 30$, we obtain a table with the code lengths and code dimensions and the dimension the information space:

system parameters		
n	k	δ
800	20	126
800	10	220
1600	20	260
1600	10	449
3200	20	529
3200	10	907

The on line redundancy is between 1/4 and 1/10 for the parameters chosen. It diminishes when the dimension of the code increases.

6. Security considerations and possible sizes.

In this section the number of codes i.e. the integer s will be studied in relation with the probability of success of the different attacks listed in paragraph 3. This is done through the values of the volume K of secret storage needed. Let P_{Had} the probability of success of the correlation attack, P_{cr} the probability of having vectors leading to the success of the Stern attack in a set of type S_{uv}, P_{nc} the probability of having vectors leading to the failure of this attack in the same set.

For $K = 1Megabyte$ choosing 1000 codes with parameters (800,10,80) we have that $P_{Had} = 2^{-270}, P_{cr} = 10^{-4}, P_{nc} > 5.10^{-2}$.

For $K = 1Megabyte$ choosing 500 codes with parameters (1600,10,80) we have that $P_{Had} = 2^{-240}, P_{cr} = 2.10^{-4}, P_{nc} > 5.10^{-2}$.

7. Conclusion.

The cryptosystem presented is a linear cryptosystem leading to very efficient implementations. The safety of the system is ensured by the introduction of redundancy. This redundancy is divided into storage requirements (the matrices G_j) and on line redundancy (limited by the rate of each G_j). There is a tradeoff between between decoding complexity at the deciphering end and on line redundancy. If the set of matrices G_j is of large size, one can lower rate random codes.

In all cases a high level of security can be reached with a moderate volume of secret storage. The determination of the matrix used, which is a time consuming operation can be accelerated by the study of the Hadamard product of a subset of vectors only and can be done in parallel.

This cryptosystem can be thought of as a public key cryptosystem in a certain way or, more adequately, as an automatic secret key cryptosystem.

In a secret key cryptosystem the user has to have some secret information in order to activate the enciphering process. In the decribed system the user does not need to have such information, if the set of matrices G_j are in a tamper proof environment. The system can be used by many users who do not need to know any secret information beforehand to operate the system. Moreover the knowledge of the sequence $\vec{r}$ used in a cryptogram by one user, which is the only quantity a user needs to keep secret, does not weaken the system for further use.

8. Bibliography

[1] R.J. McEliece "A public key cryptosytem based on algebraic coding theory" *CA May 1978*

[2] E.R. Berlekamp " On the inherent intractability of certain coding problems" *IEEE Trans. on Info. Theory* Vol. IT-22, pp. 644 - 654, May 1978.

[3] P.J. Lee and E.F. Brickell "An observation on the security of McEliece's Public Key Cryptosystem" *Eurocrypt 1988 Davos Switzerland* pp.153 - 157, May 1988.

[4] C.M.Adams and H. Meijer,"Security related comments regarding McEliece's public-key cryptosystem" to appear *Proc.Crypto '87* Aug. 87.

[5] J.Stern,"Private Conversation" *Eurocode 90 Udine Italy* Nov 1990.

PSEUDORANDOM PERMUTATIONS BASED ON THE D.E.S. SCHEME

Jacques PATARIN

INRIA Domaine de Voluceau BP 105 78153 Le Chesnay Cedex France

Abstract

We present in a new way the results of Michael Luby and Charles Rackoff *"How to construct pseudorandom permutations from pseudorandom functions"*, SIAM J. Comput., 1988, together with some new results thereon.

Two main results will be proved here:

1. A *three iterations DES Scheme* where the S-boxes are replaced by random functions (or by a pseudorandom function generator) will give us an *invertible pseudorandom function generator*. That is to say a cryptosystem which is secure against chosen plaintext attacks.

2. A *four (or more) iterations DES Scheme* where the S boxes are replaced by random functions (or by a pseudorandom function generator) will give us an invertible super pseudorandom permutation generator. That is to say a cryptosystem which is secure against chosen plaintext and chosen ciphertext attacks.

To be precise, we will explain (in Section 1 and 2) what we call a *k iteration DES Scheme* and an *invertible (super) pseudorandom permutation generator*.

The proof that we will give for item 1 will be based on the same "Main Lemma" as the one given by Luby and Rackoff. However we will prove this "Main Lemma" using another theorem (the "basic property") which will simplify both the proof and the understanding.

Key words: Cryptology, D.E.S., Pseudorandom generators.

1 Definitions of the pseudorandom notions

We will use the following

- I_n is the set $\{0,1\}^n$, $n \in \mathbb{N}$.

- If A and B are two sets, A^B will be the set of all functions $B \to A$.

Definition 1 A *pseudorandom bit generator* is a sequence of functions $f_n : I_n \to I_{l(n)}$, where $l(n)$ is a polynomial in n such that:

1. There exists a polynomial algorithm such that given n and x as input, it computes $f_n(x)$. (This is the *uniform case*. If we only have a polynomial $P(n)$ such that for all n there exists an algorithm which given x as input, computes $f_n(x)$ in less than $P(x)$ instructions, we call this is the *non-uniform case*. In this case, the algorithm depends on n).

2. For all polynomial $Q(n)$, for all polynomial algorithm T using an element A of $\{0,1\}^{l(n)}$ as input, and which gives 0 or 1 as output:

$$\exists\, n_0 \in \mathsf{N}, \forall n \geq n_0, \ |P_1^* - P_1| < \frac{1}{Q(n)},$$

where P_1^* is the probability that T gives the output 1 when A is randomly chosen in $\{0,1\}^{l(n)}$, and P_1 is the probability that T gives the output 1 when a is randomly chosen in I_n, and when we take $A = f_n(a)$.

To summarise, a pseudorandom bit generator is a polynomial technique to extend a real random number of n bits to a pseudorandom number of $l(n)$ bits.

If a is randomly chosen in I_n, then the pseudorandom $f_n(a)$ cannot be distinguished (in a time polynomial in n) from a real random of $I_{l(n)}$.

Definition 2 Let $P(n)$ be a polynomial in n. We will call an element of $I_{P(n)}$ a key. A *pseudorandom function generator* is a sequence of functions $f_n : I_{P(n)} \to I_n^{I_n}$ such that:

1. For all keys $k \in I_P(n)$, for all $\alpha \in I_n$, $f_n(k)(\alpha)$ can be computed in polynomial time in n. (To be precise, we should say here whether the algorithm used depends on n or not : this gives the non-uniform or the uniform case, as in Definition 1).

2. For all polynomials $Q(n)$, and for all polynomials in n algorithm T, T *using* a function $f : I_n \to I_n$ as input (that is to say that T can generate some numbers α and obtain the numbers $f(\alpha)$ in a time considered as unity), and T giving 0 or 1 as output:

$$\exists\, n_0 \in \mathsf{N}, \forall n \geq n_0, \ |P_1^* - P_1| < \frac{1}{Q(n)},$$

where P_1^* is the probability that T gives the output 1 when f is randomly chosen in $I_n^{I_n}$ and P_1 is the probability that T gives the output 1 when k is randomly chosen in $I_{P(n)}$, and when we take $f = f_n(k)$.

To summarise, a pseudorandom function generator is a polynomial technique to obtain a pseudo random function from a random key k.

If k is randomly chosen in $I_{P(n)}$, then the pseudo-random function $f_n(k)$ cannot be distinguished (in a polynomial time in n) from a real random function by an algorithm which can generate, when needed some numbers x and obtaining $f_n(x)$ (in a time considered as unity).

Definition 3

- A *pseudorandom permutation generator* is a pseudorandom function generator for which all the $f_n(k)$ obtained are permutations (that is $f_n(k)$ is 1–1 onto).

- A *pseudorandom invertible permutation generator* is a pseudorandom permutation generator for which all the $f_n^{-1}(k)$ are obtained (as the $f_n(k)$) from a pseudorandom permutation generator.

- A *super pseudorandom invertible permutation generator* is a pseudorandom invertible permutation generator where:
  if k is randomly chosen in $I_{P(n)}$, then the pseudo-random permutation $f_n(k)$ cannot be distinguished (in polynomial time in n) from a real random permutation by an algorithm which can generate some numbers x and obtaining $f_n(x)$ and $f_n^{-1}(x)$.

2 Definition of Ψ and the Ψ^k

Before explaining our results, we still have to define a *DES scheme technique* (Ψ).

Definition 4 Let f_1 be a function from $I_n \rightarrow I_n$, and let L, R, S, and T be elements of I_n.

Then by definition: $\Psi(f_1)$ is the function from $I_{2n} \rightarrow I_{2n}$ such that:

$$\forall (L, R) \in I_n^2, \ \Psi(f_1)[L, R] = [S, T] \iff \begin{cases} S = R \\ T = L \oplus f_1(R) \end{cases},$$

where $\oplus$ is the bitwise addition mod 2.

Schematically:

INPUT: L R

$\Psi(f_1)$:

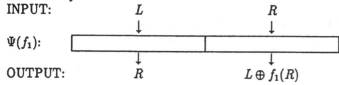

OUTPUT: R $L \oplus f_1(R)$

Note that $\Psi(f_1)$ is a permutation $I_{2n} \rightarrow I_{2n}$.

Definition 5 (Ψ^k) Let $f_1, f_2, \ldots, f_k$ be k functions from $I_n \rightarrow I_n$. $\Psi^k(f_1, \ldots, f_k)$ is the permutation from $I_{2n} \rightarrow I_{2n}$ defined by:

$$\Psi^k(f_1, \ldots, f_k) = \Psi(f_k) \circ \cdots \circ \Psi(f_2) \circ \Psi(f_1),$$

where $\circ$ is the composition of functions.

Also:

INPUT: L R

$\Psi^2(f_1, f_2)$:

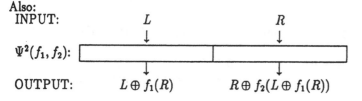

OUTPUT: $L \oplus f_1(R)$ $R \oplus f_2(L \oplus f_1(R))$

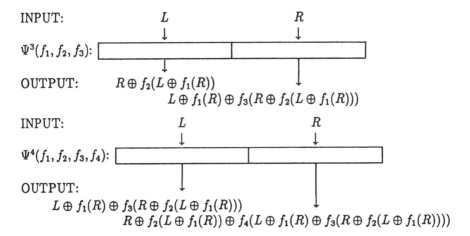

INPUT: L R

$\Psi^3(f_1, f_2, f_3)$:

OUTPUT: $R \oplus f_2(L \oplus f_1(R))$

$L \oplus f_1(R) \oplus f_3(R \oplus f_2(L \oplus f_1(R)))$

INPUT: L R

$\Psi^4(f_1, f_2, f_3, f_4)$:

OUTPUT:

$L \oplus f_1(R) \oplus f_3(R \oplus f_2(L \oplus f_1(R)))$

$R \oplus f_2(L \oplus f_1(R)) \oplus f_4(L \oplus f_1(R) \oplus f_3(R \oplus f_2(L \oplus f_1(R))))$

Remark:
$\Psi^k(f_1, \ldots, f_k)$ is in fact a k *iteration DES Scheme* where the S-boxes are replaced by the functions $f_1, \ldots, f_k$.

3 Ψ and Ψ^2 are not pseudorandom. Ψ^3 is not super pseudo-random

3.1 Ψ is not pseudorandom.

By "Ψ is not pseudorandom" we mean that a permutation generator which gives permutation of the form $\Psi(f_1)$, however f_1 is generated from a key k, is never a pseudorandom permutation generator.

This can be proved as follows : Let f be a function $I_{2n} \to I_{2n}$.
Let ϕ be this algorithm (used to test such a function f) :

1. ϕ chooses two elements L and R of I_n. (*e.g.* $L = 0$ and $R = 0$, 0 being the n-tuple $(0, \ldots, 0)$).

2. ϕ asks for the value of $f[L, R]$.

3. ϕ tests if the first n values S of $f[L, R]$ are equal to R.

4. If they are, ϕ outputs 1, otherwise, ϕ outputs 0.

Let P_1^* be the probability that ϕ outputs 1 when f is randomly chosen in $I_{2n}^{I_{2n}}$.
    Then, $P_1^* = 1/2^n$. (Because all the values for S have the same probability $1/2^n$ of appearing).
    Let P_1 be the probability that ϕ outputs 1 when f is a permutation of the form $\Psi(f_1)$. ($f_1 \in I_n^{I_n}$).
    Then $P_1 = 1$. (Because $\Psi(f_1)[L, R] = [R, L \oplus f_1(R)]$).
    So $|P_1 - P_1^*| = 1 - 1/2^n$ which does not tend to 0 quicker than any polynomial (it does not tends to 0 at all).

3.2 Ψ^2 is not pseudorandom.

We will use the same notation.

Let ϕ be this following algorithm :

1. ϕ chooses two elements L and R of I_n, and asks for the value of $f[L,R] = [S,T]$.

2. ϕ chooses an elements L' of I_n , $L' \neq L$, and asks for the value of $f[L',R] = [S',T']$.

3. ϕ tests if $S \oplus S' = L \oplus L'$.

4. If it is, ϕ outputs 1, otherwise ϕ outputs 0.

Now, $P_1^* = 1/2^n$ and $P_1 = 1$. (Because if f is a permutation $\Psi^2(f_1, f_2)$, $S = L \oplus f_1(R)$ and $S' = L' \oplus f_1(R)$).

So $|P_1 - P_1^*| = 1 - 1/2^n$ which does not tend to 0 quicker than any polynomial.

3.3 Ψ^3 is not super pseudo-random

Let ϕ be this algorithm, testing a permutation f of I_{2n} :

1. ϕ chooses two elements L and R of I_n, and asks for the value of $f[L,R] = [S,T]$.

2. ϕ chooses an elements L' of I_n, $L' \neq L$, and asks for the value of $f[L',R] = [S',T']$.

3. ϕ asks for the value $f^{-1}[S', T' \oplus L \oplus L'] = [L'', R'']$..

4. If $R'' = S' \oplus S \oplus R$, then ϕ outputs 1. If not, ϕ outputs 0.

- If f is a permutation randomly chosen, the probability P_1^{**} that ϕ outputs 1 is $\simeq 1/2^n$.

- Now assume that f is a permutation of the form :

$f = \Psi^3(f_1, f_2, f_3)$. (Where f_1, f_2, f_3 are three functions $I_n \to I_n$).

Then $f[L,R] = [S,T] \Leftrightarrow \begin{cases} S = R \oplus f_2(L \oplus f_1(R)) \\ T = L \oplus f_1(R) \oplus f_3(R \oplus f_2(L \oplus f_1(R))). \end{cases}$

And $f^{-1}[S,T] = [L,R] \Leftrightarrow \begin{cases} L = T \oplus f_3(S) \oplus f_1(S \oplus f_2(T \oplus f_3(S)))) \\ R = S \oplus f_2(T \oplus f_3(S)) \end{cases}$

So $f^{-1}[S', T' \oplus L \oplus L'] = [L'', R''] \Rightarrow R'' = S' \oplus f_2(\underbrace{T' \oplus L \oplus L' \oplus f_3(S')}_{L \oplus f_1(R)})$

$\underbrace{\qquad\qquad\qquad}_{S \oplus R}$

So $R'' = S' \oplus S \oplus R$.

So the probability P_1 that ϕ outputs 1 when f is a function $\Psi^3(f_1, f_2, f_3)$ is 1.

Then $|P_1 - P_1^{**}| \simeq 1 - 1/2^n$ which does not tend to 0.

Remark:

In Section 4 we will see how this algorithm ϕ can be found:
if we test some $f[L_i, R_i] = [S_i, T_i]$ where $i \neq j \Rightarrow S_i \neq S_j$, we will show that we cannot distinguish (in polynomial time in n) if f is a random function or a function $\Psi^3(f_1, f_2, f_3)$. This explains why here, ϕ required the value of $f^{-1}[S', -]$ in Step 3, where S' has been obtained in Step 1.

4 A basic property of Ψ^3

Let m be an integer. Let $L_i, R_i, S_i, T_i, 1 \leq i \leq m$ be elements of I_n.

Theorem 1 (Basic Property) *If $i \neq j \Rightarrow (L_i, R_i) \neq (L_j, R_j)$ (that is to say $L_i \neq L_j$ or $R_i \neq R_j$) and if $i \neq j \Rightarrow S_i \neq S_j$, then the number of 3-tuples of functions (f_1, f_2, f_3) such that :*
$$\forall i, 1 \leq i \leq m, \ \Psi^3(f_1, f_2, f_3)[L_i, R_i] = [S_i, T_i] \text{ is}$$

$$\geq \frac{F_0^3}{2^{2nm}} \left(1 - \frac{m(m-1)}{2^{n+1}} \right),$$

where $F_0 = (2^n)^{2^n} = 2^{n \cdot 2^n}$.

Proof: We want to evaluate the number of functions (f_1, f_2, f_3) such that :

$$\forall i, 1 \leq i \leq m, \begin{cases} f_2(L_i \oplus f_1(R_i)) = S_i \oplus R_i \\ f_3(S_i) = T_i \oplus L_i \oplus f_1(R_i) \end{cases}$$

First we will show that when i and j are fixed, $1 \leq i < j \leq m$, there is at most $F_0/2^n$ functions f_1 such that

$$L_i \oplus f_1(R_i) = L_j \oplus f_1(R_j). \tag{1}$$

First case $R_i = R_j$. Then $L_i \neq L_j$ by hypothesis. So no function f_1 realise (1).

Second case $R_i \neq R_j$. Whatever $f_1(R_i)$ is, the equation (1) just fixes the value of f_1 in (R_i). So exactly $F_0/2^n$ functions f_1 satisfy (1).

Thus, there are at most $\frac{F_0}{2^n} \frac{m(m-1)}{2}$ functions f_1 such that $\exists (i,j), 1 \leq i < j \leq m$ such that $L_i \oplus f_1(R_i) = L_j \oplus f_1(R_j)$. (Because there are $m(m-1)/2$ pairs (i,j) such that $1 \leq i < j \leq m$).

Let f_1 be a function which is *not* like this. For f_1, we also have at least $F_0 - F_0 \frac{m(m-1)}{2^{n+1}}$ possibilities. For such an f_1, there are exactly $F_0/2^{nm}$ functions f_2 such that: $\forall i, 1 \leq i \leq m, f_2(L_i \oplus f_1(R_i)) = S_i \oplus R_i$. (Because this is the number of functions *possible* when their values are fixed at m *distinct* points).

For such f_1 and f_2, there are exactly $F_0/2^{nm}$ functions f_3 such that : $\forall i, 1 \leq i \leq m, f_3(S_i) = T_i \oplus L_i \oplus f_1(R_i)$ for the same reason.

In conclusion, we always have at least

$$\left(F_0 - F_0 \frac{m(m-1)}{2^{n+1}} \right) \left(\frac{F_0}{2^{nm}} \right) \left(\frac{F_0}{2^{nm}} \right) \text{ solutions.}$$

This is what was claimed.

5   Demonstration of the "Main Lemma" for Ψ^3 from the Basic Property

- Let ϕ be a (deterministic) algorithm which is used to test a function f. (ϕ can test any function f from $I_{2n} \to I_{2n}$).

  ϕ can use f at most m times, that is to say that ϕ can ask for the values of some $f(C_i), C_i \in I_{2n}, 1 \leq i \leq m$. (The value C_1 is chosen by ϕ, ϕ receive $f(C_1)$, then ϕ can choose any $C_2 \neq C_1$ on receiving $f(C_2)$ etc.).
  (If $i \neq j$, C_i is always different from C_j). After a finite but unbounded amount of time, ϕ gives an output of "1" or "0".
  This output (1 or 0) is noted : $\phi(f)$

- We will denote by P_1^* the probability that ϕ gives the output 1 when f is chosen randomly in $I_{2n}^{I_{2n}}$.
  So
  $$P_1^* = \frac{\text{Number of functions } f \text{ such that } \phi(f) = 1}{\text{number of all functions from } I_{2n} \to I_{2n} = (2^{2n})^{(2^{2n})}}$$

- We will denote by P_1 the probability that ϕ give the output 1 when f is chosen as a function $\Psi^3(f_1, f_2, f_3)$, when the functions f_1, f_2, f_3 are randomly chosen in $I_n^{I_n}$.
  So
  $$P_1 = \frac{\text{Number of } (f_1, f_2, f_3)/\phi(\Psi^3(f_1, f_2, f_3)) = 1}{F_0^3 = \text{number of all possible } (f_1, f_2, f_3)}$$

We will now see the same "Main Lemma" as in the paper of M. Laby and Rackoff. And, as they showed, this "Main Lemma" will imply that Ψ^3 *is an invertible pseudo-random generator*.

Here, we will use our basic property to prove their "Main Lemma", which will give us a simple and quick proof of it.

Theorem 2 ("Main Lemma") *For all such algorithms ϕ,*

$$|P_1 - P_1^*| \leq \frac{m(m-1)}{2^n} \quad \left(so \leq \frac{m^2}{2^n}\right).$$

Proof.

Evaluation of P_1^*.

Let f be a fixed function, and let $C_1, \ldots, C_m$ be the successive values that the program ϕ will ask for the values of f (when ϕ tests the function f). We will denote by $\sigma_1 = f(C_1)$, $\ldots, \sigma_m = f(C_m)$. $\phi(f)$ depends in fact *only* of the outputs $\sigma_1, \ldots, \sigma_m$. That is to say that if g is another function of $I_{2n} \to I_{2n}$ such that $\forall i, 1 \leq i \leq m, g(C_i) = \sigma_i$, then $\phi(g) = \phi(f)$. (Since for $i < m$, the choice of C_{i+1} depends only of $\sigma_1, \ldots, \sigma_i$. Also the program ϕ cannot distinguish f from g, because ϕ will give each the same inputs, and obtain the same outputs).

Conversely, let $\sigma_1, \ldots, \sigma_m$ be m elements of I_n. Let C_1 be the first value that ϕ chooses to know $f(C_1), C_2$ the value that ϕ chooses when ϕ has obtained the answer σ_1

for $f(C_1), \ldots$ and C_m the mth value that ϕ presents to f, when ϕ has obtained $\sigma_1, \ldots, \sigma_i$ for $f(C_1), \ldots, f(C_m)$. Let $\phi(\sigma_1, \ldots, \sigma_m)$ be the output of ϕ (0 or 1).

Then

$$P_1^* = \sum_{\substack{\sigma_1, \ldots, \sigma_m \\ \phi(\sigma_1, \ldots, \sigma_m) = 1}} \frac{\text{Number of functions } f \text{ such that } \forall i, 1 \le i \le m, f(C_i) = \sigma_i}{(2^{2n})^{(2^{2n})}}$$

Since the C_i are all distinct the number of function f such that $\forall i, 1 \le i \le m$ $f(C_i) = \sigma_i$ is exactly $(2^{2n})^{(2^{2n})}/2^{2nm}$.

So

$$P_1^* = \frac{\text{Number of } m \text{ output } \sigma_1, \ldots, \sigma_m \text{ such that } \phi(\sigma_1, \ldots, \sigma_m) = 1}{2^{2nm}}$$

Let N be the numbers of outputs $\sigma_1, \ldots, \sigma_m$ such that $\phi(\sigma_1, \ldots, \sigma_m) = 1$.

So

$$P_1^* = \frac{N}{2^{2nm}}.$$

Evaluation of P_1.

With the same notations $\sigma_1, \ldots, \sigma_m$ and $C_1, \ldots, C_m$:

$$P_1 = \sum_{\substack{\sigma_1, \ldots, \sigma_m \\ \phi(\sigma_1, \ldots, \sigma_m) = 1}} \frac{\text{Number of } (f_1, f_2, f_3) \text{ such that } \forall i, 1 \le i \le m, \Psi^3(f_1, f_2, f_3)(C_i) = \sigma_i}{F_0^3}$$

There are N outputs $\sigma_1, \ldots, \sigma_m$ such that $\phi(\sigma_1, \ldots, \sigma_m) = 1$ (by definition of N). We will denote by S_i the first half of σ_i.

There are at most $\frac{m(m-1)}{2} \frac{2^{2nm}}{2^n}$ outputs $\sigma_1, \ldots, \sigma_m$ such that $\exists (i, j), 1 \le i < j \le m$ such that $S_i = S_j$. (Because there are $m(m-1)/2$ possibilities for choosing (i, j) and when i and j are fixed, there are exactly $2^{nm}/2^n$ outputs $\sigma_1, \ldots, \sigma_m$ such that $\sigma_i = \sigma_j$).

So there are at least $N - 2^{nm} \frac{m(m-1)}{2^{n+1}}$ outputs $\sigma_1, \ldots, \sigma_m$ such that:
$\phi(\sigma_1, \ldots, \sigma_m) = 1$
$\forall i, \forall j, 1 \le i < j \le m, S_i \ne S_j$.

By the Basic Property, for each of these outputs, there are at least $\frac{F_0^3}{2^{2nm}}(1 - \frac{m(m-1)}{2^{n+1}})$ possibilities for (f_1, f_2, f_3) such that $\forall i, 1 \le i \le m, \sigma_i = \Psi^3(f_1, f_2, f_3)[C_i]$.

So,

$$
\begin{aligned}
P_1 &\ge \left(N - \frac{m(m-1)}{2^{n+1}} 2^{2nm} \right) \frac{F_0^3}{2^{2nm}} \left(1 - \frac{m(m-1)}{2^{n+1}} \right) \frac{1}{F_0^3} \\
&\ge \left(\frac{N}{2^{2nm}} - \frac{m(m-1)}{2^{n+1}} \right) \left(1 - \frac{m(m-1)}{2^{n+1}} \right) \\
&\ge P_1^* - \frac{m(m-1)}{2^{n+1}} - \frac{m(m-1)}{2^{n+1}} P_1^* + \frac{m^2(m-1)^2}{2^{2n+2}}.
\end{aligned}
$$

So

$$P_1 \ge P_1^* - \frac{m(m-1)}{2^n}. \tag{2}$$

If we let P_0^* be the probability of obtaining 0 when f is a random function $I_{2n} \to I_{2n}$ ($P_0^* = 1 - P_1^*$) and P_0 the probability of obtaining 0 when f is a function $\Psi^3(f_1, f_2, f_3)$, ($P_0 = 1 - P_1$), the same reasoning gives : $P_0 \geq P_0^* - m(m-1)/2^n$.

So, $1 - P_1 \geq 1 - P_1^* - m(m-1)/2^n$, so

$$P_1 \leq P_1^* + \frac{m(m-1)}{2^n}. \tag{3}$$

Finally, (2) and (3) give : $|P_1 - P_1^*| \leq m(m-1)/2^n$. This completes the proof.

6 The Basic Property for Ψ^4

Theorem 3 (Basic Property for Ψ^4) *Let (L_i, R_i), $1 \leq i \leq m$, be distinct inputs, and let (S_i, T_i), $1 \leq i \leq m$ be distinct outputs. (Distinct means that $i \neq j \Rightarrow S_i \neq S_j$ or $T_i \neq T_j$)*

Then the number of 4-tuples of functions (f_1, f_2, f_3, f_4) such that $\forall i$, $1 \leq i \leq m$, $\Psi^4(f_1, f_2, f_3, f_4)[L_i, R_i] = [S_i, T_i]$ is $\geq \frac{F_0^4}{2^{2nm}} \left(1 - \frac{m(m-1)}{2^{n+1}}\right)^2$ where $F_0 = 2^{n \cdot 2^n}$.

Remark:
For Ψ^4 we do not need the hypothesis $S_i \neq S_j$ if $i \neq j$. (And this is why Ψ^4 will be a *super* pseudo-random permutation generator and not Ψ^3).

Proof: We require that, $\forall i$, $1 \leq i \leq m$,

$$f_3(R_i \oplus f_2(L_i \oplus f_1(R_i))) = S_i \oplus L_i \oplus f_1(R_i)$$
$$f_4(S_i) = T_i \oplus R_i \oplus f_2(L_i \oplus f_1(R_i))$$

Let k be the number of independent equalities of the form $S_i = S_j$ (k may be zero).

First we will take a function f_1 such that $\forall i, j, i \neq j$, $L_i \oplus f_1(R_i) \neq L_j \oplus f_1(R_j)$. As we did for Ψ^3 we have at least $F_0(1 - m(m-1)/2^{n+1})$ possibilities for f_1.

We take f_2 such that

1. All the $R_i \oplus f_2(L_i \oplus f_1(R_i))$ are distinct ($\forall i$, $\forall j$, $1 \leq i < j \leq m$).

2. There are k equations of the form $T_i \oplus R_i \oplus f_2(L_i \oplus f_1(R_i)) = T_j \oplus R_j \oplus f_2(L_j \oplus f_1(R_j))$ which are consequences of the k equalities $S_i = S_j$.

We have $F_0/2^{kn}$ functions f_2 which satisfy item 2.

First case. (i, j) is such that $S_i = S_j$. Then $T_i \neq T_j$. So from 2 we have : $R_i \oplus f_2(L_i \oplus f_1(R_i)) \neq R_j \oplus f_2(L_j \oplus f_1(R_j))$. So 1 is always true for these (i, j).

Second case. (i, j) is such that $S_i \neq S_j$. (We have at most $m(m-1)/2$ such (i, j), $1 \leq i < j \leq m$).

Let S be the system comprising the k equations of item 2 and the equation $R_i \oplus f_2(L_i \oplus f_1(R_i)) = R_j \oplus f_2(L_j \oplus f_1(R_j))$. S is a system of $k+1$ independent equations. So there are at most $F_0/2^{(k+1)n}$ solutions for S.

Conclusion There are at least $\frac{F_0}{2^{kn}}\left(1 - \frac{m(m-1)}{2^{n+1}}\right)$ functions f_2 which satisfy 1 and 2.

When f_1 and f_2 are chosen in this way, we take f_3 such that

$$\forall i, \ 1 \leq i \leq m, \ f_3(R_i \oplus f_2(L_i \oplus f_1(R_i))) = S_i \oplus L_i \oplus f_1(R_i).$$

(This is possible because the $R_i \oplus f_2(L_i \oplus f_1(R_i))$ are all $\neq$). There are exactly $F_0/2^{nm}$ possibilities for f_3.

Then, we write the $m - k$ independent equalities of the form $f_4(S_i) = T_i \oplus R_i \oplus f_2(L_i \oplus f_1(R_i))$. (The S_i are all different in these equalities). These are $F_0/2^{n(m-k)}$ such functions f_4.

Finally, we have at least :

$$\underbrace{F_0\left(1 - \frac{m(m-1)}{2^{n+1}}\right)}_{\text{choice of } f_1} \quad \underbrace{\frac{F_0}{2^{kn}}\left(1 - \frac{m(m-1)}{2^{n+1}}\right)}_{\text{choice of } f_2 \text{ when } f_1 \text{ is fixed}} \quad \underbrace{\frac{F_0}{2^{nm}}}_{\text{choice of } f_3} \quad \underbrace{\frac{F_0}{2^{n(m-k)}}}_{\text{choice of } f_4}$$

4-tuples (f_1, f_2, f_3, f_4) solutions.

So there are at least $\frac{F_0^4}{2^{2nm}}\left(1 - \frac{m(m-1)}{2^{n+1}}\right)^2$ solutions.

7 The main lemma for Ψ^4.

In this case, the algorithm ϕ, which is used to test a permutation f, can choose m values C_i during its execution, $C_i \in I_{2n}$ and for each $i, 1 \leq i \leq m$, ϕ can ask for the value $f(C_i)$ or $f^{-1}(C_i)$. We will denote by σ_i the value obtained.Then (after a finite but unbounded amount of time), ϕ gives the output "1" or "0". This output is denoted by $\phi(f)$.

- We will denote by P_1 the probability that $\phi(f) = 1$ when f is chosen as a function $\Psi^4(f_1, f_2, f_3, f_4)$ where the functions f_1, f_2, f_3, f_4 are randomly chosen in $I_n^{I_n}$.

So

$$P_1 = \frac{\text{Number of } (f_1, f_2, f_3, f_4) \text{ such that } \phi\left(\Psi^4(f_1, f_2, f_3, f_4)\right) = 1}{F_0^4}$$

As for ψ^3, we have

$$P_1 = \sum_{\substack{\sigma_1, \dots, \sigma_m \\ \text{outputs giving } 1}} \frac{\begin{array}{c}\text{Number of } (f_1, f_2, f_3, f_4) \text{ such that } \phi \text{ receives } \sigma_1, \dots, \sigma_m \\ \text{when it tests } \Psi^4(f_1, f_2, f_3, f_4)\end{array}}{F_0^4}. \tag{4}$$

Let P_1^{**} be the probability that ϕ gives 1 as output when f is chosen randomly in the set of all the permutations of I_{2n}.

That is to say that if N is the number of $\sigma_1, \dots, \sigma_m$ compatible with a permutation f, such that ϕ gives 1 in output when it receives successively $\sigma_1, \dots, \sigma_m$,

$$P_1^{**} = \frac{N}{2^{2n}(2^{2n} - 1)\cdots(2^{2n} - m + 1)} = \frac{N}{2^{2nm}\left(1 - \frac{1}{2^{2n}}\right)\left(1 - \frac{2}{2^{2n}}\right)\cdots\left(1 - \frac{m-1}{2^{2n}}\right)} \tag{5}$$

The object of this section is to prove

Theorem 4 ("Main lemma for Ψ^4") *For all such algorithm ϕ,*

$$|P_1 - P_1^{**}| \leq \frac{m^2}{2^n}.$$

Proof: From the "basic property for Ψ^4", for all $\sigma_1, \ldots, \sigma_m$, the number of (f_1, f_2, f_3, f_4) such that ϕ receives $\sigma_1, \ldots, \sigma_m$ when it tests $\Psi^4(f_1, f_2, f_3, f_4)$ is at least $\frac{F_0^4}{2^{2nm}}\left(1 - \frac{m(m-1)}{2^{n+1}}\right)^2$.

So, using (4), (N is the number of such $\sigma_1, \ldots, \sigma_m$) :

$$P_1 \geq N\frac{F_0^4}{2^{2nm}F_0^4}\left(1 - \frac{m(m-1)}{2^{n+1}}\right)^2 \geq P_1^{**}\left(1 - \frac{m(m-1)}{2 \cdot 2^{2n}}\right)\left(1 - \frac{m(m-1)}{2^{n+1}}\right)^2$$

So

$$P_1 \geq P_1^{**}\left(1 - \frac{m(m-1)}{2 \cdot 2^{2n}} - \frac{m(m-1)}{2^n}\right) \geq P_1^{**} - \frac{m^2}{2^n}.$$

We also find that : $P_1^{**} \geq P_1 - m^2/2^n$ when we use the output 0 instead of the output 1, applying the same argument as was used for Ψ^3.

8 Conclusion : Ψ^3 is pseudo-random, Ψ^4 is super pseudo-random, other results.

We can now precisely formulate the results of Luby and Rackoff, and the new results on the subject of pseudorandom permutation based on the DES Scheme.

Convention. Let F be a pseudorandom function generator (the existence of which is discussed farther) with key of length $P(n)$ and let k be a fixed integer. We will simply denote by Ψ^k the sequence of permutation Ψ_{2n}^k obtained from :

1. The keys K used are integers of length $k\,P(n)$.

2. f_1 is the function generated by F with the first $P(n)$ bits of K, f_2 is the function generated by F with the next $P(n)$ bits of K, ..., f_k is the function generated by F with the last $P(n)$ bits of K.

3. Ψ_{2n}^k is then the permutation $\Psi^k(f_1, \ldots, f_k)$.

Theorem 5 (Luby and Rackoff) *If F is a pseudorandom function generator, then Ψ^3 is a pseudorandom invertible permutation generator. (Further, all the Ψ^k, $k \geq 3$ are also pseudorandom permutations generators).*

Proof: Luby and Rackoff showed in [1] that this theorem is a relatively easy consequence of the Main Lemma for Ψ^3.

Theorem 6 *If F is a pseudorandom function generator, then Ψ^4 is a super pseudorandom invertible permutation generator. (And all the Ψ^k, $k \geq 4$ are also super pseudorandom permutations generators).*

Proof: Exactly the same as for Theorem 1, but using the "Main Lemma for Ψ^4" of Section 7.

We can express Theorem 5 in the following way :

Theorem 7 *If $f_1, \ldots, f_k$ are random functions from $I_n \rightarrow I_n$, or if $f_1, \ldots, f_k$ are pseudorandom functions from $I_n \rightarrow I_n$ obtained by a pseudo-random function generator, then $\Psi^k(f_1, \ldots, f_k)$ ($k \geq 3$) is a permutation which cannot be distinguished (in polynomial time) from a true random permutation.*

Other results.

- It is possible to use only two, or even one function to obtain a pseudorandom permutation generator. Indeed:

  Theorem 8 *If f and g are independently generated by a pseudorandom function generator, then $\Psi^3(f, f, g)$ is a pseudorandom permutation.*

  (The demonstration is exactly the same as for Theorem 1).

- **Theorem 9** *For all integers i, j, k, $\Psi^3(f^k, f^j, f^i)$ is not pseudorandom (where f^l is the composition of f, l times).*
  (see [3]).

- **Theorem 10** $\Psi^4(f, f, f^2)$ *is pseudorandom.*
  (see [2]).

References

[1] M. LUBY, C. RACKOFF *How to construct pseudorandom permutations from pseudorandom functions* SIAM J. Compt, 1988.

[2] J. PIERPRZYK *How to construct pseudorandom permutations from single pseudorandom Functions*, Abstract of Eurocrypt 90.

[3] Y. ZHENG, T. MATSUMOKO, H. IMAI, *Impossibility and optimality results on constructing pseudorandom permutations*, Abstract of Eurocrypt 89.

LINEAR COMPLEXITY OF TRANSFORMED SEQUENCES

Harriet J. Fell[1]

College of Computer Science, Northeastern University
Boston, Massachusetts 02115 USA

Abstract:

This paper deals with the effect of bit change errors on the linear complexity of finite sequences. Though a change in a single bit can cause a large change in linear complexity, it is shown that on the average the change will be small even when many bits, e.g. 10%, are changed. General bijections on the set of sequences of length n are studied and tight bounds are found on the average difference in linear complexity between a sequence and its image. It is also shown that to change all sequences of length n into sequences with linear complexity less than $c(n)$ where $\lim_{n\to\infty} c(n)/n = 0$, at least $\frac{n-1}{n}2^n$ of the sequences must have close to half of their bits changed.

1 Introduction

The linear complexity of a finite sequence can change drastically if a single bit is changed or deleted. For example, the sequence $0, 0, \ldots, 0, 1$ of length n has maximal linear complexity, n, while deleting the last bit or changing it to a 0 results in a sequence of linear complexity 0. A shift register that generates a given sequence can be found, with Λ^2 operations, after seeing only 2Λ bits where Λ is the linear complexity of the sequence [2], so sequences of low linear complexity are not cryptographically secure. At the *Workshop on Stream Ciphers*, at Karlsruhe, Germany, January 9-12, 1989, W. Diffie suggested that for many finite sequences, it might be possible to find small linear feedback shift registers (LFSRs) that generate nearby sequences. That is, if we can tolerate some errors, we might find it easy to generate a sequence close enough to a given sequence for cryptanalytic purposes.

In this paper, we look at functions that take sequences of length n into sequences of length n and study the average difference in the linear complexity of a sequence and its image. We then apply these results to analyze the average change in linear complexity when errors, caused by changed bits, are introduced into a sequence.

---

[1]This work was supported in part by The Institut National de Recherche en Informatique et en Automatique, Rocquencourt, France.

2 Notation

We restrict our attention to sequences over the field with two elements, Z_2. Let $S_n = \{0,1\}^n$ be the set of all sequences of zeros and ones of length n. If $s \epsilon S_n$ then $s = s_1, s_2, \ldots, s_n$ where each $s_i, \{i = 1, \ldots, n\}$ is either 0 or 1. We assume that $s_i, 1 \leq i \leq n$ are uniformly and independently distributed random variables, so that all $s \epsilon S_n$ are equiprobable.

An infinite sequence $s = \{s_i\}_{i=1\ldots\infty}$ is said to be generated by an LFSR of length k if there exist constants $a_0, \ldots, a_{k-1}$ such that

$$s_{i+k} = a_{k-1} s_{i+k-1} + \ldots + a_1 s_{i+1} + a_0 s_i \quad for\ i \geq 1$$

For $s \epsilon S_n$ we define the *linear complexity*, $\Lambda(s)$, to be the length of the smallest LFSR that generates a sequence whose first n terms are $s_1, s_2, \ldots, s_n$. If $k \leq n$, we will use $\Lambda_k(s)$ to denote the linear complexity of the sequence, $s_1, s_2, \ldots, s_k$.

If $s, t \epsilon S_n$, we define $\Delta(s,t) = |\Lambda(s) - \Lambda(t)|$ and for $1 \leq k \leq n, \Delta_k(s,t) = |\Lambda_k(s) - \Lambda_k(t)|$.

3 Changing a Bit and Other Bijections

The drastic increase in linear complexity caused by changing the last bit of a sequence of n zeros brings up the question of the general effect of a single bit change on the linear complexity of a finite sequence.

Fix an integer $k, 1 \leq k \leq n$ and for each $s \epsilon S_n$, let $\tilde{s}^k \epsilon S_n$ be the sequence obtained from s by changing the kth bit, i.e. $\tilde{s}_k^k = 1 - s_k$ and $\tilde{s}_i^k = s_i$, for $i = 1 \ldots n$, $i \neq k$. Define

$$\overset{k}{\Delta}(n) = \frac{1}{2^{n-1}} \sum_{s \epsilon S_n, s_k = 0} |\Delta(s, \tilde{s}^k)|. \tag{1}$$

This is the average change in linear complexity caused by a change in the kth bit. Although a change in the nth bit can cause a severe change in linear complexity, theorem 3.1 states that on the average, the change is close to one. We first present two lemmas that will be used in in the proof of this theorem and in later sections.

Lemma 3.1 *Let $s \epsilon S_n$ and let $\tilde{s}^n$ be defined as above, then $\Delta(s, \tilde{s}^n) = |\Lambda(s) - \Lambda(\tilde{s}^n)|$ is given by*

$$\Delta(s, \tilde{s}^n) = \begin{cases} 0 & if \Lambda_{n-1}(s) = \Lambda_{n-1}(t) \geq \frac{n}{2} \\ n - 2k & if \Lambda_{n-1}(s) = \Lambda_{n-1}(t) = k < \frac{n}{2} \end{cases}.$$

Proof: This result is due to Massey, [2].

Lemma 3.2 *The distribution of $\Lambda(s), s \epsilon S_m$ is given by*

$$card\{s : \Lambda(s) = k\} = \begin{cases} 1 & k = 0 \\ 2 \cdot 4^{k-1} & k \leq m/2 \\ 4^{m-k} & k > m/2 \end{cases}.$$

Proof: This result follows, by induction, from lemma 3.1. It appears in a slightly different form in Rueppel, [3, page 36].

Theorem 3.1 *The average change in the linear complexity of a n-bit string caused by a change in the last bit is given by*

$$\overset{n}{\triangle}(n) = \begin{cases} \frac{8}{9} + \frac{3n-4}{9 \cdot 2^{n-1}} & \longrightarrow \frac{8}{9} \quad n \text{ even} \\ \\ \frac{10}{9} + \frac{3n-4}{9 \cdot 2^{n-1}} & \longrightarrow \frac{10}{9} \quad n \text{ odd} \end{cases}.$$

Proof: From definition (1), we have

$$\overset{n}{\triangle}(n) = \frac{1}{2^{n-1}} \sum_{s \in S_n, s_n = 0} |\Delta(s, \tilde{s}^n)|.$$

Lemma 3.1 implies that $\Delta(s, \tilde{s}^n)$ depends only on $\Lambda_{n-1}(s)$ so

$$\overset{n}{\triangle}(n) = \frac{1}{2^{n-1}} \sum_{0 \le k < \frac{n}{2}} (n - 2k) \, card\{s \in S_n, s_n = 0 \mid \Lambda_{n-1}(s) = k\}$$

and applying lemma 3.2 with $m = n - 1$, yields

$$\overset{n}{\triangle}(n) = \frac{1}{2^{n-1}} \left[n \cdot 1 + \sum_{k=1}^{M} (n - 2k)(2 \cdot 4^{k-1}) \right]$$

where $M = \frac{n}{2} - 1$ when n is even and $M = \frac{n-1}{2}$ when n is odd. Observing that

$$\sum_{k=1}^{M} 4^{k-1} k = \frac{(3M - 1)4^M + 1}{9} \tag{2}$$

we have,

$$\overset{n}{\triangle}(n) = \frac{1}{2^{n-1}} \left[n + 2n \left(\frac{4^M - 1}{3} \right) - 4 \left(\frac{(3M - 1)4^M + 1}{9} \right) \right].$$

Finally, substituting the appropriate values for M gives the desired results.

□

Having found the average change in linear complexity caused by changing the last bit of a sequence of n zeros, we now look, more generally, at the average change when the kth bit is changed, $1 \le k \le n$. Changing the kth bit (or m bits in fixed positions) induces a bijection on S_n. Given a bijection, $\varphi : S_n \longrightarrow S_n$, we denote by Δ_φ, the average value of $|\Lambda(s) - \Lambda(\varphi(s))|$. We then obtain an upper bound on Δ_φ which serves, also, as an upper bound on the average change in linear complexity caused by flipping the kth bit.

Theorem 3.2 *Let $\varphi : S_n \longrightarrow S_n$ be a bijection. Then the average value, Δ_φ, of $|\Lambda(s) - \Lambda(\varphi(s))|$ is bounded above by*

$$\begin{cases} \frac{4}{3} - \frac{1}{3(2^{n-2})} & n \text{ even} \\ \\ \frac{5}{3} - \frac{1}{3(2^{n-2})} & n \text{ odd}. \end{cases}$$

For each n, there exists a bijection that attains this bound.

Proof: The average of the absolute value of the differences in linear complexity between s and $\varphi(s)$ is given by

$$\Delta_\varphi = \frac{1}{2^n} \sum_{s \epsilon Sn} | \Lambda(s) - \Lambda(\varphi(s)) |$$

$$\leq \frac{1}{2^n} \sum_{s \epsilon Sn} \left(| \Lambda(s) - \frac{n}{2} | + | \Lambda(\varphi(s)) - \frac{n}{2} | \right).$$

$$= \frac{1}{2^{n-1}} \sum_{s \epsilon Sn} | \Lambda(s) - \frac{n}{2} |$$

since φ is a bijection. So we evaluate the sum

$$S \equiv \sum_{s \epsilon Sn} | \Lambda(s) - \frac{n}{2} | .$$

From the distribution of linear complexity, (lemma 3.2), we see that if n is even,

$$S = \frac{n}{2} + \sum_{k=1}^{n/2} 2 \cdot 4^{k-1}(\frac{n}{2} - k) + \sum_{k=1+n/2}^{n} 4^{n-k}(k - \frac{n}{2})$$

and if n is odd,

$$S = \frac{n}{2} + \sum_{k=1}^{(n-1)/2} 2 \cdot 4^{k-1}(\frac{n}{2} - k) + \sum_{k=(n+1)/2}^{n} 4^{n-k}(k - \frac{n}{2}).$$

Let us first consider n even. Replacing $n - k + 1$ with k in the last sum and combining terms yields

$$S = \frac{n}{2} + \sum_{k=1}^{n/2} 4^{k-1} \left((1 + \frac{3n}{2}) - 3k \right) = \frac{2}{3}(2^n - 1).$$

and finally,

$$\Delta_\varphi \leq \frac{S}{2^{n-1}} = \frac{4}{3} - \frac{1}{3 \cdot 2^{n-2}}.$$

A similar calculation yields the result for n odd.

To construct a bijection, φ, that obtains the upper bound, start with the sequences of highest and lowest linear complexity, working inward, and always choosing for an image the sequence most distant in linear complexity and not yet used. By the above proof φ attains the maximum value for Δ_φ.

$\square$

4 Other Bit Change Functions

A bijection on S_n will take some strings to images of lower linear complexity, but others will have images with higher linear complexity. An algorithm which, given a string, $s \epsilon S_n$, tries to produce a sequence, of "low" linear complexity, that differs from s in only a small percentage of its bits, should not be a bijection. Ideally, bits will only be altered when the change results in a string of lower linear complexity. In general, such an algorithm will induce a function $\varphi : S_n \to S_n$ but φ will not be a bijection. Here, we first consider such functions φ with the restriction that φ transforms only a bounded number of strings to the same image string, e.g. functions that alter no more than k fixed bits. This leads to upper bounds similar to those in the previous section. We then consider functions on strings subject to the condition that the linear complexity of all the image strings be "low". Our final result shows that, under this condition, there must be strings that have "many" bits changed by the function.

Theorem 4.1 *Let* $0 \leq k \leq n$ *and let* φ *be a function,* $\varphi : S_n \to S_n$ *such that* $card\{\varphi^{-1}(s)\} \leq 2^k$ *for all* $s \epsilon S_n$. *Then an upper bound for* Δ_φ *is given by*

$$\Delta_\varphi \leq \frac{k}{2} - \frac{1 + 2^k}{3 \cdot 2^{n-1}} + \begin{cases} 4/3 & n \text{ even} \quad k \text{ even} \\ 3/2 & n \text{ odd} \quad k \text{ odd} \\ 3/2 & n \text{ even} \quad k \text{ odd} \\ 5/3 & n \text{ odd} \quad k \text{ even} \end{cases} .$$

For each n *and* k, *there exists a function, satisfying the conditions above such that* Δ_φ *attains this upper bound.*

Proof:

As in the analysis of bijections, we have

$$\Delta_\varphi \equiv \frac{1}{2^n} \sum_{s \epsilon S_n} |\Lambda(s) - \Lambda(\varphi(s))|$$

$$\leq \frac{1}{2^n} \sum_{s \epsilon S_n} \left(|\Lambda(s) - \frac{n}{2}| + |\Lambda(\varphi(s)) - \frac{n}{2}| \right)$$

$$= \frac{1}{2^n} \sum_{s \epsilon S_n} |\Lambda(s) - \frac{n}{2}| + \frac{1}{2^n} \sum_{s \epsilon S_n} |\Lambda(\varphi(s)) - \frac{n}{2}| .$$

The analysis for bijections gives an upper bound on the first of these sums so we have,

$$\Delta_\varphi \leq \frac{1}{2^n} \sum_{s \epsilon S_n} |\Lambda(\varphi(s)) - \frac{n}{2}| - \frac{1}{3 \cdot 2^{n-1}} + \begin{cases} 2/3 & n \text{ even} \\ 5/6 & n \text{ odd} \end{cases} . \qquad (3)$$

Now we must find an upper bound for

$$S = \sum_{s \in \mathbf{S_n}} \left| \Lambda(\varphi(s)) - \frac{n}{2} \right|$$

$$= \sum_{j=0}^{n} \left| \frac{n}{2} - j \right| card\{\varphi(s) \mid \Lambda(\varphi(s)) = j\}.$$

This will be maximal when the image values, $\Lambda(\varphi(s))$ are as far as possible from $\frac{n}{2}$. Since each s can have up to 2^k pre-images, we start with the elements of $\mathbf{S_n}$ whose linear complexity is farthest from $\frac{n}{2}$, assuming each has 2^k pre-images until we have enough pre-images to cover the 2^n elements of $\mathbf{S_n}$, so by lemma (3.2) we obtain:

$n - k$ odd:

$$S \leq 2^k \left[\frac{n}{2} + \frac{n}{2} + \sum_{j=1}^{M} (\frac{n}{2} - j)(2 \cdot 4^{j-1} + 4^j) \right]$$

with $M = \frac{n-k-1}{2}$ as for this value of M,

$$2^k \left[1 + 1 + \sum_{j=1}^{M} (2 \cdot 4^{j-1} + 4^j) \right] = 2^n$$

so that we have used all 2^n sequences in $\mathbf{S_n}$ as pre-images. Regrouping terms, we now obtain

$$S \leq 2^k \left[n + \sum_{j=1}^{M} \frac{3}{2} 4^j (\frac{n}{2} - j) \right]$$

$$= 2^k \left[n + \frac{3n}{4} \left(\frac{4^{M+1} - 4}{3} \right) - 2 \left(\frac{(3M - 1)4^M + 1}{3} \right) \right].$$

By substituting $M = \frac{n-k-1}{2}$ and reducing we obtain

$$S \leq 2^k \left[n + n(2^{n-k-1} - 1) - \frac{2}{3} \left(\left(\frac{3(n - k - 1)}{2} - 1 \right) 2^{n-k-1} + 1 \right) \right]$$

$$= 2^k \left[2^{n-k-1}(k + \frac{5}{3}) - \frac{2}{3} \right]$$

and dividing by 2^n yields

$$\frac{S}{2^n} \leq \frac{1}{2}(k + \frac{5}{3}) - \frac{2^k}{3 \cdot 2^{n-1}} = \frac{k}{2} + \frac{5}{6} - \frac{2^k}{3 \cdot 2^{n-1}}.$$

$\underline{n-k \text{ even}}$:

$$S \le 2^k \left[\frac{n}{2} + \frac{n}{2} + \left(\sum_{j=1}^{M-1} (\frac{n}{2} - j)(2 \cdot 4^{j-1} + 4^j) \right) + (\frac{n}{2} - M)\left(2 \cdot 4^{M-1} \right) \right]$$

with $M = \frac{n-k}{2}$ as for this value of M,

$$2^k \left[1 + 1 + \left(\sum_{j=1}^{\frac{n-k}{2}-1} (2 \cdot 4^{j-1} + 4^j) \right) + 2 \cdot 4^{\frac{n-k}{2}-1} \right] = 2^n.$$

Regrouping terms, we now obtain

$$S \le 2^k \left[n + \left(\sum_{j=1}^{M-1} \frac{3}{2} 4^j (\frac{n}{2} - j) \right) + 2^{2M-1}(\frac{n}{2} - M) \right]$$

$$= 2^k \left[n + 2^{2M-1}(\frac{n}{2} - M) + \frac{3n}{4}\left(\frac{4^M - 4}{3} \right) - 2\left(\frac{(3M-4)4^{M-1}+1}{3} \right) \right].$$

By substituting $M = \frac{n-k}{2}$ and reducing we obtain

$$S \le 2^k \left[2^{n-k-2}(2k + \frac{8}{3}) - \frac{2}{3} \right]$$

and dividing by 2^n yields

$$\frac{S}{2^n} \le \frac{k}{2} + \frac{2}{3} - \frac{2^k}{3 \cdot 2^{n-1}}.$$

Finally, substituting these upper bounds into the inequality (3) gives the desired results.

As in the case of a bijection, we can construct a function, $\varphi : S_n \rightarrow S_n$ that satisfies the conditions of the theorem and attains the upper bound by starting with those elements of S_n which have highest and lowest linear complexity, working inward and always choosing for an image the element of S_n which is still available and maximizes the difference in linear complexity. The only difference is that each element of S_n can be used as an image 2^k times. The following table shows how to construct this function when $n = 6$ and $k = 2$. Each string can have 4 pre-images. The table says that the unique string of linear complexity 6 and ant three of those with linear complexity 5 should be mapped to the string with linear complexity zero. The average change, $\Delta_\varphi = 2.28125$, as the theorem states.

Λ	$card\{s \mid \Lambda(s) = \Lambda\}$	Λ_*	$card\{s' \mid \Lambda(\varphi(s')) = \Lambda$ and $\Lambda(s') = \Lambda_*\}$
0	1	6	1
		5	3
1	2	5	1
		4	7
2	8	4	9
		3	23
3	32		
4	16		
5	4	3	9
		2	7
6	1	2	1
		1	2
		0	1

$\square$

Theorem 4.2 *Given $c(n), 0 < c(n) < \frac{n}{2}$, there exists a largest $\mu, 0 < \mu < 1$, such that there are at least $\frac{n-1}{n} 2^n$ sequences for which at least a fraction of μ bits must be changed to get a sequence with linear complexity less than $c(n)$. The fraction μ is a function of n and $c(n)$. If $\lim_{n\to\infty} \frac{c(n)}{n} = 0$ (e.g. if $c \leq p(\log_2(n))$ for a polynomial, p) then μ converges to $\frac{1}{2}$ as $n \to \infty$.*

Proof:
Let $M(\mu, n)$ be the number of sequences in $\mathbf{S_n}$ within Hamming distance μn of a fixed string, $s \epsilon \mathbf{S_n}$. Then, for $0 < \mu < 1/2$,

$$M(\mu, n) = \sum_{k=0}^{\lfloor \mu n \rfloor} \binom{n}{k} \leq 2^{n H_2(\mu)} \tag{4}$$

where $H_2(x) = -x \log_2(x) - (1-x) \log_2(1-x)$ is the *entropy* function, (MacWilliams and Sloane, [1, page 310]).

The number of sequences in $\mathbf{S_n}$ with linear complexity less than $c(n)$ is given by:

$$1 + \sum_{k=1}^{c(n)} 2 \cdot 4^{k-1} = 1 + 2\left(\frac{2^{2c(n)} - 1}{3}\right) = \frac{1 + 2^{2c(n)+1}}{3} \leq 2^{2c(n)+1}.$$

In order to change $\frac{2^n}{n}$ of the sequences in S_n to sequences of linear complexity less than $c(n)$ by altering at most μn bits of each sequence, we must have:

$$2^{2c(n)+1} M(\mu, n) \geq \frac{2^n}{n}.$$

Hence $M(\mu, n)$ must satisfy

$$M(\mu, n) \geq \frac{2^{n - \log_2(n)}}{2^{2c(n)+1}} = 2^{n - \log_2(n) - 2c(n) - 1}.$$

Substituting the upper bound from (4), we have

$$2^{nH_2(\mu)} \geq M(\mu, n) \geq 2^{n - \log_2(n) - 2c(n) - 1}.$$

Comparing the exponents, we see that

$$H_2(\mu) \geq 1 - \frac{\log_2(n) + 2c(n) + 1}{n}.$$

So $H_2(\mu) \to 1$ and hence, $\mu \to \frac{1}{2}$ as n tends to infinity, (MacWilliams and Sloane [1, page 308]).

$\square$

5 Conclusion and Future Work

The results of this paper tell us two things, of cryptographic importance, about the difference in linear complexity of strings and their neighbors vis-à-vis Hamming distance.

1: For large n there are strings in S_n which are cryptographically secure in the sense that they are far, in Hamming distance, from any string of "low" linear complexity.

2: There are enough such secure strings that we cannot expect to find an algorithm which for "most" strings produces nearby strings (in Hamming distance) of "low" linear complexity.

This suggests the following paths for future investigation:

1. Classify those sequences which are close (in Hamming distance) to sequences of "low" linear complexity. The results of this paper put bounds on how many such sequences there are but do not indicate what they look like.

2. Study the effect of synchronization errors on the linear complexity of sequences. We say that two sequences are $k - close$ if one can be obtained from the others by a sequence of no more then k errors where we now include added and lost bits as well as changes of bits. A sequence is k-close to many more sequences than it is within k of in Hamming distance. Theorem 6.2 does not immediately generalize in a useful way. As with Hamming distance, we should classify those sequences which are k-close to sequences of "low" linear complexity.

3. Answer these questions for "practical" sequences, e.g. those that can be generated by nonlinear registers of acceptable size.

References

[1] F. J. MacWilliams and N. J. A. Sloane. *The Theory of Error-Correcting Codes.* North-Holland Mathematical Library, Amsterdam, 1977.

[2] J. M. Massey. *Shift-Register Synthesis and BCH Decoding.* IEEE Trans. Information Theory 15, 122-127 (1969).

[3] R. A. Rueppel. *Analysis and Design of Stream Ciphers.* Springer, Berlin,1986.

Some remarks on a LFSR "disturbed" by other sequences

Guy Chassé
CNET
PAA/TIM
38-40, rue du Général Leclerc
92131 Issy les Moulineaux
FRANCE

ABSTRACT.

We study a recurring sequence obtained by disturbing (in a sense defined in the paragraph I) a linear feedback shift register (LFSR). We compute the generating series and the minimal polynomial of this sequence.

I INTRODUCTION.

Let $\mathbb{F}_q$ be the finite field with q elements (where q is a power of the prime number p). In this paper we are considering the situation where the cells of a LFSR over $\mathbb{F}_q$ are "disturbed" by sequences (of elements of $\mathbb{F}_q$).

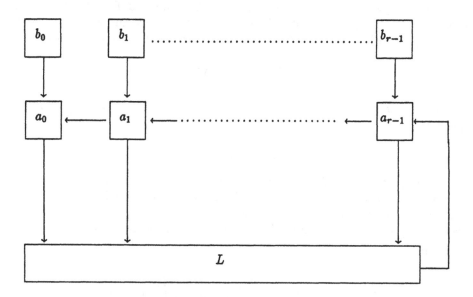

Let $a_i(t)$ (respectively $b_i(t)$) be the content of the cell a_i (respectively b_i) at the time t ($t \in \mathbb{N}$, $0 \leq i \leq r-1$). At the time $t+1$ we shall have :

$$a_i(t+1) = a_{i+1}(t) + b_i(t) \text{ for } i, \ 0 \le i \le r-2,$$
$$a_{r-1}(t+1) = L(a_0(t), a_1(t), ..., a_{r-1}(t)) + b_{r-1}(t).$$

where L is the linear feedback of our LFSR.

For the general properties of the LFSR we refer to [2] but we use results written in the style of [1].

We are interested in the properties of the sequence $(a_0(t))_{t \in \mathbb{N}}$. We reduce the problem to one where only one cell (let us say a_k, for a k, $0 \le k \le r-1$) is disturbed by a sequence $(b_k(t))_{t \in \mathbb{N}}$. Then the resulting sequence is the sum of :
- the sequence produced by the LFSR without disturbance;
- a sequence whose the generating series is essentially a product of two power series : the generating series of $(b_k(t))_{t \in \mathbb{N}}$ and the generating series of a LFSR sequence only depending on the coefficients of its feedback polynomial.

In fact, this last sequence is the one generated by the described scheme with the initial conditions $a_i(0) = 0$, $0 \le i \le r-1$.

II PREREQUISITES AND NOTATIONS.

Let $s = (s_t)_{t \in \mathbb{N}}$ be a sequence defined over the finite field $\mathbb{F}_q$. The least positive integer n such that $s_n \ne 0$ is called the order of the sequence and will be denoted by $\omega(s)$. Classically with s is associated a power series in $\mathbb{F}_q[[X]]$, the generating series of s :

$$S(s) = \sum_{t \in \mathbb{N}} s_t X^t.$$

Let $F = F(X) = \sum_{i=0}^{d} a_i X^i$ be a polynomial in $\mathbb{F}_q[X]$, of degree d. This polynomial acts on the sequence s :

$$s \mapsto F \circ s = (t \mapsto \sum_{i=0}^{d} a_i s_{t+i}).$$

This operation gives a $\mathbb{F}_q[X]$ module structure to the set of all these sequences. We say that s is a linear recurring sequence if $F \circ s = 0$ for a polynomial $F \ne 0$, the linear complexity of s is then defined to be the least degree of such an F. Such an F can be chosen monic, so it is unique and called the minimal polynomial of s.

Let us set :

$$F_{(i)} = \begin{cases} 0 \text{ for } i < 0 \text{ or } i \ge d, \\ a_{i+1} + a_{i+2}X + a_{i+3}X^2 + ... + a_d X^{d-i-1} \text{ for } 0 \le i < d. \end{cases}$$

For each sequence s and for each polynomial F we define the initial polynomial of s relatively to F by (see [1]) :

$$P(s, F) = \sum_{i \ge 0} s_i F_{(i)}.$$

It can be easily seen that $deg(P(s, F)) = deg(F) - \omega(s) - 1$.

At last the reciprocal polynomial of any polynomial F is formally defined as :

$$F^*(X) = X^d F(\frac{1}{X})$$

where d is the degree of F. The following properties are immediate and well known :
1) $deg(F^*) = deg(F_1)$ if F_1 is the polynomial such that $F = X^n.F_1$, n is an integer and $F_1(0) \neq 0$,
2) $F^{**} = (F^*)^*$ divides F and $F^{**} = F$ if $F(0) \neq 0$.

In all the sequel "gcd" will mean "greatest common divisor".

We shall need the two following results where "sequences" and "polynomials" have to be considered, for our purpose, as defined over $\mathbb{F}_q$ (see [1]).

Proposition A. *Let s be a sequence and F a polynomial cancelling s (i.e. $F \circ s = 0$). Then the minimal polynomial of s is $\frac{F}{gcd(F, P(s, F))}$.*

Proposition B. *1) Let s be a linear recurring sequence and F a monic polynomial such that $F \circ s = 0$ then the generating series of s is the rational function $S(s) = \frac{X^{\omega(s)} P^*(s, F)}{F^*(X)}$.*

2) Conversely suppose the generating series of a sequence s is a rational function $S(s) = \frac{Q(X)}{G(X)}$ where $deg(Q) < deg(G)$ and $G(0) \neq 0$. Then s is a linear recurring sequence, $G^* \circ s = 0$ and $Q(X) = X^{\omega(s)} P^*(s, G^*)$.

III REDUCTION OF THE PROBLEM.

If we consider the system introduced above and look at the sequence $(u_t)_{t \in \mathbb{N}} = (a_0(t))_{t \in \mathbb{N}}$ we easily see that we can write u_t as

$$u_t = \alpha_t^{(0)} + \alpha_t^{(1)} + ... + \alpha_t^{(r-1)} + \beta_t$$

where :
- the term $\alpha_t^{(i)}$, for $0 \leq i \leq r - 1$, can be expressed by $\sum_{0 \leq j < t} y_j^{(i)} b_j^{(i)}$, with $y_j^{(i)}$ only depending on the coefficients of the feedback polynomial F ;
- the term $(\beta_t)_{t \in \mathbb{N}}$ is the linear recurring sequence generated by the linear feedback shift register with the initial conditions $a_i(0)$, $0 \leq i \leq r - 1$) without perturbation.

It is an easy exercise to compute the linear complexity of a sequence sum involving sequences of known linear complexity. So we restrict ourselves to the case when the initial contents of all the cells A_i are 0 and only one sequence is disturbing the linear feedback shift register.

In this way, now we are no longer studying the perturbation of a LFSR by some sequence but the perturbation of a sequence going through a LFSR with zero initial cell contents.

IV EMBEDDING ONE SEQUENCE IN A LFSR WITH ZERO INITIAL CONDITIONS.

We refer to the scheme and notations of the introduction but we suppose that only the cell number k, $0 \leq k \leq r-1$ is disturbed. The disturbing sequence $(b_t^{(k)})_{t \in \mathbb{N}}$ will be denoted $b = (b_t)_{t \in \mathbb{N}}$.

We set $a_i(0) = 0$ for every i, $0 \leq i \leq r-1$.

Let $F(X) = X^r + \sum_{i=0}^{r-1} c_i X^i$ be the feedback polynomial of the register.

The following formulas are immediately obtained by applying the properties of the LFSR :

(i) for $\left\{ \begin{array}{c} i, m \in \{0, ..., k\} \\ i+j = m+n \end{array} \right\}$ and $\left\{ \begin{array}{c} i, m \in \{k+1, ..., r-1\} \\ i+j = m+n \end{array} \right\}$, $a_i(j) = a_m(n)$;

(ii) for $\left\{ \begin{array}{c} i \leq k < m \leq r-1 \\ i+j = m+n \end{array} \right\}$, $a_i(j) = a_m(n) + b_{m+n-k-1}$.

We set $a_{r-1}(t) = b_t = 0$ if $t < 0$.

Remark.

By use of the previous formulas we can link the sequences $(a_0(t))_{t \in \mathbb{N}}$ and $(a_{r-1}(t))_{t \in \mathbb{N}}$.

If $k < r-1$ we have :

$$a_0(t) = 0 \text{ if } t \leq k,$$
$$a_0(t) = b_{t-k-1} + a_{r-1}(t-r+1) \text{ if } t > k.$$

If $k = r-1$ we have :

$$a_0(t) = a_{r-1}(t-r+1) \text{ assuming } a_{r-1}(t) = 0 \text{ if } t < 0.$$

So we shall only study the sequence $a_{r-1}(t)_{t \in \mathbb{N}}$.

Proposition. *Consider the generating series of the sequence $a = (a_{r-1}(t))_{t \in \mathbb{N}}$. This series is equal to the usual product of power series : $X \cdot S(A) \cdot S(b)$ where the sequence A is only depending on the c_i, and is cancelled by F.*

Proof.

At time t, $t \geq 0$, $a_{r-1}(t)$ can be expressed as a linear combination of the b_i, $0 \leq i \leq t-1$ (and even $i < t-1$ if $k < r-1$) over $\mathbb{F}_q$. Let A_{t-1} be the coefficient of b_0. We want to compute A_t, the coefficient of b_0 in $a_{r-1}(t+1)$. For this we distinguish two cases.

1) If $0 \leq k < r-1$.

We have $A_0 = 0$ and $A_1 = c_k$. After we proceed by induction (setting $A_i = 0$ if $i < 0$) :

- for t, $0 \leq t \leq k+1$, we have $A_{t+1} = \sum_{j=0}^{r-1} c_j A_{j+t+1-r} + c_{k+1-t}$,

- for $t > k+1$ it is easy to see that $A_{t+1} = \sum_{j=0}^{r-1} c_j A_{j+t+1-r}$.

2) If $k = r - 1$.

At time $t = 1$ we have $a_{r-1}(1) = b_0$ so $A_0 = 1$. After we apply the recurrence relation and we find that for $t \geq 0$: $A_{t+1} = \sum_{j=0}^{r-1} c_j A_{j+i+1-r}$.

What is the coefficient of b_i, if $0 < i \leq t-1$ in the linear combination giving $a_{r-1}(t)$?

We immediately see that it is equal to the coefficient of b_{i-1} in the expression of $a_{r-1}(t-1)$. So we can write :

$$a_{r-1}(t) = A_{t-1}b_0 + A_{t-2}b_1 + ... + A_0 b_{t-1}$$

$$= \sum_{j=0}^{t-1} A_{t-1-j}b_j.$$

Now if we write down the generating series of $a = (a_{r-1}(t))_{t \in \mathbb{N}}$ we obtain :

$$S(a) = \sum_{t \in \mathbb{N}} a_{r-1}(t) \cdot X^t$$

$$= \sum_{t \in \mathbb{N}} \left(\sum_{j=0}^{t-1} A_{t-1-j} \cdot b_j \right) \cdot X^t$$

(by using the previous expression of $a_{r-1}(t)$)

$$= \sum_{t \in \mathbb{N}} \left(\sum_{j=0}^{t-1} A_{t-1-j} X^{t-1-j} \cdot b_j X^j \right) \cdot X$$

$$= X \cdot S(A) \cdot S(b).$$

Corollary. *Assume b is also a linear recurring sequence whose the minimal polynomial is $G = G(X)$. Then the generating series of a is :*

$$\frac{X^{\omega(A)+\omega(b)+1}.P^*(A, F).P^*(b, G)}{F^*(X).G^*(X)}.$$

Also suppose $F(0) \neq 0$ and $G(0) \neq 0$ then the minimal polynomial of the sequence a is :

$$\frac{F.G}{gcd(F.G, P(A, F).P(b, G))}.$$

Proof.

We have the relations : $S(A) = \frac{X^{\omega(A)}.P^*(A,F)}{F^*(X)}$ and $S(b) = \frac{X^{\omega(b)}.P^*(b,G)}{G^*(X)}$ following proposition B. So we obtain the generating series from our proposition.

For the end of the corollary, the hypothesis on F and G yield $deg(F^*) = deg(F)$ and $deg(G^*) = deg(G)$. Since $deg(P(s, F)) = deg(F) - \omega(s) - 1$ (for every F and s as mentioned in the part II) the degree of the numerator of $S(a)$ is less than the degree of its denominator, so from proposition B we deduce that $(F^*.G^*)^* = F.G$ cancels a. Now proposition A gives the minimal polynomial of a.

Remark.

In particular if F and G are irreducible, F does not divide $P(b, G)$ and G does not divide $P(A, F)$ then the minimal polynomial of a is $F.G$.

V CONCLUSIONS.

1) Going back to the introduction, we notice that the sequence we decided to study : $(a_0(t))_{t \in \mathbb{N}}$ with the described conditions gives an example of solution of an equation of the type :

$$F \circ \alpha = \beta$$

where $F = F(X) = \sum_{i=0}^{r-1} c_i X^i + X^r$ is the feedback polynomial of the LFSR as usual and β a fixed sequence. This type of equation generalises the linear homogeneous difference equations. To see this, using the recurring relations given in the introduction, we obtain, for an arbitrary $t \in \mathbb{N}$:

$$(*) \qquad \alpha(t + j) = a_j(t) + \sum_{i=0}^{j-1} b_i(t + j - i - 1)$$

for every j, $0 \leq j \leq r - 1$.

So we get :

$$\alpha(t + r) = \alpha((t + 1) + (r - 1))$$

$$= a_{r-1}(t + 1) + \sum_{i=0}^{r-2} b_i(t + r - i - 1)$$

$$= -\sum_{i=0}^{r-1} c_i . a_i(t) + b_{r-1}(t) + \sum_{i=0}^{r-2} b_i(t + r - i - 1)$$

Now, using the relation $(*)$ to express $a_i(t)$ we have :

$$\alpha(t + r) = -\sum_{i=0}^{r-1} c_i . \{\alpha(t + i) - \sum_{j=0}^{i-1} b_j(t + i - j - 1)\} + b_{r-1}(t)$$

$$+ \sum_{i=0}^{r-2} b_i(t + r - i - 1)$$

$$= -\sum_{i=0}^{r-1} c_i . \alpha(t + i) + \sum_{i=0}^{r-1} c_i . \{\sum_{j=0}^{i-1} b_j(t + i - j - 1)\} + b_{r-1}(t)$$

$$+ \sum_{i=0}^{r-2} b_i(t + r - i - 1)$$

Putting all the terms involving α in the left part of the equation we obtain an equation of the announced type :

$$F \circ \alpha = \beta.$$

2) Not surprisingly, this type of scheme does not improve the linear complexity of the used sequences but it shows a simple way to linearly transform a given sequence $(b_k(t))_{t \in \mathbb{N}}$ by embedding it in a LFSR. For cryptographic purposes, this suggests further study of the $\mathbb{F}_q$ linear transforms of sequences regarding other properties than linear complexity.

REFERENCES.

[1] D. FERRAND, "Suites récurrentes", Dept. of Math., University of Rennes, France, 1987.

[2] E.S. SELMER, "Linear recurrence relations over finite fields", Dept. of Math., University of Bergen, Norway, 1966.

Parameters for Complex FFTs in Finite Residue Class Rings

Reiner Creutzburg

University of Karlsruhe
Institute of Algorithms and Cognitive Systems
Am Fasanengarten 5
P. O. Box 6980
D-7500 Karlsruhe 1
phone: (+49)-721-608 4325
fax: (+49)-721-696893

Introduction

With the rapid advances in large scale integration, a growing number of digital signal processing operations becomes attractive. The convolution property of certain transforms can be used to compute the cyclic convolution of two discrete signals. One such transform is the discrete Fourier transform (DFT) in the modified version of the fast Fourier transform (FFT) which operates on signals in the complex number field [4,7]. The FFT of length N requires $(N/2)\log_2(N/2)$ complex multiplications, which take most of the computation time. In addition, the FFT implementation of cyclic convolution introduces significant amounts of round-off error, thus deteriorating the signal-to-noise-ratio.

Number-theoretic transforms (NTT) were introduced as a generalization of the DFT over residue class rings of integers in order to perform high-speed and efficient cyclic convolutions and correlations without round-off errors [1,2]. Interesting applications of the NTT are in fast coding, decoding, digital filtering, image processing, and deconvolution. A large number of transform methods were developed.

The main drawback of the NTT is a rigid relationsship between computer word length and obtainable transform length and a limited choice of posible word length. In order to enlarge the transform length of conventional NNTs, complex number-theoretic transforms (CNT) were introduced [3]. However, it is not easy to find convenient moduli m that are large enough to avoid overflow, and to find primitive Nth roots of unity modulo m with small binary weight for transform lengths N that are highly factorizable and large enough for practical applications [5]. In a recent paper [6], we have presented a solution to this problem in the ring $\mathbb{Z}$ of integers by studying cyclotomic polynomials. In this letter we present simple constructive methods for the finding of all convenient moduli m for CNTs under the assumption that a special transform length N and a special element $\alpha \in \mathbb{Z}[i]$ with small binary weight are given.

Complex number-theoretic transforms.

Let $\mathbb{Z}[i]$ be the ring of the Gaussian integers $\xi = \xi_1 + i\xi_2$, where $N(\xi) = \xi_1^2 + \xi_2^2$, $(\xi_1, \xi_2 \in \mathbb{Z})$ denotes the norm of ξ. Furthermore, let $m > 1$ be an odd integer with the prime factorization

$$m = p_1^{u_1} \dots p_s^{u_s} q_1^{v_1} \dots q_t^{v_t} \tag{1}$$

where $p_j \equiv 1 \bmod 4$, $(j = 1, \dots, s)$ and $q_k \equiv 3 \bmod 4$, $(k = 1, \dots, t)$. Then $\alpha \in \mathbb{Z}[i]$ is called a primitive N-th root of unity modulo m if

$$\alpha^N \equiv 1 \bmod m,$$
$$\text{GCD}(\alpha^n-1, m) = 1 \text{ for every } n = 1, \dots, N-1. \tag{2}$$

Note that by definition (2) the integer $m > 1$ is a so-called primitive divisor of α^N-1. The following theorem states necessary and sufficient conditions that $\alpha \in Z[i]$ is a primitive N-th root of unity modulo m. Let Φ_N be the N-th cyclotomic polynomial.

Theorem 1. Let $m > 1$ be an odd integer. An element $\alpha \in \mathbb{Z}[i]$ is a primitive N-th root of unity modulo m if and only if one of the following conditions holds:

1) $\Phi_N(\alpha) \equiv 0 \bmod m$; $\text{GCD}(N, m) = 1$; $\tag{3}$

2) $\alpha^N \equiv 1 \bmod m$; $\text{GCD}(\alpha^d-1, m) = 1$ for every prime divisor d of N with N/d prime;

3) $\alpha^N \equiv 1 \bmod m$; $\text{GCD}(N(\alpha^d-1), m) = 1$ for every prime divisor d of N with N/d prime.

A necessary and sufficient condition for the existence of such primitive N-th roots of unity modulo m is [4]

$$N \mid \text{GCD}(p_1-1, \dots, p_1-1, q_1^2-1, \dots, q_1^2-1). \tag{4}$$

Now let $x = (x_0, \dots, x_{N-1})$ and $y = (y_0, \dots, y_{N-1})$ be two N-point integer sequences. Note that the equality of such sequences x and y is explained by $x_k \equiv y_k \bmod m$ $(k = 0, \dots, N-1)$. The CNT of length N with

$\alpha \in \mathbb{Z}[i]$ as a primitive N-th root of unity modulo m and its inverse are defined as the following mappings between N-point integer sequences:

$$X_n \equiv \sum_{k=0}^{N-1} x_k \, \alpha^{nk} \bmod m, \quad (n = 0, ..., N-1),$$

(5)

$$x_k \equiv N' \sum_{k=0}^{N-1} X_n \, \alpha^{-nk} \bmod m, \quad (k = 0, ..., N-1),$$

where $NN' \equiv 1 \bmod m$. The CNT has a similar structure and properties like the DFT, particularly the cyclic convolution property. For given transform length N and given element $\alpha \in \mathbb{Z}[i]$ one has to choose the modulus m by (3) as a divisor of $\Phi_N(\alpha)$. But in general the prime factorization of $\Phi_N(\alpha)$ is difficult to find in $\mathbb{Z}[i]$. Hence we consider some special cases in which this prime factorization is easy to perform in $\mathbb{Z}$ by the help of the properties of cyclotomic polynomials. The following example shows the application of known properties [10] of $\Phi(x)$ for the Winograd-number [4] $N = 840 = 2^3 \times 3 \times 5 \times 7$:

$$\Phi_{840}(x) = \Phi_{2\times3\times5\times7}(x^4) = \Phi_{3\times5\times7}(-x^4) =$$

$$\Phi_3(-x^{140}) \, \Phi_3(-x^4) \, (\Phi_3(-x^{20}))^{-1} \, (\Phi_3(-x^{28}))^{-1}.$$

If one chooses the element $\alpha = 1 + i \in \mathbb{Z}[i]$ then the calculation gives

$$\Phi_{840}(1+i) = \Phi_{105}(4) = \Phi_{105}(2) \, \Phi_{210}(2)$$

$$= 211 \times 29\ 191 \times 106\ 681 \times 152\ 041 \times 6\ 644\ 441 \times 1\ 564\ 921.$$

For the prime factorization of such large numbers the reader is referred to [12]. Note that the calculation of $\Phi_{840}(1+i)$ was carried out by only real operations.

Construction of moduli for CNTs.

Theorem 1 states a necessary and sufficient condition for moduli m, so that an element α is a primitive N-th root of unity modulo m. But in some important cases for practical applications one can specify theorem 1 in the following way.

Theorem 2. Let $N = 2^n N_1$ ($N_1 \geq 1$ odd, $n \geq 3$) and $\alpha = \pm 2^k(1+i)$, ($k \geq 0$) be given. Then one has to choose as a modulus m for a CNT a divisor of

$$\Phi_N(\pm 2^k(1+i)) = \{ \begin{array}{l} \Phi_{N/8}(2^{4k+2}) = \Phi_{N/4}(2^{2k+1})\Phi_{N/8}(2^{2k+1}) \text{ for } n = 3, \\[2mm] \Phi_{N/4}(2^{2k+1}) \qquad \text{for } n > 3 \end{array}$$

with $GCD(m, N) = 1$. In other words, m is a primitive divisor of $2^{(2k+1)N/4}-1$, if $n = 3$ and $2^{(2k+1)N/2}-1$ if $n > 3$, respectively.

Note that the primitive divisors of 2^s-1 are listed [12].

Corollary 1. If $\alpha = 1 + i$ is a primitive Nth root of unity modulo m with $4 N$, then $\alpha^2 = 2i$ is a primitve $N/2$th root of unity modulo m.

Corollary 2. Let $N = 8p$ ($p > 3$ prime) and $\alpha = 1 + i \in \mathbb{Z}[i]$ be given. Then one has to choose as a modulus m for a CNT a divisor of

$$\Phi_{8p}(1+i) = \Phi_p(2) \ \Phi_{2p}(2) = (2^p+1)(2^p-1)/3. \tag{6}$$

The related CNT is called complex Pseudo-Mersenne transform.

Corollary 3. Let $N = 8p^2$ ($p > 3$ prime) and $\alpha = 1 + i \in \mathbb{Z}[i]$ be given. Then one has to choose as a modulus m for a CNT a divisor of

$$\Phi_{8p^2}(1+i) = [(2^{p^2} + 1)(2^p+1)^{-1}] \ [(2^{p^2} - 1)(2^p-1)^{-1}] \tag{7}$$

$$= (2^{2p^2} - 1)(2^{2p}-1)^{-1}.$$

In other words, m is a primitive divisor of $2^{2p^2}-1$.

Corollary 4. Let p, q be primes with $2 < q < p$ and $2^q \ne \mod p$. Then for $N = 8pq$ and $\alpha = 1 + i$ one has to choose for a modulus m for a CNT a divisor of

$$\Phi_{8pq}(1+i) = [3(2^{pq} + 1)((2^p+1)(2^q+1))^{-1}] \times$$
$$\times [(2^{pq} - 1)((2^p-1)(2^q-1))^{-1}]. \tag{8}$$

In other words, m is a primitive divisor of $2^{2pq}-1$.

Note that the first factor in (7)-(8) is a Pseudo-Fermat number and the second factor is a Pseudo-Mersenne number. In this direction our theorem 2 and the above corollaries generalizes known results of recent works on Pseudo-Fermat- and Pseudo-Mersenne transforms, respectively [5,8,9].

Conclusion

Using the concept of factorization of cyclotomic polynomials and the factorization of large numbers. we obtain valuable results for the determination of convenient parameters (transform length N, modules m, primitive root of unity modulo m) for complex number-theoretic transforms. Many known results of [3-6,8,9,] were generalized.

References.

[1] Pollard, J. M.: The fast Fourier transform in a finite field. Math. Comp. **25** (1971), pp. 365-374

[2] Agarwal, R. C.:, C. S. Burrus: Number theoretic transforms to implement fast digital convolution. Proc. IEEE **63** (1976), pp. 550-560

[3] Reed, I. S., T. K. Truong: Complex integer convolutions over a direct sum of Galois fields. IEEE Trans. **IT-21** (1976), pp. 169-175

[4] Nussbaumer, H. J.: Fast Fourier transform and convolution algorithms. Springer: Berlin - Heidelberg- New York 1981

[5] Dubois, E., A. N. Venetsanopoulos: The generalized Fourier transform in rings of algebraic integers. IEEE Trans. **ASSP-28** (1980), pp. 169-175

[6] Creutzburg, R., M. Tasche: Number-theoretic transforms of prescribed length. Math. Comp. **47** (1986), pp. 693-701

[7] Beth, T.: Verfahren der schnellen Fourier-Transformation. Teubner: Stuttgart 1984

[8] Nussbaumer, H. J.: Digital filtering using complex Mersenne transforms. IBM J. Res. Dev. **20** (1976), pp. 498-504

[9] Nussbaumer, H. J.: Digital filtering using pseudo Fermat transforms. IEEE Trans. **ASSP-25** (1977), pp. 79-83

[10] Lüneburg, H.: Galoisfelder, Kreisteilungskörper und Schieberegisterfolgen. Bibliographisches Institut: Mannheim 1979

[11] Creutzburg, R., M. Tasche: F-Transformation und Faltung in kommutativen Ringen. Elektron. Informationsverarb. Kybernet. **21** (1985), pp. 129-149

[12] Brillhart, J., D. H. Lehmer, J. L. Selfridge, B. Tuckerman, S. S.Wagstaff Jr.: Factorizations of $b^n \pm 1$, b = 2,3,5,6,7,10,11,12 Up to High Powers. Contemporary Math. **22** , Amer. Math. Soc., Providence, R. I., 1983

A CRYPTOGRAPHIC TOOL FOR PROGRAMS PROTECTION

Josep Domingo-Ferrer
Dpt. d'Informàtica
Universitat Autònoma de Barcelona
08193 Bellaterra, Barcelona

Llorenç Huguet-Rotger
Dpt. de Matemàtiques i Informàtica
Universitat de les Illes Balears
07071 Palma de Mallorca, Balears

Abstract. In this paper we present a cryptographic mechanism for ensuring effective software protection. A new method is featured which solves this problem by properly coding rather than encrypting, that is without adopting the lengthy and expensive solution which consists of having the whole program signed/encrypted by an authority [PaWi89].

The coding procedure is based on a one-way function F (for example, the discrete logarithm) and a public-key signature scheme (D,E) (for example, RSA or ElGamal). The mechanism follows these three basic steps, for a sequential algorithm:

1.- If the algorithm consists of machine code executable instructions $i_1, i_2,..., i_n$ then obtain the fixed-lenght normalized instructions $I_1, I_2,..., I_n$.

2.- Replace each I_j with a trace T_j obtained recursively on $F(T_{j+1})$ and I_{j+1} ($T_j = F(T_{j+1}) \oplus I_{j+1}$); with $T_n = F(I_n)$.

3.- The so called authority computes a closing trace $t_0 = D(F(T_1) \oplus I_1)$, with D the private transformation of the public-key signature.

Using two pipe-lined processors for decoding-executing, we can state the two following main results, concerning the correctness and run-time integrity of our proposal:

Correctness: Any program $i_1, i_2,..., i_n$ can be retrieved and executed from its corresponding trace sequence $t_0, T_1, T_2,..., T_n$.

Run-Time Integrity: If a program $i_1, i_2,..., i_n$ is stored as $t_0, T_1, T_2,..., T_n$, any instruction substitution, deletion or insertion will be detected at run-time, thus causing the processor to stop executing *before* the substituted, deleted or inserted instruction(s). Moreover, only the last five read traces must be kept in the internal secure memory of the processor.

Furthermore, a general use of the proposed system could practically prevent any viral attack with minimum authority operation.

1. Introduction.

Our goal is program integrity, i.e. ensuring that, given an image code no instructions are inserted or deleted before execution. This requires that the image be stored under a suitable structure, which can be almost completely worked out by the same user who wrote the program.

A *one way function F* [DiHe76], such that $F(X \oplus Y) \Leftrightarrow F(X) \oplus F(Y)$ (where $\oplus$ denotes addition modulo 2 on the binary representations of the operands), and a *public-key signature scheme* must be agreed upon before implementing the method. The signature consists of a private transformation D, exclusively owned by an authority, and a public registered inverse transformation E. Our proposal does not depend on the particular choice of F, D and E. The reader can think, by way of illustration, of F being a discrete modular exponentiation, ant the signature being and RSA [RiSA78] or ElGamal [ElGa85] one.

Also, a *normalized instruction format* must be defined. Supose an algorithm A consisting of machine-code executable instructions $i_1, i_2,...,i_n$. Each instruction i_j is first padded with a known filler to obtain a fixed-length one, by equalling its length to the maximum machine-code instruction length of our particular processor (for example, 22 bytes for MC68020, see [Moto85]). Finally, a redundance pattern, whose length depends on the desired security level, is appended and the resulting instruction is called I_j.

2. User Preparation Phase: Trace Evaluation.

When a user wishes to turn a program he has written into a trusted program, he first normalizes it as discussed above to obtain a sequence $I_1, I_2,...,I_n$, where n is the number of instructions in the program. Then he replaces each I_j with a *trace* T_j.

The traces are computed as follows, according to the assumptions on the type of instructions:

2.1.- Sequential Program.

Let $I_1, I_2, ..., I_n$ be the normalized instructions of the sequential program that we want to protect. Then we must compute:

$$
\begin{aligned}
T_0 &= F(T_1) \oplus I_1 \\
T_1 &= F(T_2) \oplus I_2 \\
T_2 &= F(T_3) \oplus I_3 \\
T_3 &= F(T_4) \oplus I_4 \\
&\quad .\ .... \\
&\quad .\ .... \\
T_{n-1} &= F(T_n) \oplus I_n \\
T_n &= F(I_n)
\end{aligned}
\tag{1}
$$

2.2.- Forward Unconditional Branch from I_k to I_j ($k<j$).

In this case, the branche is translated as:

$$
\begin{aligned}
...\quad T_{k-1} &= F(T_k) \oplus I_k \\
T_k &= F(T_j) \oplus I_j \\
T_{k+1} &= .... \\
&...... \\
T_j &= ...
\end{aligned}
\tag{2}
$$

2.3.- Forward Conditional Branch from I_k to I_j ($k<j$).

In this case, the branche is translated as:

$$
\begin{aligned}
...\quad T_{k-1} &= F(T_k) \oplus I_k \\
T_k &= F(T_{k'}) \oplus F(T_{k+1}) \oplus I_{k+1} \\
T_{k'} &= F(T_j) \oplus I_j \\
T_{k+1} &= .... \\
&...... \\
T_j &= ...
\end{aligned}
\tag{3}
$$

2.4.- Branch to Subroutine.

A branch to subroutine or procedure call at I_k is very similar to a conditional branch, in the sense that there are two target instructions, namely the first subroutine instruction and the next instruction of the calling program.

Here we must guarantee that there is no possibility for replacing the right subroutine at run-time. So, assuming that the instructions $I^*_1, I^*_2, ..., I^*_m$ of the subroutine are already encoded, by trace computation, as $T^*_1, T^*_2, ..., T^*_m$, we retrieve T^*_0 from t^*_0 (see next section about the heading trace t^*_0) and include it in the calling program as follows:

$$
\begin{aligned}
... \quad & T_{k-1} = F(T_k) \oplus I_k \\
& T_k = F(T_{k'}) \oplus F(T_{k+1}) \oplus I_{k+1} \\
& T_{k'} = T^*_0 \\
& T_{k+1} = ...
\end{aligned}
\qquad (4)
$$

Remark: Each separately stored machine-code subroutine requires its own trace computations and authority's endorsement, just as the main program does (see next section). A subroutine must be already endorsed before the trace computation (see the algorithm given in the annex) is run for a program calling it; else the computed T^*_0 will be garbage, so that a run-time error will occur when attempting to execute the subroutine call later.

2.5.- Backward Unconditional Branch from I_k to I_j (j ≤ k).

For the case of backward unconditional branch from I_k to I_j, with j≤k, the structure (2) cannot be computed, because the corresponding T_k needs the trace T_j which has not yet been computed and depends on T_k. So specific trace computing structures must be devised for backward unconditional branches. In this case we have

$$
\begin{aligned}
... \quad & T_{j-1} = F(T''_j) \oplus F(T_j) \oplus I_j \\
& T''_j = F(T_j) \oplus T''(j) \\
& T_j = ... \\
& ..... \\
& T_{k-1} = F(T_k) \oplus I_k \\
& T_k = F(T''(j)) \oplus I_j \\
& T_{k+1} = ...
\end{aligned}
\qquad (5)
$$

2.6.- Backward Conditional Branch from I_k to I_j ($j \leq k$).

For the case of backward conditional branch from I_k to I_j, with $j \leq k$, the structure (3) cannot be computed, because $T_{k'}$ needs the trace T_j which has not yet been computed and depends on $T_{k'}$. So specific trace computing structures must be devised for backward conditional branches. In this case we have:

$$
\begin{aligned}
\ldots \quad & T_{j-1} = F(T''_j) \oplus F(T_j) \oplus I_j \\
& T''_j = F(T_j) \oplus T''(j) \\
& T_j = \ldots \\
& \ldots \ldots \\
& T_{k-1} = F(T_k) \oplus I_k \\
& T_k = F(T_{k'}) \oplus F(T_{k+1}) \oplus I_{k+1} \\
& T_{k'} = F(T''(j)) \oplus I_j \\
& T_{k+1} = \ldots
\end{aligned}
\tag{6}
$$

Both in (5) and (6), $T''(j)$ is computed by applying a one-to-one function to j.

3.- Authority Endorsement Phase.

After the program instructions have been replaced with the corresponding traces by using the algorithm given in the annex, it is up to the authority owning the private transformation D to endorse the trace sequence. Actually, anyone possessing a signature recognized by the potential program users can perform a valid endorsement on the program traces, thus playing the authority role for that particular case. Now, when a user identifies himself as a valid user and supplies T_0, the authority computes a closing trace t_0:

$$
t_0 = D (T_0)
\tag{7}
$$

The same process is repeated for all T''_j in backward branch targets, so that $t''_j = D (T''_j)$ is computed for each backward target. Then T_0 is replaced with t_0, and the T''_j are replaced with the corresponding t''_j.

Notice that the whole program need not be supplied to the authority, but just T_0 and T''_j. Finally the traces are stored in an increasing order t_0, T_1, T_2,...., T_n, and will be used in lieu of the normalized instructions I_1, I_2,...., I_n.

4.- Program Execution with Controlled Instruction Flow.

We will sketch in this section the diagrams controlling the run-time operations of our scheme. The run-time setting considered consists of:

A usual processor P.

A preprocessor P', whose task is retrieving the instructions I_k and forwarding them to P.

It is assumed that P and P' are pipe-lined. The path between them must be a secure one, so that it is advisable that both processor and preprocessor be encapsulated in a single chip. By using a hybrid circuit (see [Ebel86]), it is possible to wire and encapsulate together two or more separate silicon wafers. For a sequential program, the following figure represents the operation of P and P'.

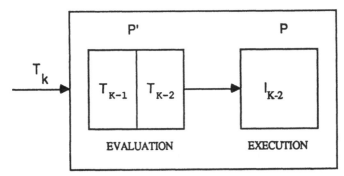

EVALUATION OF T_{k-2} => $F(T_{k-1}) \oplus T_{k-2} = I_{k-1}$ EXECUTION OF I_{k-2}

The intended operation at cycle k is:

T_k is being read by P'

T_{k-1} is available in an internal register of P'.

T_{k-2} is evaluated by P'

I_{k-2} is being executed by P (actually i_{k-2} stripped from I_{k-2} is executed, after redundance checking).

4.1.- Sequential Instruction Block.

We are ready to use induction on k to prove correctness for a sequential program. After reading t_0 and T_1 during cycles 0 and 1, at cycle 2 T_2 is read and P' evaluates t_0.

It must be pointed out that the computation $F(T_1) \oplus E(t_0) = I_1$ is feasible, from (7), because of the transformations F and E being easy and public and T_1 being available to P'. Thus the normalized instruction I_1 can be retrieved. The result follows by induction: at cycle k, I_{k-2} is executed and I_{k-1} is retrieved from T_{k-2} by $F(T_{k-1}) \oplus T_{k-2} = I_{k-1}$. Execution stops at cycle n+2 after executing I_n.

This diagram illustrates this reasoning:

cycle	P' read	evaluate	P execute
0	t_0	*	*
1	T_1	*	*
2	T_2	t_0	*
3	T_3	T_1	I_1 $(=F(T_1) \oplus E(t_0))$
4	T_4	T_2	I_2 $(=F(T_2) \oplus T_1)$
.	.	.	.
.	.	.	.
n	T_n	T_{n-2}	I_{n-2}
n+1	*	T_{n-1}	I_{n-1}
n+2	*	T_n	I_n

4.2.- Forward Unconditional Branch.

Considering 2.2 and reasoning similarly to 4.1, we have:

cycle	P' read	evaluate	P execute
.	.	.	.
k+2	T_{k+2}	T_k (fail)	I_k
k+3	T_j	*	*
k+4	T_{j+1}	T_k	*
k+5	T_{j+2}	T_j	I_j

2 idle p cycles are needed.

4.3.- Forward Conditional Branch.

Considering 2.3 and reasoning similarly to 4.1, we have:

4.3.1.- False Condition.

cycle	P'		P
	read	evaluate	execute
.	.	.	.
.	.	.	.
$k+2$	T_{k+1}	T_k *(fail)*	I_k
$k+3$	T_{k+2}	T_k	*
$k+4$	T_{k+3}	T_{k+1}	I_{k+1}

One extra trace $T_{k'}$ (unused) and 1 idle p cycle are needed.

4.3.2.- True Condition.

cycle	P'		P
	read	evaluate	execute
.	.	.	.
.	.	.	.
$k+2$	T_{k+1}	T_k *(fail)*	I_k
$k+3$	T_j	T_k	*
$k+4$	T_{j+1}	$T_{k'}$	*
$k+5$	T_{j+2}	T_j	I_j

One extra trace $T_{k'}$ and 2 idle p cycles are needed.

4.4.- Subroutine Branch

Considering 2.4 and reasoning similarly to 4.1, we have:

cycle	P'		P
	read	evaluate	execute
.	.	.	.
.	.	.	.
k+2	T_{k+1}	T_k *(fail)*	I_k
k+3	T^*_1	T_k	* *(save T_{k+1}, I_{k+1})*
k+4	T^*_2	$T_{k'}$	*
k+5	T^*_3	T^*_1	I^*_1
.	.	.	.
.	.	.	.
k+m+3	*	T^*_{m-1}	I^*_{m-1}
k+m+4	T_{k+2}	T^*_m	I^*_m *(restore T_{k+1}, I_{k+1})*
k+m+5	T_{k+3}	T_{k+1}	I_{k+1}

One extra trace $T_{k'}$ and 2 idle p cycles are needed.

4.5.- Backward Unconditional Branch.

Considering 2.5 and reasoning similarly to 4.1, we have:

cycle	P'		P
	read	evaluate	execute
.	.	.	.
.	.	.	.
k+2	T_{k+2}	T_k *(fail)*	I_k
k+3	t''_j	*	*
k+4	T_j	*	*
k+5	*	t''_j	*
k+6	T_{j+1}	T_k	*
k+7	T_{j+2}	T_j	I_j

One extra trace t''_j and 4 idle p cycles are needed.

4.6.- Backward Conditional Branch With a True Condition.
(Else see 4.3.1)

Considering 2.6 and reasoning similarly to 4.1, we have:

cycle	P'		P
	read	evaluate	execute
.	.	.	.
.	.	.	.
k+2	T_{k+2}	T_k (fail)	I_k
k+3	t''_j	T_k	*
k+4	T_j	*	*
k+5	*	t''_j	*
k+6	T_{j+1}	$T_{k'}$	*
k+7	T_{j+2}	T_j	I_j

Two extra traces t''_j and $T_{k'}$ and 4 idle p cycles are needed.

5.- Correctness and Run-Time Integrity.

Using these diagram approach it is straightforward to prove the correctness and run-time integrity of our scheme.

Theorem 1 (Correctness)

Any program $i_1, i_2,..., i_n$ can be retrieved and executed from its corresponding trace sequence $t_0, T_1, T_2,..., T_n$. ♦

Theorem 2 (Run-Time Integrity)

If a program $i_1, i_2,..., i_n$ is stored as $t_0, T_1, T_2,..., T_n$, and is evaluated as described in section 4, any instruction substitution, deletion or insertion will be detected at run-time, thus causing the processor to stop executing *before* the substituted, deleted or inserted intruction(s).

Moreover, only the last five read traces must be kept in the internal secure memory of the processor (storing five traces becomes justified when evaluating T_k in a backward unconditional branch, see 4.5) ♦

6.- Conclusions.

In order to evaluate the performance and the applicability of our controlled instruction flow system, it has to be compared with a scheme where the whole program is kept in a signed form (for example, using RSA or ElGamal signatures). Note that DES program encryption need not be considered for comparison because it involves the sharing of a secret key between the authority and each particular preprocessor P' that should be able to decrypt the program; also processing data in their encrypted form [RiAD78] has been considered so far a rather difficult issue.

The signature of a whole program by an authority has many drawbacks:

i) Signing and verifying each instruction separately is too time-consuming, and would cause the program size to increase drastically.

ii) The authority must entirely process any program having to be certified. This causes a great amount of overhead and saturation when there is a single authority available for many users. If more authorities are used, this leads to consistency and key management problems.

iii) The authority knows the contents of the program being signed. This can be undesirable if the authority is an external one.

As for the first drawback, our system allows branches and guarantees full integrity while requiring secure storage only for the last five read traces; also, decoding and execution are pipe-lined. The second handicap is smoothed as well because the work performed by the authority in our scheme is rather small, so that a great deal of users may share a single authority. Finally, regarding the privacy problem in the third drawback, our proposal requires that only T_0 and the T''_j be supplied to the authority for endorsement. The preparation phase can be carried out by the user itself, so that the authority need not know what is being signed, but just a user's valid identification.

Finally, the properties of our system ensure that if anything is wrong in the above procedure, a run-time error shall occur. Also, the ability to derive a trusted program from an existing untrusted one by just adding signed initial traces t_0 and t''_j's greatly simplifies the migration of existing executable standard software to our controlled instruction flow scheme.

Annex: Trace Computation Algorithm.

The following algorithm computes the traces for any program according to section 2.

begin (*Trace Computation Algorithm*)

    for k:=1 **to** n **do** target_back[k] := 0; *(This vector is used to indicate whether I_k is a*
                                                   target for a backward branch)

    $T_n := F(I_n)$;

    for k:=n-1 **downto** 1 **do**

   begin

      case I_k **of**

         forward unconditional branch to I_j:

           begin

               if target_back[j] = 0 **then** $T_k := F(T_j) \oplus I_j$

                            else $T_k := F(T''_j) \oplus F(T_j) \oplus I_j$

           end;

         forward conditional branch to I_j :

           begin

               if target_back[j] = 0 **then** $T_{k'} := F(T_j) \oplus I_j$

                            else $T_{k'} := F(T''_j) \oplus F(T_j) \oplus I_j$;

               if target_back[k+1] = 0 **then** $T_k := F(T_{k'}) \oplus F(T_{k+1}) \oplus I_{k+1}$

                          else $T_k := F(T_{k'}) \oplus F(T''_{k+1}) \oplus F(T_{k+1}) \oplus I_{k+1}$

           end;

         branch to subroutine:

         begin

           $T_{k'} := E(t^*_0)$;

              if target_back[k+1] = 0 **then** $T_k := F(T_{k'}) \oplus F(T_{k+1}) \oplus I_{k+1}$

                        else $T_k := F(T_{k'}) \oplus F(T''_{k+1}) \oplus F(T_{k+1}) \oplus I_{k+1}$

         end;

         backward unconditional branch to Ij:

         begin

           $T_k := F(T''(j)) \oplus I_j$;

             target_back[j] := 1

         end;

backward conditional branch to I_j:

begin

$\quad T_{k'} := F(T''(j)) \oplus I_j;$

$\quad \text{target\_back}[j] := 1;$

$\quad \text{if target\_back}[k+1] = 0 \text{ then } T_k := F(T_{k'}) \oplus F(T_{k+1}) \oplus I_{k+1}$

$\qquad\qquad\qquad\qquad \text{else } T_k := F(T_{k'}) \oplus F(T''_{k+1}) \oplus F(T_{k+1}) \oplus I_{k+1}$

end;

otherwise

begin

$\quad \text{if target\_back}[k+1] = 0 \text{ then } T_k := F(T_{k+1}) \oplus I_{k+1}$

$\qquad\qquad\qquad\qquad \text{else } T_k := F(T''_{k+1}) \oplus F(T_{k+1}) \oplus I_{k+1}$

end

end;

if target_back[k] $<> 0$ **then** $T''_k := F(T_k) \oplus T''(k)$

end;

if target_back[1] $= 0$ **then** $T_0 := F(T_1) \oplus I_1$

$\qquad\qquad\qquad\qquad \text{else } T_0 := F(T''_1) \oplus F(T_1) \oplus I_1$

end.

References

[DeDe77] D.E. Denning and P.J. Denning, "Certification of Programs for Secure Information Flow", *Communications of the ACM*, vol 20, pp. 504-513, July 1977.

[DiHe76] W. Diffie and M.E. Hellman, "New Directions in Cryptography", *IEEE Transactions on Information Theory*, vol IT-22, pp. 644-654, Nov. 1976.

[Ebel86] G.H. Ebel, "Hybrid Circuits: Thick and Thin Film", in *Handbook of Modern Microelectronics and Electrical Engineering*, Ed. C. Belove, Wiley, New York, 1986.

[ElGa85] T. ElGamal, "A Public Key Cryptosystem and a Signature Scheme Based on Discrete Logarithms", *IEEE Transactions on Information Theory* , vol, IT-31, pp. 469-472, July 1985.

[Moto85] Motorola, *MC68020, 32-bit Microprocessor User's Manual*, Prentice Hall, 1985.

[PaWi89] G. Parkin and B. Wichmann, "Intelligent Modules", in *The Protection of Computer Software - Its Technology and Applications*, Ed. Grover, Cambridge University Press, 1989.

[RiAD78] R.L. Rivest, L. Adleman and M.L. Dertouzos, "On Data Banks and Privacy Homomorphisms", in *Foundations of Secure Computation*, Ed. R.A. DeMillo et al., Academic Press, New York, 1978.

[RiSA78] R.L. Rivest, A. Shamir and L. Adleman, "A Method for Obtaining Digital Signatures and Public-Key Cryptosystems", *Communications of the ACM*, vol. 21, pp.120-126, Feb 1978.

SECTION 5

CONVOLUTIONAL CODES

DECODING TECHNIQUES FOR CONVOLUTIONAL CODES

David Haccoun
Department of Electrical and
Computer Engineering
École Polytechnique de Montréal
Montréal, Québec, Canada   H3C 3A7

ABSTRACT

The basic principles and limitations of decoding techniques for convolutional codes are presented. Regarding the decoding process as being a search procedure through the tree or trellis representation of the code, methods to circumvent the inherent shorthcomings of Viterbi and sequential decoding are presented.

1. INTRODUCTION

In modern digital communications systems the advantages and potential benefits provided by forward error correction (FEC) are widely recognized, and hence channel coding is now becoming an essential element in the design of these systems. A significant improvement in the quality of a transmission channel may be obtained if the codes are properly chosen and decoding techniques efficiently applied.   The efficiency of a digital communication system is often measured by the bit energy to noise power ratio, Eb/No, which is required to achieve a given error probability P_E. The use of error correction coding allows to reach the target error probability P_E at a smaller value of Eb/No. The difference in Eb/No values that are required for a given P_E between a coded and non coded system is called the coding gain. This coding gain depends on the code parameters (e.g. coding rate, block or constraint length, etc.) that is being used and may be quite substantial. For example for rate 1/2 convolutional codes Viterbi decoding can provide about 5 dB coding gain at $P_B = 10^{-5}$ for a constraint length 7 code, whereas sequential decoding can provide coding gains of 7 to 8 dB using more powerful codes. In practice these coding gains can be translated either as a reduction of the transmitting power for the same information data rate, or as a corresponding increase of the information data rate using the same transmitting power, or, naturally as a combination of the two. In environments where power is at a premium, the gain provided by coding may be a rather attractive and

This work was supported in part by the Natural Sciences and Engineering Research Council of Canada.

elegant alternative to brute force transmission power. Therefore error control coding may be a key element that influence the efficency and cost of digital communication systems.

In forward error correction, the principal difficulties usually reside at the decoder, and one of the problems is the development and material realization of powerful decoders that can operate at high data rates with high efficiency, deliver low error probabilities while being practical to implement.

In this paper we will concentrate on error control techniques based on convolutional coding with probabilistic decoding: Viterbi decoding, sequential decoding and their variants [1] - [7]. Systems using convolutional encoding at the transmitting end of the link and probabilistic decoding at the receiving end are among the most attractive means of approaching the performance limits predicted by the Shannon theory: these systems provide substantial coding gains while being amenable to implementations of a reasonable complexity.

After introducing the basic notions of decoding in section 2 the conventional techniques of Viterbi and sequential decoding are presented, and the problems associated with these techniques are introduced in section 3. The simplifying approach of punctured codes for encoding and decoding high rate convolutional codes is presented in section 4. The growth of the number of paths that must be searched by the decoder is examined in section 5, and different approaches and algorithms that alleviate the shortcomings of the conventional Viterbi and sequential decoding are finally presented in section 6.

We assume the reader is familiar with the tree and trellis structure of convolutional codes. Without loss of essential generality, considering binary convolutional codes of coding rate $R=1/V$ and constraint length K, two branches with V codesymbols emerge from each node, and beyond the (K-1)th level there are 2^{K-1} distinct states. A path is traced from the root node out, according to the input information sequence that specifies it, and two paths remerge after their input sequences have been identical over the last (K-1) symbols. The information sequences are assumed to be L-bit long so that the tree or trellis are finite with L levels, and the longest paths have L branches.

For high rate codes, $R=b/V$, the encoder and its tree or trellis representation become very quickly more complex. From each node of the tree or trellis emerge 2^b V-symbol branch-extensions, and the encoder may be implemented by b shift registers in parallel, connected to V modulo-2 adders. The complexity of a high rate $R=b/V$ encoder is thus about b times that of a low rate $R=1/V$ encoder.

2. PRINCIPLES OF DECODING

Decoding may be seen as the operation for determining the most likely transmitted information sequence given the received sequence. For convolutional codes, it consists essentially of finding the "best" path $U=(u_1,u_2,u_3,...)$ through a graph (tree or trellis) in which the branches are assigned "branch metric values" $\{\gamma\}$. These branch metrics are essentially the log-likelihood function $\log P(y_i|x_i^{(U)})$, between the code symbol $x_i^{(U)}$ of the branch corresponding to a possible data input u_i' and the channel output sequence y_i corresponding to the actual data input symbol u_i. Assuming a discrete memoryless channel, the branch metrics are cumulative so that, over a graph of length L branches, for both techniques the objective of the decoder is to find the path U for which the total metric

$$\Gamma_L^{(U)} = \sum_{i=1}^{L} \gamma_i^{(U)} \tag{1}$$

is maximum over all possible transmitted paths. If all input data sequences are equally likely, then the decoded path $U=(u_1,u_2,...,u_L)$ is the "best", in the sense that it minimizes the sequence error probability. Consequently decoding for convolutional codes may be seen as simply some path searching procedures through a tree or trellis. Naturally this searching may be performed along single or multiple paths. Three principal classes of search procedures may be considered: **breadth-first**, **metric-first** and **depth-first**, with each search procedure giving rise to particular decoding algorithms.

In breadth-first procedures, the tree or trellis exploration is performed simultaneously along several paths, all of the same length. No backtracking is allowed, but paths may be eliminated according to some defined discard criterion. The celebrated Viterbi decoding algorithm[5] is an example of breadth-first search procedure. In metric-first procedures, the searched paths are ranked according to their metric values and the path that is currently ranked highest is searched one step further in the tree. Backtracking is not forbidden in these procedures. The stack algorithm of sequential decoding and some of its variants [6],[7] are metric-first procedures. Finally, in depth-first procedures, the searching is performed along the currently most likely path, with backtracking and searching an alternate path should some discard criterion be not satisfied. The Fano algorithm of sequential decoding is an example of a depth-first procedure [8].

3. VITERBI AND SEQUENTIAL DECODING

The two fundamental decoding techniques for convolutional codes are Viterbi decoding and sequential decoding [5],[8]. In accordance with the discussion above, for each technique the objective of the decoder is to find with the highest reliability and the minimum computational effort and/or memory storage, the tree or trellis path that has the largest cumulative metric $\Gamma = \sum_j \gamma_j$ over all possible transmitted paths.

Viterbi and sequential decoding are different techniques with different error performances, different inherent problems, and somewhat different domains of application. They have developed independently and appear to use opposite strategies in determining the most likely information sequence given the received sequence. The Viterbi algorithm is a breadth-first procedure which exploits the path merges of the trellis structure of the code and exhaustively examines all distinct paths at every trellis level. It is an optimal procedure for trellis codes [5]. A sequential decoder operates on the tree structure of the code and follows only the single path that is currently the most likely, without searching the entire tree, making it a suboptimal procedure. This tree searching may be performed using either a depth-first procedure or metric-first procedure, corresponding respectively to either the Fano algorithm [8] or the stack algorithm [6].

As a consequence of the two different tree or trellis searching methods the computational effort is constant and equal to the number of trellis states for Viterbi decoding whereas it is on the average typically very small, but highly variable with a Pareto distribution for sequential decoding [9],[10]. The principal characteristics of Viterbi and sequential decoding are compared in Table 1, indicating somewhat opposite computational and complexity behaviours between these two techniques. However a class of "middleway" algorithms which fill the gap between these two extremes have been developed. In these algorithms, called "Generalized Stack Algorithms" the idea is to keep the search properties of sequential decoding while using some of the multiple path extensions features of Viterbi decoding. Viterbi and sequential decoding are then shown to be only particular cases of the Generalized Stack Algorithms [7].

TABLE 1: Comparison of Viterbi and sequential decoding characteristics

PARAMETER	VITERBI	SEQUENTIAL
Exploration (graph)	Exhaustive(trellis)	Partial (tree)
Searching procedure	Breadth-first	Metric-first (Stack)
		Depth-first (Fano)
Computational effort	Constant	Variable (Pareto)
Average comput./bit,C_{AV}	$C_{AV} = 2^{K-1}$	C_{AV} independ. of K
Complexity	exp(K)	independ. (K)
Decoding delai	Constant	Variable
Input buffer	No	Yes
Real-time decoding	Yes (contin. Data)	No (Data framed)
R_{comp} limitation	No	Yes $\quad (R<R_{comp})$
Constraint length K	Short, $(K \leq 7)$	Long, $(K < 60)$
Error Probability P(E)	$P(E) < e^{-KR_{comp}/R}$	$P(E) < e^{-KR_{comp}/R}$
Applications	High speed $P(E)<10^{-5}$	High speed $P(E)<10^{-7}$

The Generalized Stack Algorithms notwithstanding, the improvement in Viterbi or sequential decoding techniques can be obtained by alleviating their inherent shortcomings: reduction of the computational effort for Viterbi decoding and reduction (even elimination) of the computational variability for sequential decoding. Naturally in either case, these benefits should be obtained at no degradation of the error performance. Another problem concerns the difficulty of decoding high rate codes using Viterbi or sequential decoding. All these problems will be addressed, starting with the latter first.

4. HIGH RATE CONVOLUTIONAL CODES

By far, error control techniques using convolutional codes have been dominated by low rate $R=1/V$ codes. Optimal low rate codes providing large coding gains are available in the literature, and practical implementations of powerful decoders such as Viterbi or sequential exist for data rates in the 10-40 Mbits/s range [1]. However, as the trend for ever increasing data transmission rates and high error performance continues while conserving bandwidth, the needs arise for good high rate $R=b/V$ convolutional codes as well as practical encoding and decoding techniques for these codes. Unfortunately, a straightforward application of Viterbi or sequential decoding to high rate codes becomes very rapidly impractical as the coding rate increases. Furthermore, a conspicuous lack of good nonsystematic convolutional codes with rates R larger than 2/3 prevails in the literature.

We recall that for binary rate $R=b/V$ codes, $b<V$, there are 2^b paths entering and leaving each encoder state instead of two paths as for rate $1/V$ codes. Therefore path searching and inherent complexity of either Viterbi a sequential decoding becomes multiplied by a factor of

$2^{(b-1)}$ with rate $R=b/V$ codes. However by using the notion of punctured codes, this difficulty can be entirely circumvented. Regardless of the coding rate $R=(V-1)/V$, the encoding or decoding techniques (Viterbi or sequential) are hardly more complex than for rates $1/V$ codes. Furthermore the punctured code technique may be readily applicable to adaptive and variable-rate decoding. Finally, although the class of punctured codes of a given rate may not include the best know codes for that rate, the error performance loss is but only minimal [11]-[16].

4.1 Encoding for punctured codes

To illustrate the basic notions of high rate punctured convolutional codes, consider a rate $R=1/2$ encoder. If any one of every four code symbols corresponding to two consecutive information bits is periodically deleted from the output of this encoder, then clearly a rate $2/3$ code will result since three code symbols are produced by two information bits. By a judicious choice of the original low rate code and detection or perforation pattern, most high rate codes may thus be obtained. For example, Fig. 1 shows the trellis diagram of a rate $1/2$, memory $M=2$ code where every fourth symbol is punctured (indicated by X on the diagram). Reading this trellis two branches at a time, we see that its structure is identical to that of the rate $2/3$ code shown in Fig. 2. From every state emerge four branches, each represented by three code symbols, and on both trellises the state transitions are identical. A punctured rate $2/3$ code has therefore been generated by an original rate $1/2$ encoder. Of course, the procedure is not restricted to rate $1/2$ original codes only. Any low rate $1/V_0$ code could be used in the same manner to generate high rate punctured codes, but naturally, a larger number $S=(bV_0-V)$ symbols must be deleted from the b branches in order to obtain the rate $R=b/V$ code. However, in general, good results and ease of implementation tend to favor the use of rate $1/2$ original codes to generate good punctured codes with rates $R=(V-1)/V$ varying from $2/3$ to $16/17$ [11]-[16].

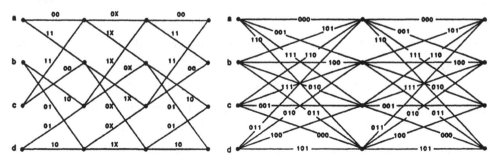

Fig. 1 Trellis for $M=1/2$, $R=1/2$ code   Fig. 2 Trellis for punctured $M=2$, $R=2/3$ code

As shown in Fig. 3 an encoder for rate $R=b/V$ punctured codes consists of an original low rate $R=1/V_0$ convolutional encoder followed by a symbol selector which deletes specific codesymbols according to a given perforation pattern. This perforation pattern is usually expressed as a matrix [P] having V_0 rows and b columns with only binary elements 0's and 1's corresponding to the deleting or keeping of the corresponding code symbols of the original encoder. Both the punctured code and punctured coding rate can be varied by suitably modifying the perforation matrix. For example, starting from an original rate 1/2 code, punctured rates 2/3 and 5/6 codes are obtained by the perforation matrices $[P_1]$ and P_2 given below:

$$[P_1] = \begin{bmatrix} 1 & 1 \\ 1 & 0 \end{bmatrix} \qquad [P_2] = \begin{bmatrix} 1 & 1 & 0 & 0 & 1 \\ 1 & 0 & 1 & 1 & 0 \end{bmatrix}$$

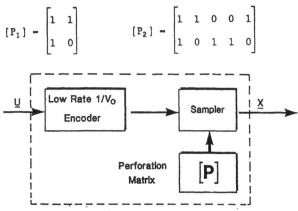

Figure 3   Encoder for punctured codes.

Variable-rate encoding is readily obtained if all punctured rates of interest are obtained from the same low rate encoder. Only the perforation matrices need to be modified as illustrated by $[P_1]$ and $[P_2]$ above. Furthermore if the restriction that all the code symbols of the higher rate punctured codes are required by a lower rate codes, then the punctured codes are said to be rate-compatible [17]. These rate compatible punctured codes may be especially attractive for some ARQ-FEC rate-adaptive applications since only the incremental redundancy needs to be transmitted for decreasing the coding rate [18]. Finally any arbitrary coding rate, $1<R<0$, may be obtained if the principle of puncturing is generalized to include the repetition of codesymbols. Considering an original low rate code with $R=1/V_0=b/bV_0$, then a coding rate $R=b/n < b/bV_0$ can be obtained by repeting $(n-bV_0)$ codesymbols out of every bV_0 original code symbols. This repetition may be conveniently represented by a repetition matrix [Q] having V_0 rows and b colums, but with some non binary elements. For example starting with a coding rate $R=5/10$, the rates $R=5/12$ and $R=5/16$ codes can be obtained by the following repetition matrices:

$$[Q_1] = \begin{bmatrix} 2 & 1 & 1 & 1 & 1 \\ 1 & 2 & 1 & 1 & 1 \end{bmatrix} \quad \text{and} \quad [Q_2] = \begin{bmatrix} 3 & 1 & 2 & 1 & 2 \\ 1 & 2 & 1 & 2 & 1 \end{bmatrix}$$

4.2 Decoding of punctured codes

For punctured high rate $R=b/V$ codes, Viterbi decoding is hardly more complex than for the original low rate $1/V_0$ code from which the punctured codes are derived. The decoding is performed on the trellis of the original low rate code where the only modification consists of discarding the metric increments corresponding to the punctured code symbols. Given the perforation pattern of the code, this can be readily performed by inserting dummy data into the positions corresponding to the deleted code symbols. In the decoding process this dummy data are discarded or erased by assigning them the same metric value (usually zero) regardless of the code symbol, 0 or 1. For either hard or soft-quantized channels, this procedure in effect inhibits the conventional metric calculation for the punctured symbols. Therefore, Viterbi codecs for high rate punctured codes involve none of the complexity of the straightforward decoding of high rate b/V codes. They can be implemented by adding relatively simple circuitry to the codecs of the original low rate $1/V_0$ code; in fact all recent Viterbi decoders implementations include now the puncturing option. Furthermore, since a given low rate $1/V_0$ code can give rise to a large number of high rate punctured codes, the punctured approach leads to a very attractive realization of variable-rate Viterbi decoding.

Just like for Viterbi decoding, the punctured approach to high-rate codes can, as easily, be applied to sequential decoding. The searching operations of decoding are performed on the tree of the original low rate code rather than on the high rate code, and any sequential decoding algorithm could be used. Given the perforation pattern, the decoder proceeds one branch at a time and again, the discarding of the punctured symbols is accomplished by inhibiting the metric evaluation corresponding to these symbols, that is, by assigning to them the same metric increment.

Long memory punctured codes present no problem in sequential decoding. The achievement of low error probabilities, fast decoding speed and good computational behaviour require selecting long memory codes having both a large free distance and a good distance profile. Lists of good rates $R=2/3$ and $R=3/4$ nonsystematic codes of memory length M varying from 9 to 23, and systematic codes with rates varying from $R=2/3$ to $R=7/8$ and memory length M ranging from 44 to 48 have been established [14]-[16].

Extensive computer simulation with punctured codes of rates varying from R=2/3 to 7/8 have shown that the distributions of the number of computations to decode one branch exhibit all a plateau before assuming the usual asymptotic Pareto behaviour (See Figure 4). These plateaux start at about 2 or 3 computations per branch for the rate 2/3 codes, extending further as the coding rate increases: up to 7 or 8 computations for the R=2/3 codes, and all the way to about 100 computations for the R=7/8 codes. These plateaux correspond to a full exploration of the tree over a length of (b+1) branches for rate R=b/V codes [19]. Now compared to straightforward decoding of high rate codes, the decoding of punctured codes involves a larger number of computations. However these computations are far simpler to execute, leading to a reduction of decoding time by a factor ranging between $2^b/2b$ and $2^b/b$ [19].

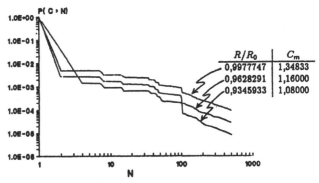

Figure 4 Number of computations per branch for M=12, R=6/7 punctured code.

In conclusion, for all applications where the use of high rate convolutional codes is indicated, and especially in digital communication systems where both the bandwidth and available power may be restricted, the use of high rate punctured convolutional codes appear to be very attractive, involving practically no additional complexity over the well established techniques for low rate codes.

5. SIMPLIFICATIONS OF THE DECODING ALGORITHMS

We have seen that with the puncturing approach, high rate codes present no more difficulty than low rate codes. We now turn our attention to the more basic problem of simplifying the two basic decoding algorithms of Viterbi and sequential decoding. From our earlier discussions, in simplifying sequential decoding the objective is essentially to reduce (or even eliminate) the computational variability without degrading the error performance. The approach of solution is to perform the search over a subset of the most likely paths (rather than along the single most likely path), discarding along any path considered to be useless. In addition limiting the backtracking will help reduce further the computational variability.

For Viterbi decoding the main objective is to reduce somewhat the computational load without degrading the error performance. Stated differently, let C_v be the complexity (in number of computations or storage) of an usual Viterbi decoder, and let $P(E)_v$ be its corresponding error probability. The objective is then to obtain a given $P(E)_v$ at a smaller complexity than C_v, or equivalently, for a given complexity C_v, obtain a superior error performance than $P(E)_v$. Here the approach of solution is simply to use more powerful codes (i.e. $K > 7$) and perform non optimal decoding on the trellis, that is, perform a nonexhaustive trellis search in a breadth-first manner. However before presenting techniques that attempt to solve these problems a closer examination of tree search characteristics is required.

5.1 Tree search characteristics

It is well known that, by far, most of the computational load (and storage) of tree search algorithms are due to path searching in the incorrect subsets, that is, in the ensemble of incorrect paths that diverge from the correct path. Now, to correct t errors, all the incorrect paths lying at Hamming distance up to t from the correct path must be searched (or stored). An important factor is thus the population N_t of these incorrect paths that must be searched (or stored) since each path searched corresponds to one computation. This number N_t is exponentially increasing with t, and an important parameter is the rate of growth $Q(R,M) = V_{t+1}/V_t$ of that computational effort, where R is the coding rate and M the memory of the code. (It may be mentionned that the problem of computational effort or storage in searching the tree for decoding purposes is equivalent to that of the computational effort for the weight spectrum determination of that code [21]). It has been shown [20]-[21] that the rate of growth decreases as the coding rate decreases and increases very slightly with the memory of the code, reaching rapidly its asymptotic value $Q(R) = (2^{1-R} - 1)^{-1}$. However the length of the incorrect paths at distance t is only linearly increasing with t. For example for rate R=1/2 codes, the number of computations to correct t errors is $(2.414)^t$ over incorrect subtrees of length about 9.1 t [22]. Given these basic facts about searching tree paths we now examine specific tree searching algorithms.

6. TREE SEARCHING ALGORITHMS

In attempting to find the correct path, the tree search algorithms may perform either an unlimited search or a limited search. With an unlimited search, the decoder attempts to decode correctly, with probability 1 any error pattern that appeared at the channel output. From our discussion above then clearly such a decoder may occasionnally perform a very large number of computations or use a correspondingly large storage. This is the case for conventional sequential decoders for which, consequently, the number of paths searched or stored has a finite expectation only if the coding rate R is smaller than R_{comp}, the cut off-rate of the channel. Clearly then, a reduction of the computational effort or storage of the conventional sequential decoder could be achieved by performing a limited search in the incorrect subsets. In addition, by performing a breadth-first searching procedure the computational variability may be drastically reduced, allowing even the use of codes with rates larger than R_{comp}. The approach therefore is to limit the path searching according to some limiting criterion, or simply, always extend a small and constant number of paths. We can therefore distinguish simplified decoding algorithms whose searching is limited by

    (i)   Error-based criterion (number of errors t or error probability P(E))

    (ii) Metric-based criterion, or

    (iii) Multi-path exploration.

We now examine some of these algorithms.

6.1 Error-based criterion algorithms

These algorithms attempt to correct only all combinations of t errors within some tree depth. The basic algorithm which uses a breadth-first procedure includes the following steps.

    i)   Search out all paths lying within Hamming distance t from the received sequence

    ii) Limit the search to some maximum length (decision depth).

    iii) Do not allow backtracking.

As a consequence the algorithm will have a finite computational effort for all coding rates, but will yield a decoding failure if more than t errors occur. In addition although all the explored paths have the same length and no backtracking is allowed, that number of paths is not constant throughout, leading to some computational variability. However this variability is not as severe as with the conventional sequential decoding. Naturally as the target error correction t is increased, a more powerful code would be required. Furthermore, the algorithm will perform many unnecessaray computations if the maximum number of error, t, is always assumed regardless of the state of the channel noise. This type of algorithm has been proposed by Foschini for reducing the intersymbol interference (ISI) in data communication [23] as well as by Anderson for error correction [22].

6.2 Metric-based criterion algorithms

These algorithms simply discard all paths that do not satisfy some given metric threshold. They usually implement a breadth-first procedure to minimize the computational variability. We briefly describe an algorithm due to Simmons [24] and a variant due to Chang and Haccoun [25]. Both algorithms may be suitable for the ISI problem as well as for error-correction. The Simmons algorithm includes the following steps:

(1) At each tree depth n, extend all paths j with metric Γ_j satisfying $(\Gamma_B - \Gamma_j) < T$, where T is a discarding threshold value and Γ_B is the highest metric value at depth n.

(2) Discard all paths not remerging back over length L. Limit the number of paths to some value N_{max}, $N_{max} << 2^{K-1}$.

(3) Do not allow backtracking in the tree (trellis).

The advantages of this algorithm is that the decoding effort is somewhat adapted to the channel noise, and hence on the average, the number of survivors is much smaller than for a conventional Viterbi decoder using the same code. In addition one could use a longer code than the maximum allowed with Viterbi decoding, yielding a smaller error probability. However some computational variability persists despite of the breadth-first procedure, but most important, should the correct path be discarded, then its recovery becomes quite difficult, leading to long error events and aborted decoding. (In the layered algorithm of Haccoun and Chang described next, this problem may be easily remedied). Nevertheless this algorithms is amenable to parallel processing implementation by VLSI, and is a good candidate for applications in error correction, ISI, source coding etc...

In the layered algorithm of Chang and Haccoun [25], the notion of a discarding threshold is kept but the survivors are then ordered by metric values and classified as follows:

(1) Classify all paths with a metric value satisfying the threshold T and insert them in bins.

(2) Order the bins in decreasing metric values.

(3) Extend all paths in the N highest bins and discard all other paths. The maximum number of paths is $M_{max} = 2^{K-1}$.

(4) Exploit the path remergers and do not allow backtracking in the trellis.

This algorithm maintains all the advantages of Simmon's algorithm but since the maximum number of extended paths is equal to the maximum number of distinct trellis states, there is no propagation of the errors; the algorithm tends to resynchronize by itself following a correct path elimination. Again, one could use somewhat longer codes than the maximum allowed by conventional Viterbi decoders, yielding an improved error performance. However the number of bins N should be judiciously determined. Computer simulations with optimal $R = 1/2$ codes having constraint lengths K varying between $K = 4$ and $K = 11$, show that the number of bins N should be increased from 3 to 5 as K increases from 4 to 11. However for a given bit error probability P_B the resulting computational effort is far smaller than Viterbi's. For example using a $R = 1/2$, $K = 10$ code over a binary symmetric channel at Eb/No $= 5.5$ dB, yields the same bit error performance P_B as Viterbi decoding using a $R = 1/2$, $K = 9$ code, at a cost of only 30 computations/bit rather than 256. Furthermore the data need not be organized in relatively short frames as in Simmon's algorithm, and can be decoded continuously in very long sequences [25].

6.3 Multi-path Algorithms

In this approach for limiting the path searching, the algorithm extends always the same number of paths, M, with M kept relatively small even for long constraint length codes, where $M << 2^{K-1}$. Using either the tree or trellis representation of the code, at every step the M paths having the largest metrics are extended into their 2M successors. Using sorting, the best M out of the 2M paths are selected for further extensions, and the other M paths are discarded. The algorithms may be implemented using a metric-first procedure as in Haccoun's M-path stack algorithms, [7] or using a breadth-first procedure as for example Anderson and Mohan's M-algorithm [26]. Variants of this algorithm have been developed for source coding, channel coding and intersymbol interference [27]-[20]. In general all these algorithms combine good performance with ease of implementation. However one of the shortcomings of the breadth-first Multi-path algorithms is that wherever the correct path is not among the M extended paths, then the decoded sequence is typically in error from the point of the correct path loss all the way to the terminal node of the trellis. The resulting bit error probability becomes therefore quite large, approaching that of random coding. Examining this with more detail, an error event may be divided in two types: (1). Loss of the correct path due to the non exhaustive trellis exploration (M too small) and (2). Elimination of the correct path in favor of an incorrect path that merged with it. This later event corresponds to a Viterbi-type error and could not be avoided even by an exhaustive trellis exploration. Therefore we may write

$$P(\text{error event}) = P(\text{error}|\text{correct path loss})\, P(\text{correct path loss}) \quad (2)$$
$$+\, P(\text{error}|\text{correct path elimin.})\, P(\text{correct path elimin.})$$

That is

$$P(E) = P_{E|L}\, P_L + P_{E|V}\, P_V \quad (3)$$

The events being disjoint we can write

$$P(E) = P_L + (1-P_L)\, P_{E|V} \quad (4)$$
$$\approx P_L + P_{E|V} \quad (5)$$

Now, for M small, $P_E >> P_{E|V}$ and hence $P(E) \approx P_L$, whereas for $M = 2^{K-1}$, then $P_L = 0$ and $P(E) = P_{E|V}$. Therefore when M is not too large, an improvement of the error performance will come from a reduction of the correct path loss event. Several approaches have been considered for the correct path recovery following a correct path loss event. In particular Lin and Anderson use quick-look-in procedures to estimate the state of the lost correct path [29]. Another approach due to Haccoun and Belzile [30],[31], considers bidirectional multi-path breadth first searching for the correct path recovery. In these decoding algorithms suited for long constraint length codes, there is no computational variability, and furthermore the regular and repetitive nature of the algorithm renders it suitable for VLSI implementation.

Assuming the data is organized in blocks terminated with a tail of (K-1) known branches, then both the starting and ending nodes of the correct path are known to the decoder. Therefore, a bidirectional multiple path exploration of the trellis may be undertaken, offering a simple and elegant solution to the bit error propagation due to the correct path loss. In this bidirectional approach the error propagation due to the correct path loss in one direction is stopped by the exploration in the other direction. By judicious sharing of the forward and reverse decoding it may be shown that bidirectional decoding restricts the extend of the decoding errors to the heavy channel noise regions that caused these errors.

Several variants of the bidirectional search algorithms have been developed, analyzed, and simulated [31]. Using a constraint length K = 20, R = 1/2 code, computer simulations show that for M = 32, 64, and 128 paths, a substantial coding gain is achieved by the bidirectional over the conventional M-algorithm using the same number of paths.

Comparing to the Viterbi decoding of the best known codes of constraint length K = 6, 7, and 8, the bidirectional algorithm using M = 32, 64, and 128 paths achieves more than 1 dB improvement in E_b/N_0 at $P_B = 10^{-5}$ and as shown in Fig. 5, the improvement increases as the code constraint length increases [31]. These results indicate a far more effective use of the computational effort over the Viterbi algorithm without suffering any computational variability. All these improvements are obtained at hardly any additional cost over the Viterbi algorithm, making the technique an attractive alternative to Viterbi decoding for moderate to high error performance.

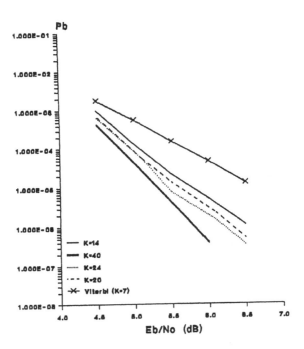

Fig. 5 Bit error performances of the bidirectional
algorithm for M=64 paths over the BSC.

7. CONCLUSION

In this paper we have considered the decoding of convolutional codes as being the problem of searching a path through an oriented graph, tree or trellis. The basic search procedures of breadth-first, metric-first and depth-first have been presented and related to the well known conventional Viterbi and sequential decoding. Analyzing the growth of the number of paths that must be searched, basic approaches to alleviate some of the shortcomings of Viterbi and sequential decoding have been examined, leading to new simplified decoding algorithms. These algorithms allow superior error performance while being suitable to efficient VLSI implementations, thus widening further the domains of applications of convolutional codes.

REFERENCES

[1] W.W. Wu, D. Haccoun, R. Peile, Y. Hirata, "Coding for Satellite Communications", IEEE Trans. Select Areas Commun., Vol. SAC 5, pp. 724-748, May 1987.

[2] I.M. Jacobs, "Practical Applications of Coding", IEEE Trans. Inf. Theory, Vol. IT20, pp. 305-310, May 1974.

[3] V.K. Bhargava, D. Haccoun, R. Matyas, P. Nuspl, "Digital Communication by Satellite", J. Wiley, New York, 1981.

[4] W.W. Wu, "Coding for Satellite and Space Channels", in Special Issue on Coding for Digital Transmission Systems, Inter. J. Electronics, Vol. 55, No 1, pp. 189-212, July 1983.

[5] A.J. Viterbi, "Convolutional Codes and their Performance in Communication Systems", IEEE Trans. on Commun, Technol., Vol. COM-19, pp. 751-772, Oct. 1971.

[6] F. Jelinek, "A Fast Sequential Decoding Algorithm Using a Stack", IBM Journ. Research Develop., Vol. 13, pp. 675-685, Nov. 1969.

[7] D. Haccoun, M. Ferguson, "Generalized Stack Algorithms for the Decoding of Convolutional Codes", IEEE Trans. Inf. Theory, Vol. IT-21, pp. 638-651, Nov. 1975.

[8] R.M. Fano, "A Heuristic Discussion of Probabilistic Decoding", IEEE Trans. Inform. Theory, Vol. IT-9, Apr. 1962.

[9] I.M. Jacobs and E.R. Berlekamp, "A Lower Bound to the Distribution of Computation for Sequential Decoding", IEEE Trans. Inform. Theory, Vol. IT-13, pp. 167-174, Apr. 1967.

[10] D. Haccoun, "Variabilité de calculs et débordements de décodeurs séquentiels à pile", Traitement du Signal (Paris, France), Vol. 3, no. 3, pp. 127-143, Dec. 1986.

[11] J.B. Cain, G.C. Clark, Jr., and J.M. Geist, "Punctured convolutional codes of rate (n-1)/n and simplified maximum likelihood decoding", IEEE Trans. Inform. Theory, Vol. IT-25, pp. 97-100, Jan. 1979.

[12] Y. Yasuda, Y. Hirata, K. Nakamura, and S. Otani, "Development of variable-rate Viterbi decoder and its performance characteristics", presented a the 6th Int. Conf. Digital Satellite Commun., Phoenix, AZ, Sept. 1983.

[13] Y. Yasuda, K. Kashiki, and Y. Hirata, "High-rate punctured convolutional codes for soft decision Viterbi decoding", IEEE Trans. Commun., Vol. COM-32, pp. 315-319, Mar. 1984.

[14] D. Haccoun and G. Bégin, "High rate punctured convolutional codes for Viterbi and sequential decoding", IEEE Trans. Commun., Vol. 37, pp. 1113-1125, Nov. 1989.

[15] G. Bégin and D. Haccoun, "High rate punctured convolutional codes: Structure properties and construction technique", IEEE Trans. commun., Vol. 37, pp. 1381-1385, Dec. 1989.

[16] G. Bégin, D. Haccoun, C. Paquin, "Further results on high-rate punctured convolutional codes for Viterbi and sequential decoding", IEEE Trans. Commun., Vol. 38, pp. 1922-1928, Nov. 1990.

[17] J. Hagenauer, "Rate-compatible punctured convolutional codes (RCPC codes) and their applications", IEEE Trans. on Commun., Vol. COM-36, pp. 389-400, Apr. 1988.

[18] S. Kallel, D. Haccoun, "Generalized Type II hybrid ARQ scheme using punctured convolutional codes", IEEE Trans. on Commun., Vol. 38, pp. 1938-1946, Nov. 1990.

[19] G. Bégin, D. Haccoun, "Sequential decoding of high rate punctured convolutional codes", to be published.

[20] Chi-Chao Chao and R.J. Mc Eliece, "On the path weight enumerators of convolutional codes", Book of Abstracts, IEEE Intern. Symp. on Inform. Theory, Kobe, Japan, June 1988.

[21] D. Haccoun, P. Montreuil, C. Paquin, "On the rate of growth of the incorrect paths of convolutional codes", to be published.

[22] J.B. Anderson, "Limited search trellis decoding of convolutional codes", IEEE Trans. on Inform. Theory, Vol. IT-35, pp. 944-955, September 1989.

[23] G.J. Foschini, "A reduced state variant of maximum likelihood sequence detection attaining optimum performance for high signal-to-noise ratios", IEEE Trans. Inform. Theory, Vol. IT-23, pp. 605-609, Sept. 1977.

[24] S.J. Simmons, "Breadth-first trellis decoding with adaptive effort", IEEE Trans. on Commun., Vol. COM-38, pp. 3-12, January 1990.

[25] F. Chang, D. Haccoun, "Layered Viterbi decoding", to be published.

[26] S. Mohan and J.B. Anderson, "Speech encoding by a stack algorithm", IEEE Trans. on Commun., Vol. COM-28, pp. 825-830, June 1980.

[27] P.R. Chevillat and E. Eleftheriou, "Decoding of trellis-encoded signals in the presence of intersymbol interference and noise", IEEE Trans. on Commun., Vol. COM-37, pp. 669-676, July 1989.

[28] D. Haccoun, S. Kallel, "Application of multiple path sequential decoding for intersymbol interference reduction problem", Proceedings of IEE, Feb. 1991.

[29] C.F. Lin, "A truncated Viterbi algorithm approach to trellis codes", Ph.D. dissertation, Elec., Comput, Syst. Eng. Dept., Rensselaer Polytech. Inst., Troy, NY, Sept. 1986.

[30] D. Haccoun, J. Belzile, "Bidirectional algorithms for the decoding of convolutional codes", Book of Abstract, IEEE International Symp. on Information Theory, San Diego, Calif., Jan. 1990.

[31] J. Belzile, D. Haccoun, "Bidirectional algorithms for decoding long convolutional codes", to be published.

A WEIGHTED-OUTPUT VARIANT OF THE VITERBI ALGORITHM
FOR CONCATENATED SCHEMES USING A CONVOLUTIONAL INNER CODE

UNE VARIANTE À SORTIE PONDÉRÉE DE L'ALGORITHME DE VITERBI
POUR DES SCHÉMAS CONCATÉNÉS À CODE INTÉRIEUR CONVOLUTIF

Rémi Sfez (*) and Gérard Battail (**)

(*) TRT, 5 avenue Réaumur, 92352 Le Plessis-Robinson Cedex, France
(**) TELECOM PARIS, 46 rue Barrault, 75634 Paris Cedex 13, France

Abstract

We describe a means for weighting the output of Viterbi decoding. At variance with previous similar algorithms, it results from simplifying the optimum weighted-output decoding rule. Simulation results over both the additive Gaussian channel and the Rayleigh fading channel are reported.

1. Introduction

Concatenation of several error-correcting codes is well known as a means to design very efficient coding schemes for noisy channels [1]. The most commonly used is composed of a convolutional inner code, combined with a Reed-Solomon outer code. The inner decoding is carried out by the Viterbi algorithm ; most of the residual error bursts from the inner decoder are corrected by the outer decoder. If we assume that the inner decoder delivers soft decisions on the decoded symbols (i.e. further than simple hard decisions, it provides an estimation of their reliability), soft decoding of the outer code becomes possible and improves the performance of the whole coding scheme.

In this connection, a variant of the Viterbi algorithm with weighted-output decisions is proposed in the following. Means for such weighted-output of the Viterbi algorithm were already published [2,3] but the method we describe in this paper does not use them as a starting point. We thus defer their discussion to section 3. Instead, we first state the problem to be solved.

For simplicity's sake, we first assume a binary PSK modulation over an ideal additive white Gaussian noise (AWGN) channel. Furthermore, although several codes may be concatenated according to a widened point of view [4], we only consider here two concatenated codes.

The inner code is assumed to be a half-rate convolutional one : each binary information symbol u_k is encoded into two binary symbols v_{2k-1} and v_{2k}. We consider the weighting of decoded symbols, and their perfect de-interleaving (in order to ensure statistical independence of the decisions provided for the outer decoder). Then, soft-decision decoding of the outer code can be achieved.

The outer code is assumed to be a binary block code : due to the weighting of binary symbols, there would actually be no advantage in using an outer burst-error correcting code, at variance with the hard decision case. For the same reason, bit-by-bit interleaving is assumed. The description above is summarized in Figure 1 where the modem was included in the channel.

With these assumptions, this paper is focused on the study of weighted-output inner decoding, and on the advantage of this technique with respect to conventional Viterbi decoding. As for the outer code, a maximum likelihood (ML) soft- (or hard-) decision decoding (according to the case considered) is assumed.

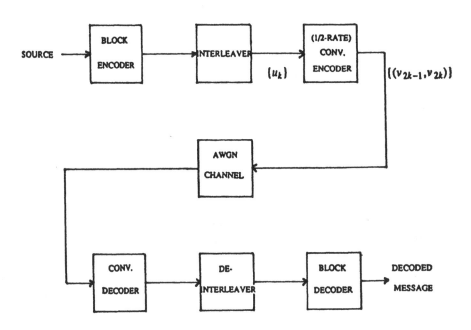

Figure 1 - Coding scheme over the AWGN channel

2. Optimum weighted-output decoding of binary symbols

In the following, we shall consider that any binary symbol is associated with one of the levels "+1" or "-1", the former corresponding to "0" and the latter to "1". With a slight abuse of language, these levels will be referred to as "symbols".

Using this convention, let u_k ($u_k = +1$ or $u_k = -1$) denote the symbol to be decoded at time $t = k$. In this case, optimum weighted-output decoding consists of computing the log-likelihood ratio

$$A_k = \ln \frac{\Pr \{u_k = +1\}}{\Pr \{u_k = -1\}} \ , \tag{1}$$

where the probabilities considered are calculated *a posteriori* (i.e. accounting for the received signals and the coding constraints [5]). It can be shown that A_k is given by :

$$A_k = \ln \frac{\sum\limits_{p_i \in S_{k,+1}} \Pr(p_i)}{\sum\limits_{p_j \in S_{k,-1}} \Pr(p_j)} \ , \tag{2}$$

where $S_{k,+1}$ (resp. $S_{k,-1}$) is the set of trellis paths whose information symbol at time $t = k$ is "+1" (resp. "−1"), and $\Pr(p_i)$ (resp. $\Pr(p_j)$) is the *a priori* probability of a particular path in this set, which can be calculated from the received signals, as shown in the following.

Let $\{v_n\}$ denote the sequence of transmitted symbols ($v_n = +1$ or $v_n = -1$). The received signal y_n, a real quantity, is available as the demodulator output :

$$y_n = \sqrt{E} \ v_n + w_n \ , \tag{3}$$

where E is the received energy per transmitted bit, and w_k is a Gaussian random variable with zero mean and variance $N_0/2$. Denoting by M_i the metric of any trellis path p_i, we have :

$$M_i = - \ln \Pr(p_i) \ , \tag{4}$$

to be minimized by the Viterbi algorithm. It is easily shown that

$$M_i = \sum_{\text{any } n} \ln (1 + e^{a_n}) - \sum_{\{n, \ v_{n,i} = +1\}} a_n \ , \tag{5}$$

where $v_{n,i}$ is the n-th encoded symbol ("+1" or "−1") from path p_i, and a_n is defined as

$$a_n = \frac{4\sqrt{E}}{N_0} \ y_n \ . \tag{6}$$

Since the first term of expression (5) is common to all trellis paths, we may use as a metric for any path p_i the quantity N_i defined as

$$N_i = \sum_{n, v_{n,i} = +1} a_n \ . \tag{7}$$

This metric will be used in the following. The most likely path has the largest one.

From relations (6) and (3), a_n is clearly a Gaussian random variable with mean :

$$E(a_n) = \pm 4 \frac{E}{N_0} \tag{8}$$

(+ or − depending on whether symbol "+1" or "−1" is transmitted), and variance :

$$\text{Var}(a_n) = 8 \frac{E}{N_0} \ . \tag{9}$$

Using relations (4) to (7), the log-likelihood ratio (2) can be derived from the sequence of received signals $\{y_n\}$, so that we can finally write :

$$A_k = \ln \frac{\sum\limits_{p_i \in S_{k,+1}} \exp(N_i)}{\sum\limits_{p_j \in S_{k,-1}} \exp(N_j)} \quad . \tag{10}$$

The hardening of the optimum weighted-output decoding rule (10) leads to a hard decision similar to the symbol-by-symbol maximum *a posteriori* (MAP) estimate derived from [6].

3. Former methods of weighted-output Viterbi decoding

From now on, without loss of generality, we shall refer to the half-rate convolutional code with constraint length $K = 3$, whose encoder is depicted by Figure 2 : at each time $t = k$, one bit u_k enters the shift register and two encoded bits v_{2k-1} and v_{2k} are output before next shifting. Defining the encoder state as the sequence of the last $K - 1$ bits in the shift register, it is clear that at a given state, a new input bit uniquely results in both a new state and a new pair of encoded bits.

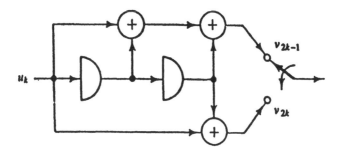

Figure 2 - Half-rate, 4-state convolutional encoder

This encoder operation is summarized on the trellis diagram of Figure 3 : the possible encoder states are labelled 0, 1, 2, 3 and correspond to the trellis nodes, while state transitions are represented by the trellis branches ; solid-line branches correspond to the input information bit "0" (i.e. symbol "+1"), and dashed ones correspond to the input information bit "1" (i.e. symbol "−1").

In the following, we denote by L the length of the path memory in the Viterbi algorithm : assuming that information symbol at time $t = k$ is to be decoded, the current time is then $t = k + L$, and $\hat{p}_{k+L}$ denotes the most likely path, estimated by the Viterbi algorithm at this time. Furthermore, at each trellis node h, there are two incoming paths, denoted by $p_{+1}^{(h)}$ and $p_{-1}^{(h)}$, corresponding to

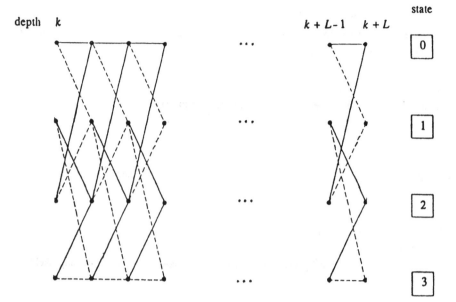

Figure 3 - 4-state trellis diagram

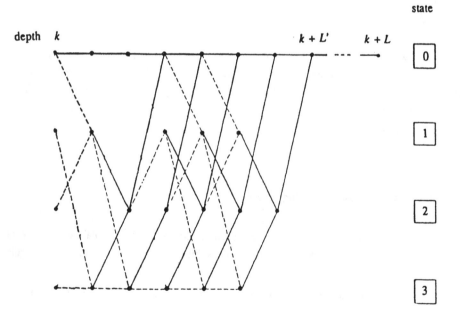

Figure 4 - Subtrellis of the 4-state trellis diagram

information symbols "+1" and "−1", respectively, leaving the encoder shift register.

In former methods intended to weight the output of Viterbi decoding [2,3], the process consisted of five steps : at current time $t = k + L$,

- step 1 : as in the Viterbi algorithm,
    - the metric of each candidate path is updated according to (7) ;
    - the most likely path $\hat{p}_{k+L}$ is identified ;

- step 2 : for each node h,
    - calculation of $A_{k+L}^{(h)}$ = difference between the metrics of paths $p_{+1}^{(h)}$ and $p_{-1}^{(h)}$;
    - storage of $A_{k+L}^{(h)}$;

- step 3 : for previous times $t = k+l$, $l = 1,.., L-1$,
    for each node h,
    the log-likelihood ratios $A_{k+l}^{(h)}$ are revised in order to account for the current log-likelihood ratios ;

- step 4 : the log-likelihood ratio of the outgoing node of path $\hat{p}_{k+L}$ is delivered ;

- step 5 : for previous times $t = k+l$, $l = 1,.., L$,
    for each node h,
    $A_{k+l-1}^{(h)} = A_{k+l}^{(h)}$.

To summarize, in formerly published methods of weighted-output Viterbi decoding [2,3] the log-likelihood ratio of each information symbol was recursively re-calculated by an updating process accounting for the log-likelihood ratios of subsequent information symbols. Clearly, there is no direct link between them and the optimum rule (10). We now propose a different approach which consists of an approximate implementation of (10).

4. Description of the proposed weighting process

Applying decoding rule (10), even when restricted to the trellis paths contained in the path memory of the Viterbi algorithm, seems unfortunately of prohibitive complexity : the use of (10) calls for an approximation.

Let p_{+1} (resp. p_{-1}) denote the most likely path from set $S_{k,+1}$ (resp. $S_{k,-1}$), whose metric is N_{+1} (resp. N_{-1}) :

$$N_{+1} = \max_{p_i \in S_{k,+1}} N_i ,$$

$$N_{-1} = \max_{p_j \in S_{k,-1}} N_j .$$

Then, the restriction of rule (10) to these two paths provides the approximation

$$A_k \approx N_{+1} - N_{-1} , \tag{11}$$

whose hardening is equivalent to the ML decision.

The Viterbi algorithm necessarily provides one of the two paths : $\hat{p}_{k+L} = p_{+1}$ or p_{-1} (depending on whether symbol "+1" or "–1" is decoded by the Viterbi algorithm). It remains the problem of identifying the other path and estimating its metric.

4.1. Example

In this section, we assume that the most likely path $\hat{p}_{k+L}$ is the one composed of symbols "+1" only : this assumption holds without loss of generality for a linear code. The path p_{-1} has then to be identified and its metric estimated.

Figure 4 depicts a simplified trellis diagram, derived from Fig. 3, but restricted to :

- the most likely path $\hat{p}_{k+L}$ (the path p_{+1} composed of transitions between nodes labelled "0" only, in this example) ;

- a subtrellis composed of all possible paths from set $S_{k,-1}$ which merge with p_{+1} at depth $t = k + L'$ ($L' = 7$ on the figure).

We assume that, at each depth, all branch and node metrics are stored. Then, identification of path p_{-1} and estimation of its metric is carried out by applying the Viterbi algorithm to the subtrellis defined above, and implementing rule (11) becomes possible.

4.2. Proposed algorithm

The algorithm used in the above example can be generally described as follows : for each current time $t = k + L$,

- step 1 : as in the Viterbi algorithm,
     - updating of the metric of each surviving path ;
     - identification of the most likely path $\hat{p}_{k+L} = p_{+1}$ or p_{-1}

- step 2 : - storage of the branch metrics of time $t = k + L$;
     - storage of the metric of all surviving paths of time $t = k + L$;

- step 3 : use of the Viterbi algorithm within the subtrellis composed of
    - all paths of $S_{k,-1}$ if $\hat{p}_{k+L} = p_{+1}$,
    - all paths of $S_{k,+1}$ if $\hat{p}_{k+L} = p_{-1}$,
    merging with $\hat{p}_{k+L}$ at depth $k + L'$;

- step 4 : implementation of weighted-output decoding rule (11) ;

- step 5 : in the memory of the algorithm,
    - shifting of all surviving paths (as in conventional Viterbi decoding) ;
    - shifting of all node and branch metrics.

L' in step 3 is an integer parameter such that $K \leq L' \leq L$. Clearly, increasing L' allows a better estimation of the metric of the path sought, but at the price of increased complexity. The maximum value $L' = L$ exactly implements rule (11).

5. Complexity analysis

The aim of this section is to evaluate the complexity of both the conventional Viterbi decoding and its weighted-output variant according to section 5.2., in order to compare them at a subsequent stage (see section 7.).

In the following, we assume $1/m$-rate convolutional coding with constraint length K. Given the m log-likelihood ratios (6) associated with the transmitted symbols v_n ($v_n = +1$ or $v_n = -1$) delivered by the demodulator, the decoding process computes a hard- or soft-decision (according to the case considered) for a single information symbol : from now on, we shall define the algorithm complexity as the number of additions or comparisons on real numbers involved in the decoding of one information symbol.

Only transmitted symbols "+1" have to be used in the calculation of branch metrics according to (7). Owing to this remark, it is easily shown that $2^K + 2^m - 2$ additions are involved in updating the metrics of all trellis paths. In addition, $2^K - 1$ comparisons are involved in selecting the most likely trellis path, so that the complexity of conventional Viterbi decoding can be measured by :

$$C_{cv} = 2^{K+1} + 2^m - 3. \tag{12}$$

Let us now refer to Figure 4 as an example, and consider weighted-output decoding as described in section 5.2. : only steps 1, 3 and 4 imply complexity (in the sense above) :

step 1 : performing the Viterbi algorithm is required, with complexity given by (12) ;

step 3 : assuming that $2K-1 \leq L'$, step 3 implies performing the same algorithm within a part of the trellis defined as follows :

- between depths $t = k$ and $t = k+K$: at depth $t = k+K$, there is a single incoming path at each node, whose symbol at depth $t = k$ is "–1" (see the example of Figure 4) ; therefore, 2^{K-1} additions are involved ;

- between depths $t = k+K$ and $t = k+L'-K+1$ (Viterbi algorithm within the complete trellis structure) : updating the path metrics occurs $(L'-2K+1)$ times the surviving path is selected at each node. This accounts for $(L'-2K+1)$ $(3 \cdot 2^{K-1})$ additions or comparisons ;

- between depths $t = k+L'-K$ and $t = k+L'$ (Viterbi algorithm restricted to those paths merging with the most likely path at depth $t = k+L'$) : this involves $3 \cdot (2^{K-1}-1)$ additions or comparisons;

step 4: decoding rule (11) involves one addition.

From the analysis above, the average complexity of weighted-output decoding is given by:

$$C_{wo} = 2^{K-1} (3L'-6K+11) + 2^m - 5. \tag{13}$$

In order to compare the amount of complexity of weighted-output decoding with respect to conventional Viterbi decoding, we shall consider the ratio

$$\rho = \frac{C_{wo}}{C_{cv}}. \tag{14}$$

A detailed analysis shows that ρ decreases as m increases and as K decreases, so that the added complexity is minimized in the case of low-rate and low-constraint length convolutional codes. Furthermore, in any case, the complexity ratio ρ is upperbounded as follows :

$$\rho < \frac{1}{4} (3L' - 6K + 11). \tag{15}$$

In the next section, BER curves resulting from computer simulations enable a comparison of weighted-output and conventional Viterbi decoding. Accounting for the results obtained, the trade-off between performance and complexity is discussed, using the complexity ratio (14).

6. Simulation results and trade-off (performance / complexity)

In this section, we first exhibit simulation results assuming the concatenation of an (outer) (23,12,7) Golay code with an (inner) 1/2-rate convolutional code, with constraint length $K=3$. A reasonable feasibility of the computer simulations is the only motivation for the choice of these particular codes.

In these simulations, an ML hard-decision decoding (carried out by the Massey-Berlekamp algorithm) of the Golay code was performed when the convolutional code was decoded by the conventional Viterbi algorithm. When using the weighted-output Viterbi algorithm, soft decoding of the Golay code was performed as follows.

Denoting by u_k the k-th symbol of a Golay codeword, by A_k the corresponding log-likelihood ratio (provided by the weighted-output convolutional decoder), and by $\hat{u}_k$ the *a priori* most likely symbol (given by the sign of A_k), the metric associated with symbol u_k was

$$Z_k(u_k) = 0 , \qquad u_k = \hat{u}_k ,$$

$$= \mid A_k \mid , \text{ elsewhere.} \tag{16}$$

The metric of a codeword is defined as the sum of the metrics (16) of its symbols and ML soft-decoding consists of finding the codeword whose metric is the lowest.

In Figure 5, BER curves after decoding of the overall concatenated scheme over the AWGN channel are drawn, in the case of conventional Viterbi decoding (dashed line), and of weighted-outptut decoding (L' = 3, 5, 7 and 15, solid lines).

For bit error rates between 10^{-3} and 10^{-4}, it is noticeable that a coding gain of about 1 dB is obtained when the most complex scheme is used (i.e., when $L' = L = 15$ in this example). In comparison, a performance impairment of about 0.15 dB is paid for a significant reduction of complexity (when $L' = 7$).

We now consider the memoryless Rayleigh fading channel. In comparison with the AWGN channel (3), the received signal available at the demodulator output can be written as :

$$y_n = |\rho_n| \sqrt{E} \, v_n + w_n , \tag{17}$$

where ρ_n is a complex normal random variable whose average norm is unity.

BER curves over the memoryless Rayleigh fading channel are drawn in Figure 6, in the case of conventional Viterbi inner decoding (dashed line), and of weighted-output decoding (L' = 6, 7, 8, 9, and 15, solid lines).

These curves show that with the most complex scheme ($L' = L = 15$) weighted-output decoding provides a coding gain higher than 1.5 dB. A performance impairment of only 0.2 dB results from using $L' = 7$.

From the simulation results above, it can be concluded that quasi-optimum BER performance can be met by using L' between $2K$ and $3K$, so that, according to (15), the complexity ratio (14) is upperbounded by a value between 11/4 and $(3K+11)/4$, which seems acceptable especially in the case of low-constraint length codes.

269

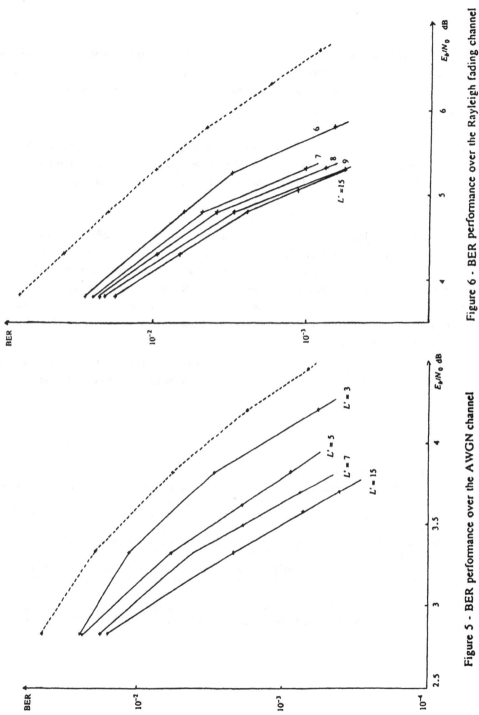

Figure 6 - BER performance over the Rayleigh fading channel

Figure 5 - BER performance over the AWGN channel

7. Conclusion

For concatenated coding schemes using an inner $1/m$-rate convolutional code, an original weighted-output variant of the Viterbi algorithm was proposed : when used for the inner decoding, it enables a trade-off between complexity and performance of the whole concatenated scheme.

Computer simulation results were presented :

- over the AWGN channel, an overall coding gain of about 1 dB is obtained by using the proposed algorithm. It is noticeable that the coding gain reached by the optimum weighted-output decoding may not exceed 2 dB, since the log-likelihood ratios delivered by the inner decoding are approximately Gaussian.

- over the Rayleigh fading channel, the coding gain amounts to more than 1.5 dB and is obtained with an acceptable amount of complexity. Since it turns out that the log-likelihood ratios delivered by the algorithm are approximately Gaussian over the Rayleigh fading channel, too, we may conclude that the reachable coding gain is still upper-bounded by 2 dB. Notice that, when applied to the mobile radio channel, a coding gain of 1.5 dB to 2 dB results in a gain in spectrum efficiency by a factor of 1.22 to 1.3 ; on the contrary, if the same complexity results from increasing the number of trellis states, no noticeable improvement is obtained.

As a conclusion, weighted-output Viterbi decoding appears as a promising technique for both satellite and mobile radio communications. Its implementation by the proposed algorithm results in a good trade-off between performance and complexity.

8. References

[1] G.D. FORNEY, Concatenated Codes, MIT Press, 1966
[2] G. BATTAIL, Pondération des Symboles Décodés par l'Algorithme de Viterbi, Annales des Télécommunications, vol. 42, n° 1-2, Jan.-Feb. 1987, pp. 31-38
[3] J. HAGUENAUER and P. HOEHER, A Viterbi Algorithm with Soft-Decision Outputs, IEEE Globecom, Dallas, 1989
[4] R. SFEZ and G. BATTAIL, Sur les Schémas de Codes Concaténés avec Décodages Pondérés, 12-th GRETSI, Juan-les-Pins, 12-16 June 1989, pp. 263-266
[5] G. BATTAIL, Le Décodage Pondéré en tant que Procédé de Réévaluation d'une Distribution de Probabilité, Annales des Télécommunications, vol. 42, n° 9-10, Sept.-Oct. 1987, pp. 499-509
[6] L.R. BAHL, J. COCKE, F. JELINEK and J. RAVIV, Optimal Decoding of Linear Codes for Minimizing Symbol Error Rate", IEEE Trans. Inf. Th., vol. 20, n° 2, March 1974, pp. 284-287

CODED MODULATION WITH CONVOLUTIONAL CODES OVER RINGS

Renato Baldini Filho
Patrick G. Farrell

Communications Research Group
Department of Electrical Engineering
The University of Manchester, Manchester, M13 9PL, U.K.

ABSTRACT

The objective of this work is to present a multilevel convolutional coding method based on rings of integers modulo-m (generally m is a power of 2) which is suitable for Coded Modulation schemes. This new method of coded modulation is proposed because of the similarities between an m-ary phase shift keying modulation and the structure of a ring of integers modulo-m. Tables of multilevel transparent systematic convolutional codes with some of their characteristics are also presented. Finally, curves of decoding performance are shown for some convolutional codes suitable for coded 4-PSK and 8-PSK.

1) INTRODUCTION

The main idea of the schemes that utilize Coded Modulation is to transmit q information bits per channel symbol by using a modulator with $m > 2^q$ waveforms (usually $m=2^{q+1}$) and thus exploit the redundancy produced by using a suitable encoding method. This procedure must establish a correspondence between binary q-tuples and symbols of the m-ary alphabet.

Ungerboeck [1] showed how it is possible to relate, monotonically, Hamming distances involved in binary convolutional coding with Euclidean distances between channel symbols. This relationship was named "mapping by set partitioning". The majority of the works developed from that date are based on this mapping by partition of the modulation signal set.

The aim of this work is to present an alternative coding method based on a ring of integers modulo-m (in practice m is a non-prime integer number, generally a power of 2) which is suitable for Coded Phase Modulation schemes. Some tables of multilevel transparent systematic convolutional codes and their performance in terms of asymptotic gain are presented. In addition, some curves of soft-decision decoding performance are also shown for some multilevel transparent convolutional codes defined over rings modulo-4 and modulo-8, which are suitable for coded 4-PSK and 8-PSK modulation, respectively.

The first author started researching this topic in 1985, and presented a paper with two colleagues in 1987 [2]. Massey (and some colleagues) is also a well-known researcher in this area, but his first publication known to us is dated 1989 [3]. It seems that this is yet another example of the parallel development of a new concept.

2) BASIC CONCEPTS [4]

The multilevel codes suitable for Coded Modulation presented in this work are defined over an algebraic entity called a *ring*. Before introducing the coding process we need to state the definition of a *ring*. To do that we state first the definition of a *group*.

Definition 1: A *group* A is a set of elements with a mathematical operation called addition (+) which satisfies the following properties:

1) *Closure*: $\forall x_1, x_2 \in A \Rightarrow x_1 + x_2 \in A$

2) *Associativity*:

$\forall x_1, x_2, x_3 \in A \Rightarrow x_1 + (x_2 + x_3) = (x_1 + x_2) + x_3$

3) *Identity*:

There exist $0 \in A$ such that $\forall x \in A \Rightarrow x + 0 = 0 + x = x$

4) *Inverse*:

There exist $-x \in A$ such that $\forall x \in A \Rightarrow x + (-x) = (-x) + x = 0$

If a group A satisfies the commutative law ($\forall x_1, x_2 \in A \Rightarrow x_1 + x_2 = x_2 + x_1$) such a group is called an *Abelian group*.

Definition 2: A *ring* R is a set of elements with two mathematical operations called addition (+) and multiplication ($\cdot$) for which the following axioms hold:

1) The set R is a Abelian group under addition.

2) *Closure*: $\forall x_1, x_2 \in R \Rightarrow x_1 \cdot x_2 \in R$

3) *Associative Law*:

$\forall x_1, x_2, x_3 \in R \Rightarrow x_1 \cdot (x_2 \cdot x_3) = (x_1 \cdot x_2) \cdot x_3$

4) *Distributive Law*:

$\forall x_1, x_2, x_3 \in R \Rightarrow \begin{cases} x_1 \cdot (x_2 + x_3) = x_1 \cdot x_2 + x_1 \cdot x_3 \\ (x_2 + x_3) \cdot x_1 = x_2 \cdot x_1 + x_3 \cdot x_1 \end{cases}$

A ring is called *commutative* if its multiplicative operation is commutative, that is,

$$\forall x_1, x_2 \in R \Rightarrow x_1 \cdot x_2 = x_2 \cdot x_1$$

3) ENCODING PROCESS

The Coded Modulation schemes proposed by Ungerboeck [1] basically utilize a binary encoder with rate $q/(q+1)$ whose $(q+1)$ output bits are mapped by the method of set partitioning into the expanded channel signal set. The alternative scheme, proposed in this work, firstly maps $(q+1)$ information bits into the expanded channel signal set and then makes a multilevel encoding of the channel symbols in order to create a redundancy which can increase the minimum Euclidean distance. Fig. I shows this alternative encoding scheme.

The q+1 parallel information bits $(b_1,..., b_{q+1})$ from the binary source are mapped into one of $m = 2^{q+1}$ channel symbols a_i belonging to the expanded modulation signal set Z_m. The set $Z_m = \{0, 1, 2,..., m-1\}$ is defined as a ring of integers modulo-m. The input of the multilevel encoder is an information sequence $a = (a_1, a_2, ... , a_k)$ with elements belonging to the ring Z_m. The multilevel encoder then sends to the channel a coded sequence $x = (x_1, x_2, ... , x_{k+1})$ the elements of which belong to the same ring Z_m.

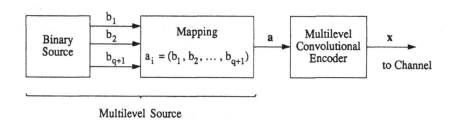

Fig. I Encoder Structure

The modulation constellations more suitable for the multilevel coding of sequences with elements belonging to a ring of integers modulo-m are those which have their modulation symbols distributed on a circumference, that is, the modulations named m-PSK. The spatial distribution of the symbols of m-PSK modulation obeys the same rules which define the rings of integers modulo–m.

4) EUCLIDEAN DISTANCES BETWEEN M-PSK SIGNALS

The Euclidean distance between two signal elements of a given m-PSK modulation is directly related to the corresponding symbols from the ring of integers modulo-m assigned to those modulation signals. The m-PSK modulation signals are represented by $s_i = \exp\left(j\frac{2\pi i}{m}\right)$, $i = 0, 1, ... , m-1$, where the corresponding elements from the ring are given by the subscripts of s_i.

The squared Euclidean distance D_E^2 between two modulation signals can be evaluated from

$$D_E^2(s_i, s_r) = |s_i - s_r|^2 \tag{1}$$

5) MULTILEVEL TRANSPARENT CONVOLUTIONAL CODES

The proposed rate k/(k+1) multilevel convolutional codes are defined by the generator matrix G(D) where D is a delay element and the entries of the matrix are polynomials with coefficients belonging to the ring of integers modulo-m. The codes are chosen to be systematic in order to avoid catastrophicity and, therefore, G(D) can be expressed by

$$G(D) = \begin{bmatrix} & \vdots & g^{(1)}(D)/f(D) \\ & \vdots & g^{(2)}(D)/f(D) \\ I_k & \vdots & \vdots \\ & \vdots & g^{(k)}(D)/f(D) \end{bmatrix} \qquad (2)$$

where I_k is a $k \times k$ identity matrix and $g^{(i)}(D) = g_0^{(i)} + g_1^{(i)}D + g_2^{(i)}D^2 + ... + g_s^{(i)}D^s$ and $f(D) = f_1D + f_2D^2 + ... + f_sD^s$ are the polynomials reresponsible for the feedforward and feedback connections, respectively, as shown in fig. II. Notice that the problem is to find the matrix $G(D)$ which maximizes the minimum Euclidean distance between all pairs of codewords. It is possible to obtain the check parity matrix $H(D)$ from the generator matrix $G(D)$ for a code defined over a ring of integers modulo–m.

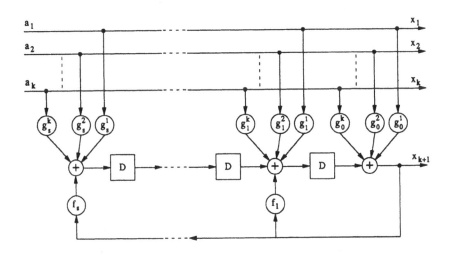

Fig. II General Structure of the Convolutional Encoder

In order to avoid error propagation problems, it is useful to have codes over ring of integers modulo-m which are invariant to multiples of $2\pi/m$ phase rotation. It is also desirable that multilevel codes have as many symmetries as possible in order to facilitate carrier phase synchronization.

Multilevel codes which have $2\pi/m$ phase invariance are generally called transparent codes and can always be utilized together with a differential coding process in order to eliminate errors which can be caused by phase-rotation produced by the channel in the input of the demodulator. Fig. III shows the differential encoder and the differential decoder which can remove phase-rotation produced by the channel from the sequence of information symbols defined over a ring of integers modulo-m.

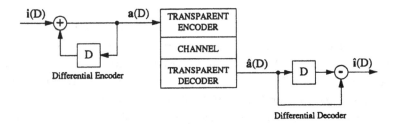

Fig. III Differential Encoder and Decoder

The symbol D inside the squares in fig.III means a time delay of one symbol. The information sequence i(D) is differentially encoded producing the sequence a(D) = i(D) ⊕ a(D)·D. The sequence a(D) is then encoded and sent through the channel which introduces in the sequence a phase-rotation of $2\pi r/m$ (r is integer, $0 \leq r < m$). Because of the transparent nature of the multilevel encoder and decoder, the input sequence â(D) of the differential decoder is a phase-rotation version of a(D) provided all possible errors have been corrected by the multilevel encoding scheme. Note that a phase-rotation of $2\pi r/m$ in the sequence a(D) is equivalent to the sum of it and the sequence r.u(D), where u(D) is the all-one sequence, therefore â(D) = a(D) ⊕ r.u(D). The output of the differential decoder is the sequence î(D) = â(D).D ⊖ â(D) which must be equal to i(D) if â(D) is error free. The symbol ⊖ means subtraction modulo-m.

In this section, we present some multilevel systematic convolutional codes defined over Z_4, Z_8 and Z_{16} which are invariant to multiples of $\pi/2$, $\pi/4$ and $\pi/8$ phase rotation, respectively. Those multilevel systematic convolutional codes were obtained by computer search. The restriction in finding convolutional codes with large k is associated with a large time-consuming computer search.

A multilevel convolutional code has $2\pi/m$ phase-invariance if and only if it has the all-one sequence as a codeword. It is straightforward to show that if a code has the all-one codeword then it has a transparent encoder. Note that a phase rotation of $2\pi r/m$ ($0 \leq r < m$) in any codeword c of the code C corresponds to the sum of the codeword c with the all-one n-tuple u multiplied by the scalar r. Therefore, if u belongs to the code C then r.u is a codeword and the rotation version of codeword c is also a codeword.

The increase in performance of the coded modulation schemes over the reference uncoded modulation is obtained from the expression for the asymptotic coding gain given below:

$$g_\infty = 10 \cdot \log_{10}\left[\frac{\log M_c}{\log M_u} \cdot R_c \cdot \frac{D^2_{Free}}{D^2_{E_u}}\right] \quad [dB] \tag{3}$$

where M_c and M_u are the number of signal symbol elements of the coded and uncoded scheme, respectively; R_c is the coding rate and D^2_{Free} and $D^2_{E_u}$ are the minimum squared Euclidean distances of the coded and uncoded scheme, respectively.

The number N_{Free} is the total number of paths in the code trellis with D_{Free} that diverge from any given state and reach it after a finite number of transitions. The number N_{Free} is an important parameter in the evaluation of the error-event probability.

Tables I, II and III show some multilevel systematic linear feedforward convolutional codes suitable for coded 4-PSK, 8-PSK and 16-PSK modulation, respectively. Feedforward convolutional codes are obtained by eliminating the feedback connections of the encoder given in fig. II, i. e., the polynomial f(D) is omitted in eq. 2 of G(D). Tables III, IV and V show some multilevel systematic linear feedback convolutional codes suitable for coded 4-PSK, 8-PSK and 16-PSK modulation, respectively.

Observe that the second column of the tables gives the coefficients of the generator polynomials $g^{(i)}(D)$ between brackets and f(D) is given after the slashes. For instance, a 16-state feedback convolutional code over Z_{16} and defined by the generator matrix:

$$G(D) = \begin{bmatrix} 1 & 0 & 0 & (10D+5)/(8D^2+14D) \\ 0 & 1 & 0 & (8D^2+8D+3)/(8D^2+14D) \\ 0 & 0 & 1 & (8D^2+12D+9)/(8D^2+14D) \end{bmatrix}$$

(4)

is given in the table VI as (8, 8, 0) (12, 8, 10) (9, 3, 5) / 8 14.

Table I: Transparent Feedforward Multilevel Convolutional
Codes over Ring of Integers Modulo-4 (4-PSK)

States	Generator Polynomial	N_{free}	D^2_{free}	g_{∞} [dB] over 2-PSK
2	2 3	5	8	3.01
4	2 2 1	1	8	3.01
8	2 2 0 1	1	8	3.01
16	2 2 1 1 3	2	12	4.77
32	3 1 1 2 0 2	16	16	6.02
64	2 0 1 1 2 3	7	16	6.02
128	2 0 0 0 1 1 2 3	3	16	6.02
256	2 2 2 1 0 1 2 3	2	16	6.02
1024	2 0 1 1 1 1 2 1	12	20	6.99

Table II: Transparent Feedforward Multilevel Convolutional
Codes over Ring of Integers Modulo-8 (8-PSK)

States	Generator Polynomial	N_{free}	D^2_{free}	g_{∞} [dB] over 4-PSK
2	(0, 4) (3, 2)	2	2.93	0.17
4	(0, 4) (0, 2) (5, 6)	3	4.00	3.01
8	(0, 4) (0, 0) (0, 6) (5, 2)	3	4.00	3.01
16	(0, 4) (0, 4) (2, 7) (5, 3)	2	4.93	3.92
32	(4, 0) (2, 6) (1, 5) (5, 2)	2	5.76	4.59
128	(4, 0) (0, 6) (4, 2) (7, 7) (6, 5)	1	6.93	5.40
256	(4, 0) (0, 2) (5, 7) (0, 6) (6, 4)	1	7.17	5.55

Table III: Transparent Feedforward Multilevel Convolutional
Codes over Ring of Integers Modulo-16 (16-PSK)

States	Generator Polynomial	N_{free}	D^2_{free}	g_∞ [dB] over 8-PSK
2	(8, 8, 8) (4, 7, 14)	4	0.89	1.81
4	(8, 8, 8) (8, 8, 0) (12, 7, 6)	2	1.04	2.49
8	(12, 8, 10) (12, 14, 9)	2	1.32	3.54
16	(0, 8, 0) (2, 2, 4) (9, 11, 13)	4	1.48	4.01
32	(8, 8, 0) (8, 0, 8) (8, 10, 10)(14, 11, 12)	2	1.63	4.44
64	(6, 0, 4)(14, 12, 2)(5, 9, 3)	10	1.78	4.83

Table IV: Transparent Feedback Multilevel Convolutional
Codes over Ring of Integers Modulo-4 (4-PSK)

States	Generator Polynomial / Feedback Constants	N_{free}	D^2_{free}	g_∞ [dB] over 2-PSK
2	3 3 3 / 2 1 1	5	8	3.01
4	1 2 / 2	1	8	3.01
8	1 2 / 2 0	1	8	3.01
16	3 1 1 / 2 2	2	12	4.77
32	1 3 3 / 2 2 2	16	16	6.02
64	3 1 3 2 / 2 0 2	2	16	6.02
128	1 3 2 3 / 2 0 2 0	1	16	6.02
1024	3 0 3 1 3 3 / 2 2 2 0 2	21	24	7.78

Table V: Transparent Feedback Multilevel Convolutional
Codes over Ring of Integers Modulo-8 (8-PSK)

States	Generator Polynomial / Feedback Constants	N_{free}	D^2_{free}	g_∞ [dB] over 4-PSK
4	(2, 2) (5, 3) / 6	3	4.00	3.01
8	(3, 1) (2, 5) / 6	2	4.59	3.60
16	(4, 0)(7, 3) (3, 6) / 2	1	5.17	4.13
32	(3, 1)(1, 4) (6, 5) / 6 7	2	6.00	4.77
64	(3, 7) (6, 2) (5, 6) / 2 2	1	6.34	5.01
128	(5, 6) (0, 3) (5, 4) / 4 2 4	1	6.93	5.55
256	(4, 0) (3, 3) (4, 3) (4, 6) / 2 6 6	1	7.51	5.75
512	(1, 4) (6, 1) (7, 3) (5, 2) / 6 0 6	2	8.00	6.02

Table VI: Transparent Feedback Multilevel Convolutional
Codes over Ring of Integers Modulo-16 (16-PSK)

States	Generator Polynomial / Feedback Constants	N_{free}	D^2_{free}	g_∞ [dB] over 8-PSK
2	(8, 8, 0) (9, 5, 3) / 8	8	0.89	1.82
4	(8, 8, 8) (13, 6, 5) / 12	1	1.04	2.50
8	(8, 8, 12) (7, 14, 13) / 14	1	1.32	3.54
16	(8, 8, 0) (12, 8, 10) (9, 3, 5) / 8 14	4	1.48	4.01
32	(8, 0, 8)(0, 11, 9) (6, 3, 2) / 2	1	1.65	4.50
64	(14, 12, 0)(6, 8, 4)(13, 9, 11) / 10 0	4	1.78	4.83
128	(10, 6, 4) (13, 2, 0) (11, 6, 8) / 2 6	4	1.80	4.88
256	(2, 8, 15) (10, 14, 2) (14, 7, 3) / 12 8	1	2.08	5.50

6) DECODING PROCESS

As for the binary convolutional codes, the Viterbi Algorithm is considered one of the most suitable decoding process for the convolutional codes defined over a ring of integers modulo-m. This algorithm produces a maximum likelihood estimation of the transmitted information sequence.

Fig. IV shows the soft-decision performance, in terms of BER (bit error rate) versus SNR (signal energy to noise spectral density ratio), for coded 4-PSK modulation using the multilevel feedback convolutional codes 1 2 / 2 and 3 1 1 / 2 2. These codes have an asymptotic coding gains of 3.01 dB and 4.77 dB over uncoded 2-PSK, respectively. Notice from fig. IV that, for the first code, the asymptotic coding gain is almost completely reached for a BER = 10^{-5} while, at the same BER, the second code reaches ~ 4.0 dB.

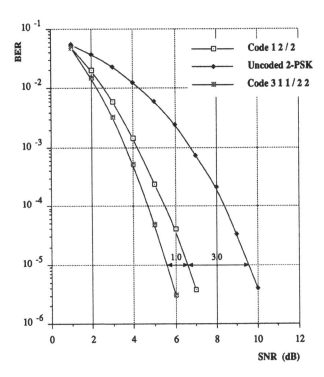

Fig. IV Performance of Coded 4-PSK Modulation

Fig. V shows the soft-decision performance for coded 8-PSK modulation using the multilevel convolutional code (3, 1)(2, 5) / 6. Observe that this code has an asymptotic coding gain of 3.6 dB and at BER = 10^{-4} it reaches a coding gain of 2.4 dB over uncoded 4-PSK.

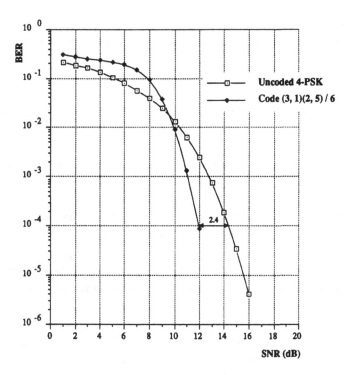

Fig. V Performance of Coded 8-PSK Modulation

7) CONCLUSION

The preliminary results presented in this work show that multilevel convolutional codes defined over a ring of integers modulo-m are an alternative coding scheme for achieving good coding gains for coded modulation. These codes give more flexibility for 2-dimensional coded modulation in terms of coding rates than the equivalent Ungerboeck schemes. Although this work only discussed multilevel convolutional codes, the theory presented here is also applicable to block codes. In addition, it is possible to devise such multilevel block and convolutional coded modulation schemes for QAM and other bandwidth-efficient signal constellations.

8) REFERENCES

[1] Ungerboeck, G., "Channel Coding with Multilevel/Phase Signals", IEEE Trans. on Inform. Theory, pp. 55-67, Jan. 1982.

[2] Baldini Filho, R., Pessoa, A. C. F. & Arantes, D. S., "Systematic Linear Codes over a Ring for Encoded Phase Modulation", presented at ISICT'87, International Symposium on Information and Coding Theory, Campinas - SP - Brazil.

[3] Massey, J. L. & Mittelholzer, T., "Convolutional Codes over Rings", Fourth Joint Swedish-Soviet International Workshop on Information Theory, August 27 - September 1, 1989. Gotland, Sweden.

[4] Peterson, W. W. & Weldon, E. J., "Error-Correcting Codes", Cambridge, Massachusetts, MIT Press, 1972

[5] Baldini Filho, R., "Códigos Pseudo-Cíclicos Multiníveis", Annals of $7^{\underline{o}}$ Simpósio Brasileiro de Telecomunicações, Sept. 1989, Florianópolis - SC - Brazil, (in Portuguese).

[6] Baldini Filho, R. & Farrell, P. G., "Codes Over A Ring Modulo-m", IEE Colloquium on "Multi-level Modulation Techniques for Point-to-Point and Mobile Radio", London - England, 19 March 1990.

[7] Baldini Filho, R. & Farrell, P. G., "Coded Modulation with Convolutional over Rings", Second Bangor Symposium on Communications, Bangor - Wales, 23 - 24 May 1990.

SECTION 6

INFORMATION THEORY

A SHANNON-THEORETIC CODING THEOREM
IN AUTHENTICATION THEORY

Andrea Sgarro

Dipartimento di Matematica e Informatica
Università di Udine, 33100 Udine (Italy)
and: Dipartimento di Scienze Matematiche
Università di Trieste, 34100 Trieste (Italy)

Abstract. What is a coding theorem in authentication theory? This seemingly obvious question has been answered unsatisfactorily in the literature. We try to mend this flaw by giving a genuine, even if naive, Shannon-theoretic coding theorem for authentication codes.

1. Authentication and impersonation.

A Shannon-theoretic frame for authentication theory has been put forward by G. Simmons (as general references for Simmons' approach to authentication we refer to [1,2,3]). A *multicode* is a finite random triple XYZ (*message, codeword, key*). Under each key (encoding rule), encoding and decoding are assumed to be deterministic. Key and message are independent r.v.'s (random variables); we rule out keys and messages of zero probability. Below we give an example of an encoding matrix and of the corresponding authentication matrix χ; in the latter one has $\chi(z,y)=1$ iff key z authenticates codeword y, that is iff there exists a message x which is encoded to y under key z, that is iff $\text{Prob}\{Z=z,Y=y\}\neq0$.

	x1	x2	x3
z1	y1	y2	y3
z2	y3	y4	y1

	y1	y2	y3	y4
z1	1	1	1	0
z2	1	0	1	1

Examples of multicodes are source codes (Z has one value), ciphers (then Y is rather called the cryptogram), and *authentication codes.* Simmons prescribes a zero-error decoding scheme and so each codeword can appear at most once in each row of the encoding matrix.

So far, nothing new. Authentication theory begins for good when the attacks are described against which an authentication code must be resistant. *Impersonation attacks* are the simplest possible attacks against an authentication code. In this case the mischievous opposer chooses a codeword Y^* hoping it to be authenticated by the current key Z, which he ignores. Formally, Y^*, which describes the random strategy of the opposer, is a r.v. with the same alphabet as Y and independent of XYZ. In the literature only the optimal attack strategy has been considered; we consider also an "aping strategy", which is truly random, by taking Y^* with the same (marginal) distribution as Y. The probabilities of fraud (successful attack) are, respectively:

$$P_I = \max_y \ \text{Prob}\{\chi(Z,y)=1\}$$
$$P_A = E_{Y^*} \ \text{Prob}\{\chi(Z,Y^*)=1\}$$

Clearly $P_I \geq P_A$. (Note that in the aping strategy the opposer "apes" the random behaviour of the true codeword Y; unlike Y^*, however, Y is correlated with Z; note also that in accordance to Kerckhoffs' principle we pessimistically assume that the opposer knows the distribution law of XYZ). A well-known lower bound for P_I is Simmons bound, (cf [1]), which involves the mutual information $I(Y;Z)$:

$$P_I \geq 2^{-I(Y;Z)}$$

We show below that this is a bound also to the aping strategy:

$$P_I \geq P_A \geq 2^{-I(Y;Z)}$$

To prove the bound for P_A we shall give (in Appendix A) a "slightly slower version" of the simple proof of Simmons bound given in [4].

Since the aping strategy is only suboptimal, it is not at all surprising that R. Johannesson and this author were able to strenghten Simmons bound; they obtained the following JS-bound:

$$P_I \geq 2^{-R(0)}$$

R(0) being the rate distortion function at zero distortion for source Z, primary alphabet equal to the key-alphabet, secondary alphabet equal to the codeword-alphabet, and distortion measure χ (cf. [5] and [6]; to enhance self-readability the proof given in [6] is sketched in Appendix B). At the moment, unfortunately, we are not able to explain why a rate-distortion function pops out here; this is a question worth further investigation. Unlike Simmons bound, the JS-bound implies the simple and well-known combinatorial bound

$$P_I \geq \frac{|X|}{|Y|}$$

(To see this, recall that R(0) is defined as a minimum mutual information I(Z;Y') subject to the constraint that the distortion between Z and Y' is zero: for ZY'=ZY one re-obtains Simmons bound, for ZY' such that the conditional probabilities codeword-given-key are constant whenever they are not constrained to be zero one re-obtains the combinatorial bound; cf [6]). In a way, Simmons bound is a "natural" lower bound for the aping strategy, while the JS-bound is a "natural" lower bound for the optimal stategy.

Equality criteria follow, based on conditions:

a) $\Pr\{Z{=}z|Y{=}y\} = \Pr\{Z{=}z|\chi(Z,y){=}1\}$

b) $\Pr\{\chi(Z,y){=}1\}$ is constant in y

i) $P_A = 2^{-I(Y;Z)}$ iff   a) and b)

ii) $P_I = 2^{-I(Y;Z)}$ iff   a) and b)

iii) $P_I {=} P_A$ iff   b)

iv) $P_I = 2^{-R(0)}$ iff   b)

v) $P_I = \frac{|X|}{|Y|}$ iff   b)

Proofs are given in appendix A and B; ii) and iv), which is new, are equality criteria for Simmons bound and the JS-bound, respectively.

For "good" impersonation codes, that is for codes which achieve the JS-bound with equality, $\Pr\{\chi(Z,y)=1\}$ has to be independent of y, (cf iv)); however, iv) holds iff v) holds, and so, rather surprisingly, the naive combinatorial bound is a *tight* bound for good codes! This is the deep reason why the naive code constructions given below, which are based on unsophisticated listing, work so well.

In the literature Simmons bound has been passed off as a coding theorem in authentication theory; mutual information was to be a sort of "capacity" of the "authentication channel". Below we give genuine coding theorems in authentication theory. *It will turn out that the relevant information measure is good old Shannon entropy, rather than mutual information or rate-distortion functions.*

2. The zero-error case.

As a preliminary step we give a (trivial) zero-error coding theorem dealing with optimal impersonation coding; it is an asymptotic result, as is usual in the Shannon-theoretic approach. For a given random message source X, let us fix the "resources" $|Z|$ and $|Y|$ (we assume $|X| \leq |Y| \leq |X||Z|$ to avoid trivial specifications) and try to make P_I as low as possible. The inverse part of the coding theorem is soon provided by the combinatorial bound $P_I \geq \dfrac{|X|}{|Y|}$. As for the direct part we use the following code construction: take Z uniform, in the encoding matrix write down the codewords cyclically, right to left, top to bottom.

If $|Y|$ divides $|X||Z|$, $P_I = \dfrac{|X|}{|Y|}$ and the construction is optimal (the computations are trivial); else $P_I < \dfrac{|X|}{|Y|} + \dfrac{1}{|Z|}$, and so the construction is "asymptotically" optimal:

$$\text{for } |Z| \gg 1 \qquad P_I \approx \frac{|X|}{|Y|}$$

An example proves that our construction, however, need not be optimal for finite lengths (in a way, we are dealing with a Shannon-Fano-type code, rather than with a Huffman-type code): for the code with 2 messages, 3 keys and 4 codewords constructed as above one has: $\dfrac{1}{2} < P_I = \dfrac{2}{3} < \dfrac{1}{2} + \dfrac{1}{3}$. Instead, with $\text{distr}(Z) = (\dfrac{1}{4}, \dfrac{1}{2}, \dfrac{1}{4})$, one obtains the theoretical optimum $P_I = \dfrac{1}{2}$.

Actually, constructing finite-length optimal codes appears to be an (apparently) non-trivial set-combinatorial problem. We recall that an m-covering of a set (of the key-space in our case) is a family of not necessarily distinct subsets such that each set-element (each key) belongs to exactly m subsets; the size of the covering is the number of subsets which make it up. Now, we shall identify codeword y with subset $A=\{z: \chi(z,y)=1\}$: we can do this because the "names" of codewords are irrelevant. Since we are dealing with deterministic codes, each key z has to belong to exactly $|X|$ subsets, that is it has to authenticate exactly $|X|$ codewords. Let us fix Z, $|X|$, $|Y|$, (for the moment being, we insist that the key should have a given distribution). Then, to minimize P_I with the given constraints, we must look for an $|X|$-covering of the key-space such as to achieve

$$\min \max \Pr\{Z \in A\}$$

the minimum being taken w.r. to $|X|$-coverings of size $|Y|$, the maximum w.r. to the elements A of the covering. If only the size of Z is fixed, we have to further minimize over all key-distributions with the prescribed size.

3. The coding theorem.

In this section we insist on deterministic decoding, *but we do not insist that it should be deterministically successful!* In other words (again as is usual in the Shannon-theoretic approach) we allow for a "small" decoding error probability P_{err}. We fix the resource Z (we insist that the key should have a given distribution; however this will turn out to be rather irrelevant). For given R (R>0) set $|Y| \approx 2^{nR}$ (more precisely, set $|Y| = \lfloor 2^{nR} \rfloor$). In a Shannon-theoretic spirit we impose that the code should have reliability given by $P_{err} \leq \varepsilon$, $P_I \leq \delta$ ($0<\varepsilon,\delta<1$). How small can R be? We shall find an asymptotic result for stationary memoryless message-key sources $X^n Z^n$ (keys and messages are n-tuples).

In the direct part of the theorem we achieve an asymptotic code-rate $R \approx H(X)$; more precisely, for any τ, $\tau>0$, we achieve an asymptotic rate $R=H(X)+\tau$. This is proved by means of an explicit code construction: write down y's cyclically in the

287

small rectangle of the encoding matrix formed by typical keys and typical messages; fill the matrix in any way but don't use extra y's in each row corresponding to a typical key. If the key is untypical decode as you like it, if the key is typical decode to the unique typical message which might have produced the received codeword; note that decoding is correct whenever both the key and the message are typical. With this construction one achieves: $P_{err}(n) \to 0 < \varepsilon$, $P_I(n) \to 0 < \delta$. The first limit is a straightforward consequence of the properties of typical sequences (cf [7]; roughly, a typical sequence is one for which the relative frequencies of letters are approximately equal to their probabilities; we just recall that the overall probability of typical sequences goes to 1, while n^{-1} log their number goes to the source entropy; typical sequences tend to become equiprobable with increasing n). As for the second limit, repeat the calculations of section 2, discarding however untypical keys and untypical messages (the overall probability of typical key-message couples goes to 1; their joint distribution becomes uniform). We omit the easy details; we just emphasize that the number of typical messages is less than $2^{n[H(X)+\tau/2]}$ for n high enough, and so this number divided by the number of codewords goes to zero with increasing n (cf the approximation $P_I \approx \frac{|X|}{|Y|}$ of section 2).

We go to the inverse part of the theorem. Let a code be given with $P_{err} \leq \varepsilon$, $P_I \leq \delta$ (actually, $P_I \leq \delta$ is never used!). Then, asymptotically, $R \geq H(X)$. We sketch the easy proof. By taking an optimal source block-code for Z^n, and juxtaposing codewords (key-codeword followed by the authentication codeword) one has a reliable source code for the joint key-message source $Z^n X^n$. Then, asymptotically: $H(Z)+R \geq H(XZ) = H(X)+H(Z)$, and so $R \geq H(X)$ (asymptotically, the rate of the optimal key-code yielding the prefixes is $H(Z)$; the rate $H(Z)+R$ of the juxtaposed code cannot be less than the entropy of the joint source; X and Z are independent).

The direct and the inverse parts of the theorem prove that the optimal asymptotic rate is precisely Shannon entropy. More formally, let $R^*(n) = R^*(n,Z,X,\varepsilon,\delta)$ be the lowest, or optimal, rate of authentication codes for $X^n Z^n$ which obey the given constraints $P_{err}(n) \leq \varepsilon$ and $P_I(n) \leq \delta$. Then:

<u>Theorem</u>: $\lim_n R^*(n) = H(X)$

We remark that, again asymptotically: 1) once $P_{err} \le \varepsilon$ is "paid for", $P_I \le \delta$ is "for free" (the same rates which allow to achieve $P_{err} \le \varepsilon$ alone allow to achieve both $P_{err} \le \varepsilon$ and $P_I \le \delta$; 2) ε, δ, the key-rate, the excess rate R-H(X) influence only the speed of convergence; in particular the key-distribution is irrelevant, provided there are exponentially many keys (provided Z is not singular: in this case we would be dealing with a usual source code which of course offers no protection against impersonation).

4. <u>Open problems.</u>

In our ongoing research, we are considering coding theorems for attacks different from impersonation. A list of possible authentication-theoretic, cryptanalytic and "mixed" attacks follows:

Substitution:	$P_S = \sum_c Pr\{Y=c\}\ max_{y \ne c} Pr\{\chi(Z,y)=1 | Y=c\}$
Deception:	$P_D = max\ (P_I, P_S)$
Decryption:	$P_X = \sum_y Pr\{Y=y\}\ max_x\ Pr\{X=x | Y=y\}$
Mixed:	$max\ (P_I, P_X)$, or: $max\ (P_D, P_X)$, etc.

In the substitution attack the mischievous opposer intercepts the legal codeword c and sends the fake codeword y in its place; in the case of deception the code is expected to be resistant both against impersonation and substitution; decryption is a purely cryptanalytic attack; finally, we have considered also codes which have to offer both authentication and secrecy.

Another relevant question: is splitting, that is probabilistic encoding, asymptotically useful? (In the case of impersonation it is easy to see that splitting does never help; instead, it is well-known that, at least for finite lengths, splitting does help in the case of substitution; cf e.g. [6]. Note also that the substitution attack has to be specified more carefully when splitting is used, since a codeword $y \ne c$ is decoded to the correct message x whenever both y and c are legal codewords for x under the given key. We also stress that in Section 1 deterministic encoding has been used only in the proofs of iv) and v); cf Appendix B below).

Acknowledgment. We gratefully acknowledge helpful discussions with J. Körner.

Appendix A. Here we sketch a proof of Simmons bound for P_I (for more details cf [4,5]); we prove that it is a bound also for P_A; we prove criteria i), ii), iii). Below we use the fact that the conditional informational divergence (cross-entropy) $D(Z;W|Y)$ is non-negative and is zero iff distr(ZY)=distr(WY) (for this and other basic fact on information measures cf e.g. [7]). In our case W is defined by

$$\text{Prob}\{W=z|Y=y\} = \text{Pr}\{Z=z|\chi(Z,y)=1\}$$

A mechanical computation shows that $D(Z;W|Y) \geq 0$ becomes

$$E_{Y*} \log \text{Prob}\{\chi(Z,Y^*)=1\} \geq -I(Y;Z)$$

with equality iff a) holds (Y^* is defined as in the aping strategy). In [4], passing from expectation to maximum, and interchanging max and log, one jumps soon to the conclusion that Simmons bound holds for P_I, with equality iff both a) and b) hold. Let us go slower, and use first Jensen's inequality to interchange log and E so as to obtain:

$$\log P_A \geq E_{Y*} \log \text{Prob}\{\chi(Z,Y^*)=1\} \geq -I(Y;Z)$$

the first inequality being an equality iff b) holds. We have proved i) and ii); iii) is trivial.

Appendix B. Here we sketch the proof of the JS-bound given in [6]; we prove criteria iv) and v). In the sequel the key distribution is fixed throughout. We begin by observing that to define P_I and to prove Simmons bound for it one needs couples ZY, rather than triples XYZ; ZY does not even have to be derivable from a triple XYZ. We shall say that ZY is an *abstract* code; we shall say that two abstract codes are *equivalent* when they have the same key Z and the same authentication matrix χ. Now, the point comes: P_I is invariant w.r. to equivalent abstract codes! This gives soon the JS-bound in the form:

$$P_I \geq \sup 2^{-I(Y';Z)} = 2^{-\inf I(Y';Z)}$$

the sup (the inf, respectively) being taken w.r. to all abstract codes ZY' equivalent to the given code XYZ. (The form given in the body of the paper is soon obtained by closing the minimization set, recalling that mutual information is a continuous function, and recalling the expression for the rate-distortion function). Clearly, the JS-bound holds with equality iff there is an abstract code ZY' equivalent to XYZ, for which *Simmons* bound holds with equality. Now, assume the JS-bound holds with equality; then the Simmons bound holds with equality for a certain ZY' equivalent to XYZ; then a) and b) hold for ZY'; however, b) is invariant w.r. to equivalent codes, and so b) holds also for XYZ. Assume conversely that b) holds for XYZ. We shall construct an abstract code ZY' equivalent to XYZ for which Simmons bound holds with equality. XYZ being deterministic, each row of the encoding matrix χ has exactly $|X|$ entries which are not constrained to be zero; in correspondence to these entries set $\text{Prob}\{Y=y|Z=z\}=\frac{1}{|X|}$; an application of Bayes rule shows that a) holds for ZY'. To prove v) it is enough to inspect the elementary derivation of the combinatorial bound:

$$|Y|\, P_I \geq \sum_y \text{Prob}\{\chi(Z,y)=1\} = \sum_y \sum_{z:\chi(z,y)=1}\text{Prob}\{Z=z\} =$$

$$= \sum_z\sum_{y:\chi(z,y)=1}\text{Prob}\{Z=z\} \geq \sum_z|X|\,\text{Prob}\{Z=z\} = |X|$$

In the last inequality we have used the fact that the size of the set $\{y:\chi(z,y)=1\}$ is at least $|X|$ (exactly $|X|$ if XYZ is deterministic). To have equality in the first inequality, $\text{Prob}\{\chi(Z,y)=1\}$ must be constant in y, so as to be equal to its maximum, that is b) must hold. Note that the combinatorial bound never holds with equality for a non-deterministic code.

References.

[1]   G.J. Simmons "A survey of information authentication", Proceedings of the IEEE, May 1988, 603-620

[2]   J. Massey "An introduction to contemporary cryptology", Proceedings of the IEEE, May 1988, 533-549

[3]   A. Sgarro "Unconditional secrecy and authentication", in *Geometries, codes and cryptography*, ed. by G. Longo, M. Marchi, A. Sgarro, CISM Courses and Lectures 313, Springer Verlag, 1990, 131-160

[4]   A. Sgarro "Informational-divergence bounds for authentication codes", in *Advances in Cryptology - Eurocrypt '89*, ed. by J.-J. Quisquater and J. Vandewalle, Lecture Notes in Computer Science 434, Springer Verlag, 1990, 93-101

[5]   R. Johannesson, A. Sgarro "Strengthening Simmons' bound on impersonation", to be published on IEEE Transactions on Information Theory

[6]   A. Sgarro "Lower bounds for authentication codes with splitting", Proceedings of Eurocrypt 90, Aarhus, May 1990 (to be printed in the Lecture Notes in Computer Science by Springer Verlag)

[7]   I. Csiszár, J. Körner, *Information theory*, Academic Press, New York, 1982

JOINT FRACTIONAL ENTROPY

Anna Fioretto *, Andrea Sgarro *,°

* Dipartimento di Scienze Matematiche

Università di Trieste, 34100 Trieste (Italy)

° Dipartimento di Matematica e Informatica

Università di Udine, 33100 Udine (Italy)

Abstract. The fractional entropy $S(X)$ has been introduced by the second author as a pragmatic measure of the "fractional uncertainty" contained in the random variable X. We prove new properties for $S(X)$. In particular we prove that the joint fractional entropy is superadditive under independence of its random arguments.

Introduction.

This paper deals with a certain rate-distortion function $R_X(0)=S(X)$ computed at zero distortion. More precisely, the r.v. (random variable) X takes its values in a primary alphabet of K letters; the secondary alphabet is made up of all the subsets of M primary letters, M being a fixed integer (M≥1); the distortion between a primary letter and a secondary letter (a subset of M primary letters) is zero if and only if the former belongs to the latter (if M>K, we agree that $S(X)$ is zero). For M=1 one re-finds Shannon entropy $H(X)$; in the sequel we assume M≥2. The second author has proposed $S(X)$ to serve as a measure of "fractional uncertainty" (cf /1/): the pragmatic user is not interested in the actual outcome of the experiment X, but rather is completely satisfied when he is given an M-set of possible outcomes where the actual outcome does belong (for applications to cryptology cf /1/; for philosophical comments cf /2/; for standard notions in rate-distortion theory cf /3/).

S(X) is called *fractional entropy*; more specifically, S(X) is called *semientropy* for M=2. In previous works several adequacy properties have been proved for S(X). This is a continuation. The idea is to establish a set of computational rules similar to those available for Shannon entropy.

Being a rate-distortion function, S(X) is expressed as (cf /3/):

$$S(X) = \min I(X;\Xi)$$

the minimum of the mutual information $I(X;\Xi)=H(X)+H(\Xi)-H(X\Xi)$ being taken w.r. to all conditional distributions $\Xi|X$, such that $\text{Prob}\{\Xi=\xi|X=a\}$ is zero whenever letter a does not belong to subset ξ. In /2/ it has been proved that if one enlarges the secondary alphabet so as to comprise also the primary subsets of size less than M, S(X) does not change. The following properties have been proved in /1/; there their adequacy is commented upon:

1) $0 \leq S(X) \leq \log \frac{K}{M}$

2) $H(X) - \log M \leq S(X) \leq H(X)$

3)   S(X) is (weakly) concave

The relevance of having *weak* concavity, rather than strict concavity as for entropy, is discussed in /2/. We list criteria for equality (cf /1/):

left of 1): X is M-ary   right of 1): X is uniform

left of 2): $p \leq \frac{1}{M}$ right of 2): X is unary

Here and below p is the highest letter-probability for X; an n-ary r.v. is one which takes at most n values with positive probability; unary r.v.'s are also called deterministic, singular or degenerate; (note that in our terminology a binary r.v. can have a range of more that two values, provided at most two are assumed with positive probability; in the same spirit, when we say that r.v. Y is a (deterministic) function of r.v. Z, Y=f(Z), say, or that function f is one-to-one, we are contented that this happens with probability 1).

The joint entropy S(XY) has been first mentioned in /2/, where the following property is proved (conditions for equality are discussed below):

4) $S(XY) \geq S(X)$

New results.

i) $S(X) \geq S(f(X))$

ii) $S(X) = S(X,X) = S(X,f(X))$

iii)   if X and Y are independent $S(XY) \geq S(X) + S(Y)$

By induction iii) soon generalizes to:

iv)   if $X_1, ..., X_n$ are independent $S(X_1 ... X_n) \geq S(X_1) + ... + S(X_n)$

Proofs. Properties i) and iii) will be proved by an application of the data-processing lemma (cf /3/, lemma 1.3.11): if f(A) and g(B) are functions of r.v.'s A and B, respectively, $I(f(A);g(B)) \leq I(A;B)$; ii) soon follows from 4) and i). Proof of i): for a suitable Ξ, one has: $S(X) = I(X;\Xi) \geq I(f(X);f(\Xi)) \geq S(f(X))$; (the notation $f(\Xi)$ is self-explaining: note that a value for $f(\Xi)$ is in general a subset of *less* than M letters, but this is irrelevant in view of the observation above). Proof of iii): For a suitable Ξ, $S(XY)=I(XY;\Xi) =I(X;\Xi)+I(Y;\Xi)$ (we have used the independence of X and Y). Let $\Phi=g(\Xi)$ be the X-projection of Ξ (the values taken by Φ can be subsets of less than M letters, again); then, by tha data-processing lemma, $I(X;\Xi) \geq I(X;\Phi) \geq S(X)$; in the same way one proceeds for $I(Y;\Xi)$. QED

Remark 1. Without independence the joint fractional entropy can be strictly subadditive: just think of S(X,X).

Remark 2. Below we list *sufficient* conditions for equality in 4), i), iii) and iv) (proofs are obvious). For 4): Y is a function of X, or S(XY)=0 (e.g.: X is unary and Y is binary; or, for $M \geq 4$, X and Y are both binary, so that XY is quaternary); for i): f is one-to-one, or S(X)=0 (e.g.: X is M-ary, and f is many-to-one); for iii), and, more generally, for iv): $S(X_1 ... X_n)=0$, or at least n-1 of the r.v.'s in question are unary. Unfortunately, we are able to prove that these conditions are also *necessary* only in the case of semientropy (M=2). In this case the following explicit formula is available (cf /1/; we recall that p is the highest probability for X; X* is a binary r.v. of distribution (p,1-p) obtained by engrossment of X):

$$S(X) = H(X) - \log 2 \quad \text{for } p \leq \tfrac{1}{2}$$
$$S(X) = H(X)-H(X^*) = H(X|X^*) \quad \text{for } p \geq \tfrac{1}{2}$$

(We shall write, in general: $S(X) = H(X)-h^*(p)$, which implicitely defines h^*). This expression allows to solve the problem of equality by a brute force attack: actually, we end up by giving also an alternative proof of 4), i), iii) for $M=2$. The (helas! boring) computations are sketched in the appendix.

Open problems (for $M\geq 3$). Find criteria for equality in 4), i) and iii). Find an explicit formula for $S(X)$.

Comment. The adequacy of i) and ii) deserves no comment. We go to iii). Assume the fractional uncertainties about independent experiments X and Y are removed separately by giving $S(X)+S(Y)$ units of information. From a subset of X-letters and a subset of Y-letters, however, one can construct more than $M!$ subsets of XY-couples (subsets have size M): so, more information is needed to remove the fractional uncertainty about the joint experiment XY. As a matter of fact, one expects that equality should hold only in the very degenerate cases when either there is no fractional uncertainty around ($S(XY)=0$), or when X, say, is deterministic, so that only Y "is actually there".

Appendix. We begin by 4). Let q be the highest probability for XY; of course, $q\leq p$. The case $p\leq\frac{1}{2}$ is trivial. For notational ease set $Z=XY$. For $p\geq\frac{1}{2}$, one has $S(Z)-S(X) = H(Y|X) + H(X^*) - h^*(q)$. First, we assume that one has also $q\geq\frac{1}{2}$, so that $h^*(q)=H(Z^*)$ (Z^* is obtained from Z in the same way as X^* is obtained from X); $q\geq\frac{1}{2}$ implies that the element a of probability p is the first component of the couple ab of probability q. Engross Z to a ternary r.v. T which takes values ab, $aN(b)$, $N(a)$ (N denotes negation); we stress that both X^* and Z^* are functions of T. One has $H(Y|X) = H(YT|X) = H(T|X)+H(Y|TX) = H(T|X^*)+H(Y|TX)$ and so $S(Z)-S(X)$ becomes $H(T)+H(Y|TX)-H(Z^*) = H(T|Z^*)+H(Y|TX)$, which clearly is not negative. To have $S(X)=S(Z)$ one must have both $H(Y|TX)=0$ and $H(T|Z^*)=0$; the first implies that Y is a function of X, except possibly for $X=a$, in which case Y can take two values; the second implies that, unless $p=1$, Y is a deterministic function of X also for $X=a$. For $q\leq\frac{1}{2}\leq p$ ($h^*(q)=\log 2$) we are left with the problem of finding the absolute minimum of the concave function $H(Y|X)+H(X^*)$ with the given constraints (for the concavity properties of entropies cf, e.g., lemma 1.3.5 in [3]). If X takes two values the problem trivializes to minimizing the *strictly* concave function $H(XY)$ over a polyhedron (it is enough to check the values of $H(XY)$ on the vertices of the polyhedron, which are either in common with other cases, or correspond to degenerate distributions: the minimum turns out to be log 2 and is taken on vertices in common with other cases). If the range of X has size 3 or more, we can go back to the preceding case: simply write X as a convex combination of binary r.v.'s X_2, ..., X_K taking value a with probability p, and keep fixed the conditional distribution of Y; then use the fact that $H(Y|X)+H(X^*)$ is linear on such a combination (cf again lemma 1.3.5 in [3]).

We go to i). Now q will denote the highest probability for $f(X)$; one has $p \leq q$. To avoid trivialities, we can assume: X is at least ternary; $q \gneqq \frac{1}{2}$. Without real restriction we also assume that f joins precisely two elements of X, a and b, and leaves the rest unchanged (else write f as a composition of functions of this type, followed by a one-to-one function). We can assume that q is the probability of the engrossed element $\{a,b\}$; else p=q, and things become trivial. For notational ease set $Z=f(X)$. Engross X to a ternary r.v. T which takes values a, b, and joins all the other values. One has $S(X)-S(Z) = H(X|Z)+H(Z^*)-h^*(p)$. But $H(X|Z)=H(T|Z^*)$, and so $S(X)-S(Z)=H(T)-h^*(p)$ (we have used the fact that Z^* is a function of T). If $p \leq \frac{1}{2}$ ($h^*(p)=\log 2$), we are again left with the easy problem of minimizing the strictly concave function $H(T)$ over a polyhedron; the minimum turns out to be precisely log 2, and is taken on vertices in common with other cases. Assume then $\frac{1}{2} \leq p \leq q$, so that $S(X)-S(Z) = H(T)-H(X^*)$; $p \gneqq \frac{1}{2}$ implies that one of the element joined by f, a, say, has probability p; in turn, this implies that also X^* is a function of T, so that the difference $H(T)-H(X^*) = H(T|X^*)$ is non-negative. Conditions for equality in i) are now obvious, $S(X)-S(Z)$ being zero when T is a function of X^*, i.e. when b has zero probability, so that f is one-to-one with probability 1.

Finally, we go to iii). X and Y are independent, and so the highest probability for XY, q, say, is obtained by product of the highest probabilities for X and Y, q=pr,say. Set $Z=XY$. One has $S(Z)-S(X)-S(Y)=h^*(p)+h^*(r)-h^*(pr)$. Things are trivial unless one has both $p\geq\frac{1}{2}$ and $r\geq\frac{1}{2}$. If $pr\leq\frac{1}{2}$ ($h^*(pr)=\log 2$) we are brought back to the easy problem of minimizing a strictly concave function (in this case $H(X^*)+H(Y^*)$), which is dealt with as above. In the case $pr\geq\frac{1}{2}$, we have $S(Z)-S(X)-S(Y) = H(X^*)+H(Y^*)-H(Z^*) = H(X^*,Y^*)-H(Z^*)$ (X and Y being independent, so are X^* and Y^*). This difference is non-negative because Z^* is a function of the couple X^*,Y^*: $S(Z)-S(X)-S(Y)=H(X^*Y^*|Z^*)$. Conditions to have equality in iv) are now obvious, Z^* having to be an *invertible* function of X^*Y^*.

References.

[1]   A. Sgarro "A measure of semiequivocation", in "Advances in Cryptology", ed. by Ch. Günther, Springer Verlag, Lecture Notes in Computer Science 330 (1988) 375-387

[2]   G. Longo, A. Sgarro "A pragmatic way-out of the maze of uncertainty measures", Proceedings of IPMU 1990, Paris, July 1990

[3]   I. Csiszár, J. Körner, *Information Thory*, Academic Press, New York, 1982

ON THE ADAPTIVE SOURCE CODING

SUR LE CODAGE DE SOURCE ADAPTATIF

Gérard Battail

Ecole Nationale Supérieure des Télécommunications,

Département Communications (also URA 820 of C.N.R.S.),

46 rue Barrault, 75634 PARIS CEDEX 13, France

and

Mauro Guazzo

CODEWORK,

Corso Cairoli 32, 10123 TURIN, Italy

Abstract

We compare adaptive source coding algorithms. Description of the Huffman-Gallager algorithm can use a transition diagram were the tree leaves merge with its root. We show that the Lempel-Ziv algorithm can be used as a means for constructing a diagram of same shape. When the probability of each branch conditioned on the node it departs from is constant and known, the Guazzo algorithm is a means for describing any path in such a diagram, hence any source output. It results in a coded message whose average length approaches the theoretical limit as a closeness criterion is made arbitrarily tight. Its adaptive version replaces the conditional probabilities by frequencies. We consider its use on a diagram constructed by the Lempel-Ziv algorithm, preferably transformed by the Huffman-Gallager algorithm.

INTRODUCTION

Source coding is a reversible transformation of any message delivered by a discrete source which aims at minimizing the average message length after encoding. Its practical interest is obvious since it achieves a saving in all physical resources that are needed to transmit or store information (transmission time, bandwidth, energy, storage space, etc).

Despite these advantages and the considerable compression it may achieve, source coding is not applied in a generalized way in computer and transmission applications. The authors feel that two facts hinder its ubiquitous use :

a) the encoded message is much more vulnerable to channel errors than the original one. Even the corruption of a single symbol in the encoded message may cause the entire recovered message to be garbled ;

b) the conventional source coding algorithms need a preliminary description of the source statistics, which most often lacks in practice.

The first drawback seems basically unavoidable. One cannot decrease redundancy without at the same time increasing to some extent the vulnerability to errors. Powerful countermeasures however are available in the form of error detecting and correcting codes and, in the case of transmission, protocols that handle requests to retransmit.

The second point is less intrinsic to source coding and there exist *adaptive* coding techniques which acquire by themselves the statistical description of the source. Not only an *a priori* knowledge of the source is no longer necessary but the algorithm can follow the evolution of a nonstationary source provided it is slow enough. Economy in message length can thus be obtained even when encoding an unknown and variable source, which fact is highly favourable to a practical use.

This paper is devoted to a comparison of adaptive source coding algorithms, namely :
- the adaptive version of the classical Huffman algorithm [1] proposed by Gallager [2] ;
- the Guazzo algorithm [3] in Battail's adaptive version [4] ;
- the Lempel and Ziv algorithm [5].

We shall begin with a short description of the Huffman algorithm and its adaptive version by Gallager, which will give us the opportunity to introduce a graphical representation of a source to be useful later. Then, after a short description of the Lempel-Ziv algorithm, we shall show that it can be interpreted as a means to construct an equivalent secondary source which possesses a graphical representation of same shape. The Guazzo algorithm can then be interpreted as a means for describing any path in such a graph.

We shall emphasize the interest of using the adaptive Guazzo algorithm in order to encode a secondary source constructed according to the Lempel-Ziv algorithm either directly or, preferably, after its encoding by the Huffman-Gallager algorithm.

BRIEF DESCRIPTION OF THE ALGORITHMS

Huffman-Gallager coding

We first assume the source to be memoryless. Let q_s and q_c denote the source and code alphabet sizes, respectively. Let us assume that $q_s > q_c$, which can always result from using a source extension of a high enough order (i.e., dealing with a block of k source symbols as a single symbol from a source of alphabet size q_s^k referred to as the extension of order k of the original source). Now consider a tree such that q_c branches depart from each of its inner nodes and which has q_s terminal nodes or "leaves". The branches departing from an inner node are labelled with the code symbols and the leaves with the source symbols. Encoding of a source symbol is performed by moving a marker from the tree root to the leaf associated with this symbol and emitting the codeword which consists of the sequence of code symbols on the path of the marker. This setup guarantees that no codeword is the prefix of another one, so that codewords can be concatenated and still uniquely separated from each others.

To represent the encoding of successive source symbols we may transform the tree into the graph which results from merging its leaves with its root (Figure 1). We will refer to this graph as the *transition diagram* (the botanic metaphor would be a tree with stolons). In this graph, the source symbols serve as labels to the branches that revert to the root since the leaves disappeared.

Figure 1

The source alphabet size is assumed to be $q_s = 4$, and its symbols are denoted by α_1, α_2, α_3 and α_4. The code alphabet size is $q_c = 2$. A possible representation of this source is a tree (solid line). Its origin is referred to as its "root" and its terminal nodes as its "leaves". A leaf is associated with each source symbol. An irreducible encoding consists of describing the path from the root to a leaf by telling at each node the direction to be taken. For instance, the binary symbol "0" means "go to left" and "1", "go to right".

The same source may be represented by a transition diagram which results from the tree by merging its leaves with its root, as shown by dashed branches. The source symbols now label the branches which revert to the root. A sequence of binary symbols "0" and "1" describes a path in this graph and, if this path begins and ends at the root, this sequence describes a message from the source.

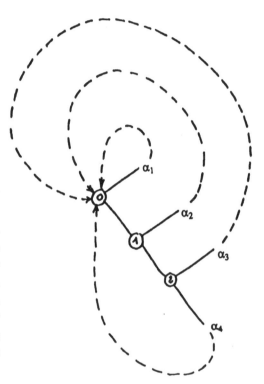

The Huffman algorithm constructs the optimum tree i.e., which minimizes the average length of the marker path in the transition diagram which describes the source output. In this case as well as in the more general case of a stationary source, the Huffman algorithm fails however to reach the theoretical limit i.e., the entropy of the source expressed with logarithms to the base q_c, except in the very particular case where all source symbol probabilities happen to be integer powers of $1/q_c$. If it is so, the conditional probability of the marker taking a given branch in the Huffman tree turns out to be $1/q_c$. When this is not the case, the only way for improving the compression rate of the Huffman algorithm consists of using an extension of the source of higher order. It can be shown that the conditional probabilities of the branches in the tree corresponding to the extension get closer and closer to $1/q_c$ as the extension order k increases, but at the same time the tree complexity grows exponentially with k. The theoretical limit of the average codeword length is asymptotically approached as the extension order k goes to infinity.

In the case of a source with memory, the use of a higher order extension of the original source has a further advantage. Clearly, the algorithm deals with each source symbol independently of the previous ones. On the other hand, the probabilities of the extension source symbols do fully take into account the dependence of the original symbols inside a block of length k. So, the larger k, the better is the compression as better account is taken of the source memory. The acceptable complexity still sets a limit on the tree size which can be used.

Gallager showed that the Huffman tree has the necessary and sufficient *sibling property* : if all the tree nodes are ordered by non-increasing (or non-decreasing) probability, then sibling nodes (i.e., having the same node as their ascendant) are adjacent in this ordered list. In Gallager's version of the algorithm, adaptivity is obtained by providing each node with a counter which is incremented each time the marker reaches it during the encoding process. Counter overflow is avoided, for instance, by halving the contents of all counters whenever one of them is about to overflow [2].

The nodes are ordered and the Huffman tree reconstructed on the basis of the counters contents. The sibling property is kept satisfied by rearranging the tree (cutting at some node and grafting at another one) as it becomes necessary. The number of source symbols being given, the number of nodes in the tree remains constant in such grafting process although the tree structure varies.

Lempel and Ziv coding

In marked constrast with the Huffman algorithm which initially assumed the source statistics to be known and was much later made adaptive, the Lempel and Ziv algorithm was conceived to deal with individual sequences, thus avoiding any *a priori* assumption on the source except knowledge of its alphabet [5].

The key idea is that a *sample* of the source output is stored in both encoder and decoder so that a reference to a previous output may save its retransmission. It is assumed that certain symbol strings identically repeat themselves in the source output : the idea of redundancy is thus literally taken.

The sample has been parsed into *template strings* such that, if one of them is of length $l > 0$, then the sample, as a set of template strings, also contains its prefix of length $l - 1$ (we assume that the 0-length string λ belongs to the sample ; it can be considered as a prefix of any other string). Newly delivered source output is observed until a string which is not contained in the sample has been emitted. Encoding consists of transmitting the ordinal number of the largest template string that matches the source output plus an innovation or extra symbol. The new source output, including the innovation symbol, can itself act as a template string to be appended to the sample. In a practical application, however, the sample size cannot indefinitely increase, so that a policy is required for limiting the allowed sample size.

The sample can be given a graphical interpretation by means of a tree having at most q_s branches, labelled by source symbols, departing from each node. Every unsaturated node (i.e., having less than q_s branches leaving it) represents one of the template strings which have to be indicated, together with an innovation symbol, in order to represent some source output. An example of such tree construction from a given source output is provided by Figure 2.

Using such a tree is a convenient way to parse new source output since the source symbols serve as steering indications at each node, except when the branch corresponding to a source symbol does not exist in which case it is transmitted as an innovation symbol.

Figure 2

Graph associated with the Lempel-Ziv algorithm. We assume that $q_s = 2$ and the sample used in order to construct the tree is as follows :
$$0^2 10^9 10^4 10^2 10^7 1^2 0^4 10^2 10^6 10^{11} 10^5 1010^4 10^{10} 1^3 010^2.$$
Its incremental parsing into template strings (separated by commas) results in :
$$0,01,0^2,0^3,0^4,1,0^4 1,0^2 1,0^5,0^2 1^2,0^4 10,010,0^5 1,0^6,0^5 10,0^4 101,0^4 10^2,0^7,01^2,10,10^2.$$

The corresponding tree was drawn on the figure. The up branches represent symbol "0" from the source, the down branches symbol "1". Using this sample, encoding according to Lempel and Ziv consists of indicating the "unsaturated" node (i.e., from which less than $q = 2$ branches depart) corresponding to the largest template string together with the following source symbol.

We obtain a transition diagram similar to that of Fig. 1 by restoring and connecting to the root all the missing branches from the unsaturated nodes (the dashed branches ; they are assumed to be connected to the root, but this was not actually drawn, for clarity's sake). Then, a source output is described by telling what path is followed in this diagram using the adaptive Guazzo algorithm.

We may also interpret the Lempel-Ziv algorithm as defining a secondary source equivalent to the original one, whose symbols are associated with the paths departing from and going back to the origin. Efficient compression results from encoding it by the Huffman-Gallager algorithm in order to construct a tree like that of Fig. 1 and then using the adaptive Guazzo algorithm.

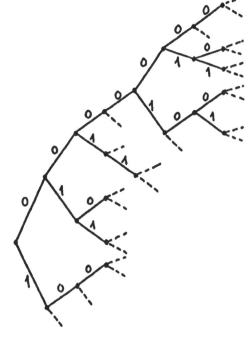

The Lempel-Ziv tree can be transformed into a transition diagram by simply completing the unsaturated nodes with branches reverting to the root (Fig. 2). We may then define a new secondary source with $q_c > q_s$ symbols by labelling the branches that revert to the root not only with a source symbol but also with a secondary one.

In this setup, encoding is performed by using the source symbols as steering indications to move a marker on the transition diagram and transmitting the secondary symbols of branches that revert to the root.

Similarly to the Huffman algorithm, the compression performance of the Lempel-Ziv algorithm for a stationary ergodic source approaches the theoretical limit as the sample size, hence the algorithm complexity, goes to infinity [5,6]. The Huffman and Lempel-Ziv algorithms differ in several important features, however, although they can be represented by transition diagrams of same shape. The most striking difference between them is that the branches are labelled by code symbols in the diagram which depicts the Huffman algorithm but by source symbols in the Lempel-Ziv case. The former thus achieves compression by transmitting the branch labels instead of those of the leaves (or branches reverting to the root) but the latter by doing the reverse.

This is not as puzzling as it seems at first, since in both cases the aim is to transmit symbols whose probabilities differ as little as possible and the key point is to construct the tree which is the most appropriate in this respect.

The Huffman algorithm is a fixed-to-variable coding technique requiring previous knowledge of the source statistics and using alphabet extension to improve the performance, especially (but not only) in order to take into account the source memory. The Lempel-Ziv algorithm implements a variable-length tailor-made extension of the source, thus fitted to its memory as it manifests itself in the sample used to construct the graph. Therefore, it efficiently takes into account the source memory whithout a prohibitive increase of the graph complexity, at variance with the Huffman algorithm.

Adaptive Guazzo algorithm

The Guazzo algorithm represents a different technique for source coding which includes Huffman coding as a very special case [3]. This algorithm can be regarded as still another way to encode steering indications in a tree or a graph.

A peculiarity of this method is that – at a given stage of the encoding process – the encoded string is not unique in the sense that any string in a set of s can represent the source output encoded so far. Subsequent encoding can thus take advantage of an alphabet of size s by using the information associated with the choice of a string in a set of s. Moreover, by appending n more symbols to the encoded string ($n \geq 0$), the number of strings that are available to encode the source output becomes $s \, q_c^n$.

To encode the next branching indication (at a node from which q branches depart), this set of cardinality $s \, q_c^n$ is partitioned into q nonempty subsets, each associated with a branch. The algorithm requires further that the cardinality of these subsets be approximately proportional to the conditional probabilities of the branches, where "approximately" means here that some precise criterion is satisfied. This criterion guarantees sufficient closeness of the proportion of strings alloted to each branch with respect to its conditional probability. The cardinality of the subset associated with the chosen branch sets the new value of s to be used in subsequent encoding.

The compression performance depends basically on the tightness of the closeness criteron, so that the theoretical limit can be approached as closely as desired while keeping the graph size finite. More precisely, in the stationary case the average length after coding $\bar{l}$ can be developed in series. If as an approximation we keep only its first two terms we get :

$$\bar{l} \approx H_{q_c} + A\ \eta^2 \tag{1}$$

q_c-ary symbols, where H_{q_c} is the source entropy in q_c-ary units, where η specifies how tight is the closeness criterion and where A depends both on the graph and on the transition probabilities at its nodes. We showed that, if the graph is drawn from a tree as in Fig. 1, A is minimum for the Huffman one [3,4].

This algorithm applies to any choice of q_c, q_s and q, irrespective of the matching between these parameters and the probability values. It works on any q-transition diagram, which we may define in general as an oriented graph having the following properties : (i) exactly q branches depart from each node ; (ii) no transition probability equals 0 or 1 ; (iii) there exists a path in the diagram from every node to any other one. Figs 1 and 2 are examples of such graphs. More generally, such a diagram is used as the model of a stationary ergodic Markovian source. We may also consider the case where the transition probabilities are initially unknown and/or slowly varying.

Battail has shown [4] that the Guazzo algorithm is easily made adaptive without any change in the graph (at variance with the Huffman-Gallager algorithm). This is achieved by providing each branch with a counter which is incremented each times the marker travels that branch. The algorithm then uses the counter contents to estimate the probabilities. In order to both follow slowly varying sources and prevent the counters from overflowing, the contents of all counters associated to branches leaving a given node should be reduced from time to time according to some policy. As a rule, this policy will set a trade-off between speed of adaptation and precision in estimating the probabilities. The simplest of these policies consists of halving all counters at a given node whenever one of them is about to overflow. In this case the counter size controls the trade-off between the ability of adaptation and the compression performance.

The adaptive Guazzo algorithm was implemented on a computer to deal with a wide variety of simulated sources, both stationary and nonstationary. Even for a moderately tight closeness criterion, results very close to the theoretical limit were obtained when using stationary memoryless sources. The ability to follow changes in the source statistics was also checked ; due to the possible coarseness of the criterion, fairly fast changes resulted in moderate degradation with respect to the stationary case. The natural binary tree, a reasonable choice in the absence of preliminary knowledge of the source, was generally used. The main shortcomings were met with sources having memory since taking it into account would have led to use a prohibitively complex tree.

COMPARISON AND POSSIBLE COMBINATION OF THE ALGORITHMS

Roles of the three algorithms

We now summarize the respective roles of the three algorithms which have been described above.

For a given source e.g., the k-order extension of the original source, the Huffman-Gallager algorithm constructs the q_c-transition diagram such that the path of the marker which represents its output is the shortest one in the average. The source memory is fully taken into account inside each block of k symbols of the original source alphabet, but not at all outside such a block.

Given the original source, the Lempel-Ziv algorithm defines a variable-length extension of it which takes account of the source memory within limits set by the accepted complexity. This extension is a *secondary source* better fitted to further processing than the original one. Moreover, parsing the output sample can be interpreted as constructing a q_s-transition diagram, as depicted in Fig. 2.

On the other hand, the adaptive Guazzo algorithm appears as a means for describing the path of a marker in a q-transition diagram of finite size with an economy of symbols which approaches the theoretical limit as the closeness criterion is made tighter and tighter.

Thus, both the Huffman-Gallager and Lempel-Ziv algorithms may be interpreted as constructing a q-transition diagram. Coding being considered as a means for describing the path of a marker in such a graph, the adaptive Guazzo algorithm appears as providing a further economy with respect to merely telling the branch to be taken at each node (Huffman) or the relevant secondary source symbol (Lempel-Ziv). According to this point of view, the adaptive Guazzo algorithm is complementary to the other two, which provide it with a graph on which to work.

Such combinations of the algorithms were simulated on stationary sources, especially memoryless ones or modelled as Markov chains, and nonstationary ones.

Using first the Huffman algorithm and then the adaptive Guazzo algorithm was very successful for memoryless sources. As already said, the graph which results from the Huffman algorithm minimizes the coefficient A in (1). However, the performance mainly depends on η, so that substitution of the Huffman tree for an arbitrary one (e.g., corresponding to the natural q-ary representation) results in little improvement.

We also tried to use the adaptive Guazzo algorithm on the graph associated with the Lempel-Ziv algorithm to represent the parsing of the sample, as in Fig. 2. A source with memory, modelled by a Markov chain, was used. Although an improvement was achieved with respect to the Lempel-Ziv algorithm alone, the compression achieved for simulated ergodic stationary sources remained significantly less than the theoretical limit. This can be understood if we remember that, although the performance of the Guazzo algorithm does not strongly depend on the tree which describes the source, the best tree results from Huffman coding [3,4]. According to a remark above, the tree which results from the Lempel-Ziv algorithm is strongly mismatched in this respect, so that the constant A in (1) is unduly large.

This drawback can be avoided, however, if we think of the Lempel-Ziv algorithm as defining a secondary source which can be given a tree representation e.g., the natural tree or that constructed using the Huffman-Gallager algorithm. The adaptive Guazzo algorithm then takes into account the actual frequencies of the Lempel-Ziv extension symbols in order to improve compression. We are thus led to consider a *symbiosis* of the three algorithms.

Symbiosis of the algorithms

The three algorithms appear as *complementary* so we may think of making them collaborate in the following way, as depicted in Figure 3.

The Lempel-Ziv algorithm will be used first in order to define a secondary source which best takes into account the source memory for a given accepted complexity.

Secondly, since the symbols of the secondary source are generally far from being equiprobable (although it tends to be so for an ergodic source as the sample size increases indefinitely), the Huffman-Gallager algorithm will result in a graph better fitted to the secondary source than that which describes the parsing of the sample into template strings.

Finally, the adaptive Guazzo algorithm will provide a description of the marker path in this diagram which approaches the shortest possible one as the closeness criterion becomes arbitrarily tight.

Figure 3. Possible combination of the three algorithms.

The *Lempel-Ziv* algorithm (LZ) is used first in order to obtain a *variable-length extension* of the source (VLE) whose size depends on the accepted complexity, to be dealt with as memoryless. Then, the *Huffman-Gallager* algorithm (HG) may operate on this source thus resulting in the *best tree* (of shortest average length) (BT). Finally, the *adaptive Guazzo* algorithm (AGA) provides in the average the *asymptotically shortest description* (ASD) of any path in the transition diagram corresponding to this tree, as its closeness criterion gets arbitrarily tight.

Of course, it may happen in certain cases that one of the algorithms is almost useless and thus may be omitted. For instance, in the case of a memorylesss source, the Lempel-Ziv algorithm is of no use. In many cases, the Huffman-Gallager will result in a very small improvement and the simpler natural tree may be used in order to represent the secondary source provided by the Lempel-Ziv algorithm. On the other hand, if this secondary source or the result of its Huffman-Gallager encoding happens to be fairly equiprobable, the Guazzo algorithm will provide little benefit. We think however that properly sharing the available overall complexity amongst the three combined algorithms may result in a data compression system which can efficiently handle a wide variety of real-world sources.

On the other hand, the adaptive Guazzo algorithm may directly work on the graph which describes a source as an ergodic Markov chain, provided of course that this model is actually available. If a Markovian model of the source is completely known (number of states, connection between them and transition probabilities), then the Guazzo algorithm can work on its transition diagram. If its structure is known but not the transition probabilities, the adaptive Guazzo algorithm can be used. If no Markovian model of the source is known, we still may think of it as a *hidden Markov chain* where we can observe only the symbols associated with its transitions (e.g., see [7]). Assuming it to be ergodic and state-calculable, a proper algorithm may be used in order to identify a Markovian model of the source, thus resulting in a graph on which the adaptive Guazzo algorithm can work.

CONCLUSION

The Guazzo algorithm appears as a means for describing the transition diagram of a stationary ergodic Markov chain, with an economy of symbols which depends only on the tightness of the criterion it uses. Its adaptive version avoids the necessity of a preliminary knowledge of the transition probabilities and can follow their possible variations if they are slow enough.

On the other hand, the Huffman-Gallager algorithm is a means for constructing the best tree which represents a given source assumed to be memoryless, while the Lempel-Ziv algorithm defines the best variable-length extension of a source with memory without requiring any preliminary knowledge of it. The three algorithms thus appear as performing not only different tasks, but also complementary ones : significantly, near optimum results are obtained for a broad class of sources by making them collaborate.

A more general approach would consist of dealing with a given source as a hidden Markov chain, of identifying its model and then applying the adaptive Guazzo algorithm on the transition diagram thus obtained.

References

[1]   D.A. HUFFMAN, A method for the construction of minimum redundancy codes, Proc. IRE, vol. **40**, 1952, pp. 1098-1101

[2]   R.G. GALLAGER, Variations on a theme by Huffman, IEEE Trans. on Inf. Th., vol. **IT-24**, n° 6, Nov. 1978, pp. 668-674

[3]   M. GUAZZO, A general minimum-redundancy source-coding algorithm, IEEE Trans. on Inf. Th., vol. **IT-26**, n° 1, Jan. 1980, pp. 15-25

[4]   G. BATTAIL, Codage de source adaptatif par l'algorithme de Guazzo, Annales des Télécommunic., to be published

[5]   J. ZIV and A. LEMPEL, Compression of individual sequences via variable-rate coding, IEEE Trans. on Inf. Th., vol. **IT-24**, n° 5, Sept. 1978, pp. 530-536

[6]   A.D. WYNER and J. ZIV, Some asymptotic properties of the entropy of a stationary ergodic data source with applications to data compression, IEEE Trans. on Inf. Th., vol. **35**, n° 6, Nov. 1989, pp. 1250-1258

[7]   L.R. RABINER and B.H. JUANG, An introduction to hidden Markov models, IEEE ASSP Mag., Jan. 1989, pp. 4-16

Minimum codeword length
and redundancy of Huffman codes*

Renato M. Capocelli

Dipartimento di Matematica

Università di Roma "La Sapienza"

00185 Roma, Italy

Alfredo De Santis

Dipartimento di Informatica ed Applicazioni

Università di Salerno

84081 Baronissi (Salerno), Italy

Abstract

A tight upper bound on the redundancy r of Huffman codes, in terms of the minimum codeword length l, $l \geq 1$, is provided. The bound is a strictly decreasing function of l. For large l it yields $r \leq l - \log(2^{l+1} - 1) + 1 + \beta + O(2^{-2l})$, where $\beta \approx 0.0860$.

By using this result we improve Gallager's bound on the redundancy when only the most likely source probability p_1 is known.

1 Introduction

Let A be a discrete source with N, $2 \leq N < \infty$, letters and let p_k denote the probability of letter a_k, $1 \leq k \leq N$. Let $x_1, x_2, \ldots, x_N$ be a set of binary codewords for encoding letters of source A and let $n_1, n_2, \ldots, n_N$ be the codeword lengths. The *Huffman* encoding algorithm provides an *optimal prefix code* C for the source A. The encoding is optimal in the sense that codeword lengths minimize the *redundancy* r, defined as the difference between the *average codeword length* E of the code and the *entropy* $H(p_1, p_2, \ldots, p_N)$ of the source:

$$r = E - H(p_1, p_2, \ldots, p_N) = \sum_{i=1}^{N} p_i n_i + \sum_{i=1}^{N} p_i \log p_i$$

where $\log$ denotes the logarithm to base 2.

---

*This work was partially supported by the National Council of Research (C.N.R.) under grant 90.01552.CT12.

According to Shannon's first theorem the redundancy of any Huffman code is always nonnegative and less than or equal to one.

If only the probability p_1 of the most likely source letter is known, it has been proved [1], [2], [3], [8] that

$$r \leq \begin{cases} 2 - \mathcal{H}(p_1) - p_1, & \text{if } 0.5 \leq p_1 < 1 \\ 3 - 5p_1 - \mathcal{H}(2p_1), & \text{if } \delta \leq p_1 < 0.5 \\ 1 + 0.5(1 - p_1) - \mathcal{H}(p_1), & \text{if } 1/3 \leq p_1 < \delta \\ 3 - 3(1 + \log 3)p_1 - \mathcal{H}(3p_1), & \text{if } \theta \leq p_1 < 1/3 \\ 3/4 + 5p_1/4 - \mathcal{H}(p_1), & \text{if } 2/9 < p_1 < \theta \\ \beta + p_1, & \text{if } 1/(2^l - 1) \leq p_1 \leq 2/(2^l + 1),\ l \geq 3 \\ 4 - 7(1 + \log 7)p_1 - \mathcal{H}(7p_1), & \text{if } \gamma \leq p_1 < 1/7 \\ 3 - \mathcal{H}(p_1) - (1 - p_1)(2\log 3 - 5/9), & \text{if } 2/17 < p_1 < \gamma \\ l - \mathcal{H}(p_1) - (1 - p_1)A_{2^l - 1}, & \text{if } 2/(2^{l+1} + 1) < p_1 < 1/(2^l - 1),\ l \geq 4 \end{cases}$$

where $\mathcal{H}(p) = -p \log p - (1 - p) \log(1 - p)$; $\delta \approx 0.4505$, $\beta = 1 - \log e + \log(\log e) \approx 0.0860$, $\theta \approx 0.3138$, $\gamma \approx 0.1422$, $A_j = \min_{\{w_i \in W_j\}}\{H(w_1, w_2, ..., w_j) - w_j\}$ and W_j, $j \geq 2$, is the set of positive real numbers $w_1, w_2, ..., w_j$ that satisfy $w_1 \geq w_2 \geq ... \geq w_j \geq w_1/2$ along with $\sum w_h = 1$. All upper bounds, but $r \leq \beta + p_1$, are tight.

It has been also proved [4], [10] that the following tight lower bound holds whatever p_1

$$r \geq m - \mathcal{H}(p_1) - (1 - p_1) \log(2^m - 1);$$

m is determined according to $\alpha_m < p_1 \leq \alpha_{m-1}$, where $\alpha_0 = 1$ and $\alpha_m = 1 - \left(\log \frac{2^{m+1}-1}{2^m-1}\right)^{-1}$, if $m \geq 1$.

Capocelli and De Santis [6] proved tight upper and tight lower bounds for the redundancy of binary Huffman codes when the least likely source letter probability is known.

Capocelli and De Santis have also analyzed the D-ary case [5]. They proved an upper bound and a tight lower bound on the redundancy of D-ary Huffman codes when the most likely source letter probability is known.

In this paper we prove that the redundancy r, as a function of the minimum codeword length l of the Huffman code, is upper bounded by

$$r \leq l - M_{2^l}$$

where $M_j = \min_{\{x_i \in X_j\}}\{H(x_1, x_2, ..., x_j) - x_j\}$ and X_j, $j \geq 2$, is the set of nonnegative real numbers $x_1, x_2, ..., x_j$ that satisfy $x_1 \geq x_2 \geq ... \geq x_j \geq x_1 - x_{j-1}$ along with $\sum x_i = 1$. The bound is the best possible bound on r expressed as a function only of l. An asymptotic estimate of M_j is

$$M_j = \log(2j - 1) - 1 - \beta + O\left(\frac{1}{j^2}\right).$$

We use the upper bound $r \leq l - M_{2^l}$ to improve Gallager's bound [1] $r \leq p_1 + \beta$ for $1/(2^l - 1) \leq p_1 \leq 2/(2^l + 1)$, $l \geq 4$, where p_1 is the most likely source probability.

2 The Bound

Define X_j, $j \geq 2$, as the set of nonnegative real numbers $x_1, x_2, ..., x_j$ that satisfy $x_1 \geq x_2 \geq ... \geq x_j \geq x_1 - x_{j-1}$ along with $\sum x_i = 1$.

Lemma 1 *The minimum*

$$M_j = \min_{\{x_i \in X_j\}} \left\{ H(x_1, x_2, ..., x_j) - x_j \right\} \tag{1}$$

is equal to $\log(j - 1)$ for $2 \leq j \leq 9$. Moreover for $j \geq 10$ it satisfies

$$0 < M_j - \log(2j - 1) + 1 + \beta < \frac{(\log e)^3}{(2j - 1)(2j - 1 + \log e)} \tag{2}$$

where $\beta = 1 - \log e + \log(\log e) \approx 0.0860$.

Proof. Since the function $H(x_1, x_2, ..., x_j) - x_j$ is convex $\cap$, the minimum must occur at an extreme point of X_j. The extreme points of X_j are exactly j, and are those for which for some k, $k \in [1, j]$, $x_i = x_i$, if $i \leq k$ and $x_i = x_1 - x_{j-1}$, if $j \geq i > k$. Thus, for $k = j$ the extreme point is $x_i = 1/j$, $i = 1, ..., j$; for $k = j - 1$ it is $x_i = 1/(j - 1)$, $i = 1, ..., j - 1$, and $x_j = 0$; and for $k \leq j - 2$ the extreme points are $x_i = x_1$ for $i = 1, ..., k$ and $x_i = x_1/2$ for $i = k + 1, ..., j$. Therefore

$$M_j = \min \left\{ \log j - \frac{1}{j}, \log(j - 1), \min_{1 \leq k \leq j-2} \left\{ \log(j + k) - \frac{1 + 2k}{j + k} \right\} \right\}. \tag{3}$$

Let $f(j)$ be the difference between $\log j - 1/j$ and $\log(j - 1)$. Since the derivative of f is negative for $j \geq 3$ and $f(j)$ approaches 0 as $j \to \infty$, it follows that f is positive for $j \geq 3$. Moreover $f(2) = 1/2$. Thus $\log j - 1/j > \log(j - 1)$ for $j \geq 2$.

The minimum M_j is then given by $M_j = \min\{\log(j-1), A_j\}$, where $A_j = \min_{1 \leq k \leq j-2}\{\log(j + k) - (1 + 2k)/(j + k)\}$. A numerical computation shows that $M_j = \log(j - 1)$ for $j = 2, ..., 9$.

To compute M_j for $j \geq 10$, we make use of the same computation performed in [3] for computing A_j.

Consider the function $\psi_j(k) = \log(j + k) - (1 + 2k)/(j + k)$. $\psi_j(k)$ is a convex $\cup$ function of k. Allowing k to take nonintegral values one finds, by elementary calculus, that $\psi_j(k)$ takes its minimum value at

$$k_j^* = (2/\log e - 1)j - 1/\log e. \tag{4}$$

Thus $A_j \geq \psi(k_j^*) = \log(2j - 1) - 1 - \beta$. Define $\epsilon_j = A_j - (\log(2j - 1) - 1 - \beta)$. From the convexity, $\psi_j(k)$ takes its minimum among the integral values either at $\lfloor k_j^* \rfloor$ or at $\lfloor k_j^* \rfloor + 1$. Thus

$$
\begin{aligned}
\epsilon_j &= \min\{\psi_j(\lfloor k_j^* \rfloor) - \psi_j(k_j^*), \; \psi_j(\lfloor k_j^* \rfloor + 1) - \psi_j(k_j^*)\} \\
&\leq \min\{\psi_j(k_j^* - 1) - \psi_j(k_j^*), \; \psi_j(k_j^* + 1) - \psi_j(k_j^*)\}
\end{aligned}
$$

Recalling the definition of $\psi_j(k)$, we obtain

$$\epsilon_j \leq \min\left\{\log\left(1 - \frac{1}{j + k_j^*}\right) + \frac{2j - 1}{(j + k_j^* - 1)(j + k_j^*)}, \; \log\left(1 + \frac{1}{j + k_j^*}\right) + \frac{1 - 2j}{(j + k_j^* + 1)(j + k_j^*)}\right\}$$

that, from (4) and by making use of the inequality $\log y \leq (y - 1)\log e$ gives

$$
\begin{aligned}
\epsilon_j &\leq \min\left\{\frac{(\log e)^3}{(2j - 1)(2j - 1 - \log e)}, \; \frac{(\log e)^3}{(2j - 1)(2j - 1 + \log e)}\right\} \\
&= \frac{(\log e)^3}{(2j - 1)(2j - 1 + \log e)}.
\end{aligned}
$$

Now, consider $A_j - \log(j - 1) = \epsilon_j + (\log(2j - 1) - 1 - \beta) - \log(j - 1) \leq \phi(j)$, where $\phi(j) = \log(2j - 1) - \log(j - 1) - 1 - \beta + (\log e)^3/((2j - 1)(2j - 1 + \log e))$. Since $\phi(j)$ is a decreasing function of j and $\phi(10) \approx -0.0002$, we have $\phi(j) < 0$ for $j \geq 10$. Hence, for $j \geq 10$, $A_j \leq \log(j - 1)$. Consequently $M_j = A_j$ and the bound (2) follows. $\quad\square$

For illustrative purposes, values of M_j and of $\log(2j - 1) - 1 - \beta$ up to $j = 511$, together with their difference ϵ_j are presented in Table 1.

j	A_j	$\log(2j-1)-1-\beta$	ϵ_j
10	$\log 13 - 7/13$	3.1618561813876512869	0.0001219982919024119
11	$\log 15 - 9/15$	3.3062460907228260820	0.0006445048856924473
12	$55/16$	3.4374906240010786654	0.0000093759989213346
13	$\log 17 - 9/17$	3.5577848577187904888	0.0002662188256665665
14	$\log 19 - 11/19$	3.6688161701075343374	0.0001639749149985248
15	$\log 5 + 29/20$	3.7719096630716379138	0.0000184318157244340
31	$\log 21 + 19/42$	4.8446660055069520695	0.0000323696527606003
63	$\log 87 - 49/87$	5.8797129526061528367	0.0000121524379777643
127	$\log 175 - 97/175$	6.8969222426383759216	0.0000031549082385958
255	$\log 353 - 197/353$	7.9054505140197610885	0.0000002048604843575
511	$\log 177 + 1021/708$	8.9096958188218672804	0.0000001267413561485

Table 1.

Theorem 1 *Let l be the minimum length of the Huffman code C. Then the redundancy of C is upper bounded by*

$$r \leq l - M_{2^l}. \tag{5}$$

The bound is the best possible bound expressed only in terms of l.

Proof. Let $q_1 \geq q_2 \geq \dots \geq q_L$ be the $L = 2^l$ probabilities at level l in the Huffman code. Let us recall from Gallager [1] that

$$r \leq l - H(q_1, q_2, \dots, q_L) + q_L.$$

Minimizing over all possible choices of q_i's we obtain

$$r \leq l - \min_{q_i \in X_L} \left\{ H(q_1, q_2, \dots, q_L) - q_L \right\}$$

that, by Lemma 1, yields

$$r \leq l - M_{2^l}.$$

We now prove that the bound is the best possible bound expressed only in terms of l.

First, suppose $l \leq 3$. Consider the source consisting of 2^l letters, one of them with probability ϵ, one of them with probability $(2^l - 1)^{-1} - \epsilon$, and each of the remaining letters with probability $1/(2^l - 1)$. The Huffman code has 2^l codewords of length l. Its redundancy approaches $l - \log(2^l - 1)$ as $\epsilon \to 0$.

Finally, suppose $l \geq 4$. Let k'_j be the value for which A_j reaches its minimum, i.e. $A_j = \log(j + k'_j) - (1 + 2k'_j)/(j + k'_j)$. Consider the source consisting of $2^l + k'_{2^l} + 1$ letters, two of them with probabilities ϵ and $(2^l + k'_{2^l})^{-1} - \epsilon$, and $2^l + k'_{2^l} - 1$ letters each with probability $(2^l + k'_{2^l})^{-1}$. The Huffman code has $2^l - k'_{2^l} - 1$ codewords of length l and $2k'_{2^l} + 2$ of length $l + 1$. Its redundancy approaches $l - M_{2^l}$ as $\epsilon \to 0$. $\square$

The first nine values of $l - M_{2^l}$ are presented in Table 2.

l	$l - M_{2^l}$
1	1
2	$2 - \log 3$
3	$3 - \log 7$
4	$95/21 - \log 21$
5	$157/44 - \log 11$
6	$313/88 - \log 11$
7	$1338/177 - \log 177$
8	$2675/354 - \log 177$
9	$6776/709 - \log 709$

Table 2.

A property of the bound stated in Theorem 1 is that it is decreasing with l.

Theorem 2 *The function $l - M_{2^l}$ is strictly decreasing with l.*

Proof. From Table 2, the theorem is true for the first few values. We prove that $l - 1 - M_{2^{l-1}} > l - M_{2^l}$, for $l \geq 5$. From (2) we have $M_{2^l} \geq \log(2^{l+1} - 1) - 1 - \beta$ and $M_{2^{l-1}} \leq \log(2^l - 1) - 1 - \beta + (\log e)^3/((2^l - 1)(2^l - 1 + \log e))$. Thus, for proving the theorem it is enough to prove that, for $l \geq 5$,

$$\log(2^{l+1} - 1) - 1 - \beta > 1 + \log(2^l - 1) - 1 - \beta + \frac{(\log e)^3}{(2^l - 1)(2^l - 1 + \log e)}. \tag{6}$$

To prove above inequality, denote by $\omega(l)$ the function $\log(2^{l+1} - 1) - 1 - \log(2^l - 1)$. The function $\omega(l)$ can be rewritten as $\log((2^{l+1} - 1)/(2^{l+1} - 2))$ which is $\geq (\log e)/(2^{l+1} - 1)$, because of the inequality $\ln x \geq 1 - 1/x$. In turn, $(\log e)/(2^{l+1} - 1)$ is $\geq 0.2/(2^l - 1)$, for $l \geq 5$. Since $(\log e)^3/(2^l - 1 + \log e)$ is a decreasing function of $l \geq 4$, it reaches its maximum at $l = 4$ where it is equal to $0.18262 < 0.2$. Hence, $\omega(l) \geq (\log e)^3/((2^l - 1)(2^l - 1 + \log e))$, which proves (6) and concludes the proof of the lemma. $\square$

Remark. Since the difference between M_j and $\log(2j - 1) + 1 + \beta$ becomes negligible as $j \to 0$, the upper bound (5) approaches $l - \log(2^{l+1} - 1) + 1 + \beta$, for large l. Thus, as $l \to \infty$ the upper bound approaches β.

3 Improved bounds when only p_1 is known

In this section we use the tight upper bound given by Theorem 1 to improve the Gallager's bound [1] $r \leq p_1 + \beta$ for $1/(2^l - 1) \leq p_1 \leq 2/(2^l + 1)$, $l \geq 4$.

For that, notice that in any Huffman code, the probability p_1 of the most likely source letter is related to the length of its codeword by the following lemma, due to Montgomery and Kumar [9].

Lemma 2 *Let p_1 be the probability of the most likely source letter of a discrete source A. If for some $l \geq 1$*

$$\frac{2}{2^{l+1} + 1} < p_1 < \frac{1}{2^l - 1},$$

then any optimal binary code for A must have a minimum codeword length of l. Furthermore, if for some $l \geq 1$,

$$\frac{1}{2^{l+1} - 1} \leq p_1 \leq \frac{2}{2^{l+1} + 1},$$

then any optimal code for A must have a minimum codeword length of either l or $l + 1$.

This result is an extension of Johnsen's Theorem 1 [8] and Capocelli et al.'s Lemma 1 [2]. Recently, Capocelli and De Santis [5] have extended it to the D-ary case.

Combining Theorems 1 and 2, and Lemma 2, we can prove the following theorem.

Theorem 3 *Let p_1 be the probability of the most likely source letter of a discrete source A. The redundancy of the corresponding Huffman code is upper bounded by*

$$r \leq p_1 + \beta, \quad \text{if } 1/(2^l - 1) \leq p_1 < l - 1 - M_{2^{l-1}} - \beta, l \geq 4, \tag{7}$$

$$r \leq l - 1 - M_{2^{l-1}}, \quad \text{if } l - 1 - M_{2^{l-1}} - \beta \leq p_1 \leq 2/(2^l + 1), l \geq 4. \tag{8}$$

Proof. Gallager [1] proved that $r \leq p_1 + \beta$ for $1/(2^l - 1) \leq p_1 \leq 2/(2^l + 1)$, $l \geq 4$. From Lemma 2 we know that for $1/(2^l - 1) \leq p_1 \leq 2/(2^l + 1)$, $l \geq 4$, the Huffman code has a minimum codeword length of either l or $l - 1$. Because of Theorems 1 and

2, for $1/(2^l - 1) \leq p_1 \leq 2/(2^l + 1)$, $l \geq 4$, one has $r \leq l - 1 - M_{2^{l-1}}$. Therefore, $r \leq \min\{p_1 + \beta, l - 1 - M_{2^{l-1}}\}$, for $1/(2^l - 1) \leq p_1 \leq 2/(2^l + 1)$, $l \geq 4$. Let us now compute this minimum.

First, notice that from (2) we get $l - 1 - M_{2^{l-1}} < l - \log(2^l - 1) + \beta$. Making use of the inequality $\ln x \leq x - 1$, it follows $l - \log(2^l - 1) + \beta = \beta + \log(2^l/(2^l - 1)) \leq \beta + (\log e)/(2^l - 1)$ which in turn results $< \beta + 2/(2^l + 1)$. Thus, $l - 1 - M_{2^{l-1}} < \beta + 2/(2^l + 1)$.

Next, notice that from (2) we also have $l - 1 - M_{2^{l-1}} > l - \log(2^l - 1) + \beta - (\log e)^3/((2^l - 1)(2^l - 1 + \log e))$. Because of the inequality $\ln x \geq 1 - 1/x$, it follows that $l - \log(2^l - 1) = \log(2^l/(2^l - 1)) \geq (\log e)/2^l$. Since $(\log e)^3/(2^l - 1 + \log e)$ is a decreasing function of $l \geq 4$ it reaches its maximum at $l = 4$ where it equals $0.18262 < 0.2$. Hence, $l - 1 - M_{2^{l-1}} > (\log e)/2^l + \beta - 0.2/(2^l - 1)$. Finally, since $(\log e)/2^l \geq 1.2/(2^l - 1)$, we have $l - 1 - M_{2^{l-1}} > \beta + 1/(2^l - 1)$.

Since $\beta + 1/(2^l - 1) < l - 1 - M_{2^{l-1}} < \beta + 2/(2^l + 1)$, then the two bounds $p_1 + \beta$ and $l - 1 - M_{2^{l-1}}$ intersect in exactly one point in the interval $[1/(2^l - 1), 2/(2^l + 1)]$, $l \geq 4$. Thus, $p_1 + \beta < l - 1 - M_{2^{l-1}}$ if $1/(2^l - 1) \leq p_1 < l - 1 - M_{2^{l-1}} - \beta$, $l \geq 4$, and $p_1 + \beta > l - 1 - M_{2^{l-1}}$ if $l - 1 - M_{2^{l-1}} - \beta \leq p_1 \leq 2/(2^l + 1)$, $l \geq 4$.

Bound $r \leq l - 1 - M_{2^{l-1}}$ improves Gallager's bound for $l \geq 4$ when $p_1 \in [l - 1 - M_{2^{l-1}} - \beta, 2/(2^l + 1)]$. $\Box$

References

[1] R. G. Gallager, "Variation on a theme by Huffman," *IEEE Trans. on Inform. Theory*, vol. IT-24, no. 6, pp. 668–674, Nov. 1978.

[2] R. M. Capocelli, R. Giancarlo and I. J. Taneja, "Bounds on the redundancy of Huffman codes," *IEEE Trans. on Inform. Theory*, vol. IT-32, no. 6, pp. 854–857, Nov. 1986.

[3] R. M. Capocelli and A. De Santis, "Tight upper bounds on the redundancy of Huffman codes," *IEEE Trans. Inform. Theory*, vol. 35, n. 5, pp. 1084–1091, Sept. 1989.

[4] R. M. Capocelli and A. De Santis, "Tight bounds on the redundancy of Huffman codes," IBM Research Report RC–14154.

[5] R. M. Capocelli and A. De Santis, "A note on D-ary Huffman codes," to appear on *IEEE Trans. Inform. Theory*, Jan. 1991.

[6] R. M. Capocelli and A. De Santis, "New bounds on the redundancy of Huffman codes," to appear on *IEEE Trans. Inform. Theory*.

[7] D. A. Huffman, "A method for the construction of minimum redundancy codes," *Proc. IRE*, vol. 40, pp. 1098–1101, 1952.

[8] O. Johnsen, "On the redundancy of Huffman codes," *IEEE Trans. on Inform. Theory*, vol. IT-26, no. 2, pp. 220–222, Mar. 1980.

[9] B. L. Montgomery and B. V. K. Vijaya Kumar, "On the average codeword length of optimal binary codes for extended sources," *IEEE Trans. on Inform. Theory*, vol. IT-33, no. 2, pp. 293–296, Mar. 1987.

[10] B. L. Montgomery and J. Abrahams, "On the redundancy of optimal binary prefix-condition codes for finite and infinite sources," *IEEE Trans. Inform. Theory*, vol. IT-33, no. 1, pp. 156–160, Jan. 1987.

SECTION 7

MODULATION

BINARY COVERING CODES AND HIGH SPEED DATA TRANSMISSION

A. R. Calderbank
Mathematical Sciences Research Center
AT&T Bell Laboratories, Murray Hill, NJ 07974

Abstract. There has been a great deal of recent interest in the covering radius of binary codes. We shall describe how good covering codes can be used to make high speed data transmission more reliable.

1. Introduction

We consider the problem of designing signal constellations for the Gaussian channel. A signal is simply a vector in N-dimensional Euclidean space, and the coordinate entries correspond to voltage levels on a transmission line. An encoder transforms a binary data stream into a sequence of signals, and the channel adds noise, which may cause errors. The reliability of the transmission scheme is determined by the minimum squared Euclidean distance between distinct signal sequences, and the cost of achieving that reliability is the average signal power (which is proportional to the square of the voltage). We start by choosing a lattice Λ in real N-dimensional space $\mathbb{R}^N$. The signal constellation Ω consists of all lattice points within a region $\mathcal{R}$. The reason we consider signal constellations drawn from lattices is that signal points are distributed regularly throughout N-dimensional space. This means that the average signal power P of the constellation Ω is approximately the average power $P(\mathcal{R})$ of a probability distribution that is uniform within $\mathcal{R}$ and zero elsewhere. This approximation is called the *continuous approximation*. It was introduced by Forney [5], and we shall use it extensively. If we fix the size of the signal constellation, then the average signal power depends on the choice of lattice and on the shape of the region that bounds the constellation. The continuous approximation allows us to quantify these dependencies.

We shall focus on savings in average power that are obtained by manipulating the shape of the region $\mathcal{R}$ that bounds the signal constellation. If the region $\mathcal{R}$ is an N-cube with faces parallel to the coordinate axes then the induced probability distribution on an arbitrary M-dimensional projection is uniform. Changing the shape of the region $\mathcal{R}$ induces a non-uniform probability distribution on the M-dimensional projection. Thus gains derived from shaping an N-dimensional constellation can be achieved in lower dimensional space by non-equiprobable signaling.

We shall describe how binary covering codes can be used to design non-equiprobable signaling schemes for use in high speed modems (data rates of between 9.6 kb/s and 24 kb/s).

2. The Continuous Approximation

A *lattice* Λ in real N-dimensional space is an additive subgroup of $\mathbb{R}^N$. A basis for the lattice Λ is a set of m vectors $v_1, ..., v_m$ such that

$$\Lambda = \left\{ \sum_{i=1}^{m} \lambda_i v_i \,\Big|\, \lambda_i \in \mathbb{Z}, \, i = 1, ..., m \right\}.$$

The lattice Λ is said to be m-dimensional and usually we have $m = N$. If $w_1, ..., w_m$ is another choice of basis then there exists a unimodular integral matrix Q such that $w_i = Q v_i$ for all $i = 1, ..., m$.

Definition. A *fundamental region* $\mathcal{R}$ for a lattice Λ is a region of $\mathbb{R}^N$ that contains one and only one point from each equivalence class modulo Λ. In other words $\mathcal{R}$ is a complete system of coset representatives for Λ in $\mathbb{R}^N$.

If $v_1, ..., v_m$ are a basis for a lattice Λ then the parallelotope consisting of the points

$$\mu_1 v_1 + \cdots + \mu_m v_m \quad (0 \le \mu_i < 1)$$

is an example of a fundamental region of Λ. This region is called a *fundamental parallelotope*.

Example 1. The hexagonal lattice A_2 is given by

$$A_2 = \left\{ a(1,0) + b(1/2, \sqrt{3}/2) \,\Big|\, a, b \in \mathbb{Z} \right\}.$$

Here $v_1 = (1,0)$ and $v_2 = (1/2, \sqrt{3}/2)$ are a basis for A_2 and Fig. 1 shows the fundamental parallelotope $\mathcal{R}$ determined by the basis v_1, v_2. The translates $\mathcal{R} + y$, $y \in A_2$, of the fundamental parallelotope by points in the lattice A_2 tile 2-dimensional space.

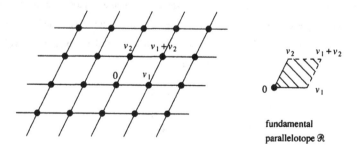

fundamental
parallelotope $\mathcal{R}$

Fig. 1. The plane tiled by translates $\mathcal{R} + y$, $y \in A_2$ of the fundamental parallelotope.

Definition. If $\Lambda \subseteq \mathbb{R}^N$ is a lattice, and $y \in \Lambda$ is a lattice point, then the *Voronoi region* $\mathcal{R}(y)$ consists of those points in $\mathbb{R}^N$ that are at least as close to y as to any other $y' \in \Lambda$. Thus

$$\mathcal{R}(y) = \left\{ x \in \mathbb{R}^N \mid \|x - y\| \leq \|x - y'\| \text{ for all } y' \in \Lambda \right\}.$$

The interiors of different Voronoi regions are disjoint though two neighboring Voronoi regions may share a face. These faces lie in the hyperplanes midway between two neighboring lattice points. Translation by $y \in \Lambda$ maps the Voronoi region $\mathcal{R}(w)$ to the Voronoi region $\mathcal{R}(w + y)$, so that all Voronoi regions are congruent.

A minimum distance decoding algorithm for the lattice Λ finds the Voronoi region $\mathcal{R}(y)$ that contains the received vector $v \in \mathbb{R}^N$. The Voronoi regions $\mathcal{R}(y)$ are the decision regions for this algorithm. We may create a fundamental region for the lattice Λ by deleting faces from a Voronoi region. Different ways of deleting faces correspond to different rules for resolving ties in a minimum distance decoding algorithm. Figure 2 shows a fundamental region for the integer lattice $\mathbb{Z}^2$ that is obtained by deleting faces from a Voronoi region. The corresponding rule for resolving ties in the first quadrant is that ½-integers are rounded down.

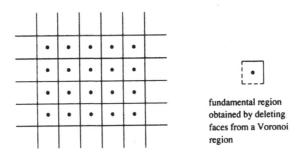

fundamental region
obtained by deleting
faces from a Voronoi
region

Fig. 2. The plane tiled by Voronoi regions of the integer lattice $\mathbb{Z}^2$.

Given a lattice $\Lambda \subseteq \mathbb{R}^N$, there are many ways to choose a fundamental region. But the volume of the fundamental region is uniquely determined by the lattice Λ. This volume is called the *fundamental volume* and we denote it by $V(\Lambda)$. There is a simple formula for the fundamental volume. Let $v_i = (v_{i1}, \ldots, v_{iN}), i = 1, \ldots, m$ be a basis for $V(\Lambda)$, and let

$$A = \begin{bmatrix} v_{11} & \cdots & v_{1N} \\ \vdots & & \vdots \\ v_{m1} & \cdots & v_{mN} \end{bmatrix}.$$

Then the fundamental volume $V(\Lambda)$ is given by

$$V(\Lambda)^2 = \det(AA^T),$$

and $V(\Lambda)^2$ is called the *determinant* or *discriminant* of the lattice. Note that the fundamental volume $V(\Lambda)$ is independent of the basis v_i, $i = 1, ..., m$, since any other basis w_i, $i = 1, ..., m$ is of the form $w_i = Qv_i$, where Q is a unimodular integral matrix. The matrix A is called *a generator matrix* for Λ since every lattice point is an integer linear combination of the rows of A.

Example 2. The lattice E_8 was discovered in the last third of the nineteenth century by the Russian mathematicians A. N. Korkin and E. I. Zolotaroff, and by the English lawyer and amateur mathematician Thorold Gosset;

$$E_8 = \{(z_1, ..., z_8) \mid z_i \in \mathbb{Z}, i = 1, ..., 8 \text{ or } z_i \in \mathbb{Z} + \tfrac{1}{2}, i = 1, ..., 8,$$

$$\text{and } z_1 + \cdots + z_8 \in 2\mathbb{Z}\}.$$

To see that the matrix A given by

$$A = \begin{bmatrix} 2 & 0 & 0 & 0 & 0 & 0 & 0 & 0 \\ -1 & 1 & 0 & 0 & 0 & 0 & 0 & 0 \\ 0 & -1 & 1 & 0 & 0 & 0 & 0 & 0 \\ 0 & 0 & -1 & 1 & 0 & 0 & 0 & 0 \\ 0 & 0 & 0 & -1 & 1 & 0 & 0 & 0 \\ 0 & 0 & 0 & 0 & -1 & 1 & 0 & 0 \\ 0 & 0 & 0 & 0 & 0 & -1 & 1 & 0 \\ \tfrac{1}{2} & \tfrac{1}{2} & \tfrac{1}{2} & \tfrac{1}{2} & \tfrac{1}{2} & \tfrac{1}{2} & \tfrac{1}{2} & \tfrac{1}{2} \end{bmatrix}$$

generates the lattice E_8, let $v_1, ..., v_8$ be the rows of A. We have to prove that any lattice point $z = (z_1, ..., z_8)$ in E_8 is an integral linear combination of these rows. If

$$w = z - 2z_8 v_8 = (w_1, w_2, ..., w_8)$$

then $w_8 = 0$, and

$$w = (w_7 + w_8)v_7 + (w_6 + w_7 + w_8)v_6 + \cdots + (w_2 + \cdots + w_8)v_2 +$$

$$\frac{1}{2}(w_1 + \cdots + w_8)v_1,$$

as required. The fundamental volume $V(E_8)$ is given by

$$V(E_8)^2 = \det(AA^T) = (\det A)^2 = 1,$$

so that $V(E_8) = 1$.

We shall say that a lattice $\Lambda \subseteq \mathbb{R}^N$ is *integral* if the inner product of any two lattice points is an integer. An integral lattice Λ with determinant $\det(\Lambda) = 1$ is called *unimodular*. If Λ is an integral lattice, and if the inner product $(x, x) \in 2\mathbb{Z}$ for all $x \in \Lambda$ then Λ is said to be *even*. It happens that E_8 is the unique even unimodular lattice (see Conway and Sloane [4], for more details).

The problem of designing an N-dimensional signal constellation Ω based on a lattice Λ is that of choosing the region $\mathcal{R}$ that bounds the signal constellation. The fact that a lattice has a fundamental volume means that we can specify the size of the signal constellation by specifying the volume of the region $\mathcal{R}$. The volume $V(\mathcal{R})$ of the region $\mathcal{R}$ is given by

$$V(\mathcal{R}) = |\Omega| \, V(\Lambda) .\tag{1}$$

Definition. Given an N-dimensional signal constellation Ω, the *average signal power P* (normalized per dimension) is given by

$$P = \frac{1}{N} \sum_{y \in \Omega} q(y) \, \|y\|^2 ,$$

where $q(y)$ is the probability of transmitting the signal y.

Points in a lattice Λ are distributed regularly throughout N-dimensional space. The average signal power P of the constellation Ω is approximately the average power $P(\mathcal{R})$ of a probability distribution that is uniform within $\mathcal{R}$ and zero elsewhere. We introduce this approximation principle with an example.

Example 3. Let $\Lambda = \mathbb{Z}$. Then the fundamental volume $V(\Lambda) = 1$ and any interval $[y, y+1)$ is a fundamental region. The signal constellation Ω consists of the $2L$ points $\pm(2i+1)/2$, $i = 0, 1, \dots, L-1$ drawn from the translate $\mathbb{Z} + 1/2$ of the integer lattice. The translate is chosen to minimize average signal power. The region $\mathcal{R}$ is the interval $[-L, L]$. If we assume that signals are equiprobable, then the true average signal power P is given by

$$P = \frac{1}{4L} \sum_{i=0}^{L-1} (2i+1)^2 = \frac{4L^2 - 1}{12} .\tag{2}$$

The approximation $P([L,L])$ is given by

$$P([-L,L]) = \frac{1}{2L} \int_{-L}^{L} t^2 \, dt = L^2/3 ,\tag{3}$$

since the probability distribution takes the value $1/2L$ throughout $[-L,L]$ (the volume of the interval $[-L,L]$ equals $2L$). The agreement between (2) and (3) is quite good.

The Continuous Approximation. Let Ω be an N-dimensional signal constellation consisting of all points from a lattice Λ that lie within a region $\mathcal{R}$, with centroid the origin. If signals are equiprobable, then the average signal power P is approximately the average power $P(\mathcal{R})$ of a continuous distribution that is uniform within $\mathcal{R}$ and zero elsewhere. Thus

$$P \approx P(\mathcal{R}) = \frac{1}{N \, V(\mathcal{R})} \int_{\mathcal{R}} \|x\|^2 \, dv ,\tag{4}$$

where

$$V(\mathfrak{R}) = \int_{\mathfrak{R}} dv$$

is the volume of the region $\mathfrak{R}$.

We rewrite (4) as

$$P \approx G(\mathfrak{R}) \, V(\mathfrak{R})^{2/N} , \qquad (5)$$

where

$$G(\mathfrak{R}) = \frac{\int_{\mathfrak{R}} \|x\|^2 dv}{N \, V(\mathfrak{R})^{1+2/N}} \qquad (6)$$

is the normalized or dimensionless second moment. The second moment $G(\mathfrak{R})$ results from taking the average squared distance from a point in $\mathfrak{R}$ to the centroid, and normalizing to obtain a dimensionless quantity.

We see that the average signal power P depends on the choice of lattice, and on the shape of the region that bounds the signal constellation. The formula $P \approx G(\mathfrak{R}) V(\mathfrak{R})^{2/N}$ separates these two contributions.

Equation (1) gives $V(\mathfrak{R}) = |\Omega| V(\Lambda)$, so that the second factor is determined by the choice of lattice. Since different lattices require different volumes to enclose the same number of signal points, it is possible to save on signal power by choosing the lattice appropriately.

Since the second moment $G(\mathfrak{R})$ is dimensionless, it is not changed by scaling the region $\mathfrak{R}$. Therefore the first factor $G(\mathfrak{R})$ measures the effect of the shape of the region $\mathfrak{R}$ on average signal power.

3. Biasing Gain

Calderbank and Ozarow [2] introduced a shaping method that applies to a signal constellation drawn from an N-dimensional lattice in $\mathbb{R}^N$. As in Section 2, the signal constellation consists of all lattice points within a region $\mathfrak{R}$. This region is partitioned into T subregions so as to obtain T equal subconstellations with increasing average power. A shaping code then specifies sequences of subregions, and it is designed so that subconstellations with lower average power are more frequent.

We start with a basic region $\mathfrak{R}$, and by scaling we obtain a nested sequence $\mathfrak{R} = \alpha_0 \mathfrak{R}$, $\alpha_1 \mathfrak{R}, ..., \alpha_{T-1} \mathfrak{R}$ of copies of $\mathfrak{R}$. This allows us to partition a signal constellation Ω bounded by $\alpha_{T-1} \mathfrak{R}$ into subconstellations $\Omega_0, ..., \Omega_{T-1}$, where $\Omega_0 = \Omega \cap \mathfrak{R}$, and

$$\Omega_i = \Omega \cap (\alpha_i \mathfrak{R} \setminus \alpha_{i-1} \mathfrak{R}) , \qquad (7)$$

for $i = 1, 2, ..., T-1$. Figure 3 shows the decomposition of a 2-dimensional constellation Ω into annular subconstellations. Recall from Section 2 that constellation size is proportional to the volume of the

326

region bounding the constellation, the constant of proportionality being the reciprocal of the fundamental volume of the lattice. Since we require subconstellations of equal size, we need to have

$$V(\alpha_i \mathcal{R}) = (i+1) V(\mathcal{R}) \,, \tag{8}$$

and it follows that

$$\alpha_i = (i+1)^{1/N} \,. \tag{9}$$

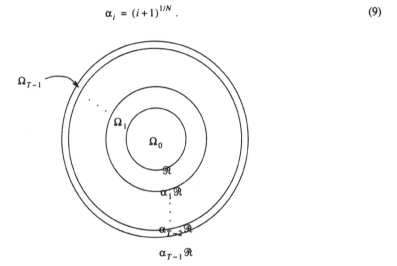

Fig. 3. A 2-dimensional constellation partitioned into annular subconstellations.

Signal points in the same subconstellation are to be equiprobable, and the subconstellation Ω_i is to be used with probability f_i. If P_i is the average signal power of subconstellation Ω_i, then the average signal power P of the non-equiprobable signaling scheme is given by

$$P = \sum_{i=0}^{T-1} f_i P_i \,. \tag{10}$$

Lemma 1. The average signal power P is given by

$$P = \sum_{i=0}^{T-1} \left((i+1)^{2/N+1} - i^{2/N+1} \right) f_i P_0 \,, \tag{11}$$

where P_0 is the average signal power of subconstellation Ω_0.

Proof. In Section 2 we argued that the average signal power of a constellation drawn from a lattice is well approximated by the average power of a probability distribution that is uniform within the region bounding the constellation, and zero elsewhere. This continuous approximation $(P \approx G(\mathcal{R}) V(\mathcal{R})^{2/N})$ predicts that scaling a basic region $\mathcal{R}$ so that the volume is increased by a factor l will increase average signal power by a factor $l^{2/N}$. If $P_i = t_i P_0$, then

$$\frac{1}{(i+1)} \sum_{j=0}^{i} t_j P_0 = (i+1)^{2/N} P_0 . \tag{12}$$

Thus $t_1 = 2^{2/N+1} - 1^{2/N+1}$, and for $i \geq 2$,

$$
\begin{aligned}
t_i &= (i+1)^{2/N+1} - \sum_{j=0}^{i-1} t_j \\
&= (i+1)^{2/N+1} - t_{i-1} - \sum_{j=0}^{i-2} t_j \\
&= (i+1)^{2/N+1} - \left(i^{2/N+1} - \sum_{j=0}^{i-2} t_j \right) - \sum_{j=0}^{i-2} t_j \\
&= (i+1)^{2/N+1} - i^{2/N+1} .
\end{aligned}
$$

The required result now follows directly from (10). ∎

If we use subconstellations $\Omega_0, ..., \Omega_{T-1}$ with probabilities $f_0, ..., f_{T-1}$ respectively, then it is possible to transmit up to $H(f_0, ..., f_{T-1})$ bits of information, where

$$H(f_0, ..., f_{T-1}) = - \sum_{i=0}^{T-1} f_i \ln f_i \tag{13}$$

is the binary entropy function (and ln denotes logarithm to base 2). To transmit this information by equiprobable signaling requires a signal constellation of size $2^{H(f_0, ..., f_{T-1})} |\Omega_0|$. If this signal constellation were drawn from the same underlying lattice, and if it were bounded by a scaled version of the region $\mathcal{R}$, then to the accuracy of the continuous approximation, the average signal power P' would be given by

$$P' = 2^{2H(f_0, ..., f_{T-1})/N} P_0 . \tag{14}$$

Definition. The *biasing gain* $\gamma_N(f_0, ..., f_{T-1})$ is the ratio of P' to P and is given by

$$\gamma_N(f_0, ..., f_{T-1}) = \frac{2^{2H(f_0, ..., f_{T-1})/N}}{\sum\limits_{i=0}^{T-1} \left((i+1)^{2/N+1} - i^{2/N+1} \right) f_i} . \tag{15}$$

It is the potential saving in average signal power that can result from non-equiprobable signaling with probabilities $f_0, ..., f_{T-1}$. It assumes that the constellations for equiprobable and non-equiprobable signaling are drawn from a common underlying lattice, and are bounded by regions with the same shape.

Note that with equiprobable signaling $f_0 = f_1 = \cdots = f_{T-1} = 1/T$ and $\gamma_N(1/T, ..., 1/T) = 1$.

Definition. A shaping code C_s is the set of allowed subregion (or subconstellation) sequences. If the subconstellations $\Omega_0, \Omega_1, ..., \Omega_{T-1}$ appear in C_s with probabilities $f_0, ..., f_{T-1}$, then the biasing gain $\gamma_N(C_s)$ is given by

$$\gamma_N(C_s) = \frac{2^{2R(C_s)/N}}{\sum\limits_{i=0}^{T-1}\left((i+1)^{2N+1}-i^{2/N+1}\right)f_i} , \tag{16}$$

where $R(C_s)$ is the information rate of the code C_s. Since the rate $R(C_s)$ is at most $H(f_0, ..., f_{T-1})$, the biasing gain $\gamma(C_s)$ is at most $\gamma_N(f_0, ..., f_{T-1})$. By increasing the complexity of C_s the biasing gain $\gamma_N(C_s)$ can be made to approach $\gamma_N(f_0, ..., f_{T-1})$.

The shaping method introduced by Calderbank and Ozarow makes use of an N-dimensional constellation with more signal points than are required by equiprobable signaling. Since this redundancy is used to drive down the average transmitted signal power, it follows that the ratio of peak to average power (PAR) must increase. However it is of considerable practical importance to limit this increase since signaling schemes are quite sensitive to non-linear perturbations. Thus our objective is to achieve significant biasing gain while minimizing the increase in the ratio of peak to average power. Since we are particularly interested in QAM (quadrature amplitude modulation) modems, we shall focus on 2-dimensional signaling.

Given an N-dimensional signal constellation Ω partitioned into T annular subconstellations $\Omega_0, ..., \Omega_{T-1}$, let C_s be a shaping code that uses subconstellation Ω_i with probability f_i. Then non-equiprobable transmission requires a signal constellation of size $|\Omega_0| T$, and equiprobable transmission of the same information requires a signal constellation of size $|\Omega_0|2^{R(C_s)}$, where $R(C_s)$ is the information rate of the shaping code C_s.

Definition. Given a shaping code C_s, the constellation expansion ratio $CER(C_s)$ is the factor by which the size of the signal constellation is magnified through non-equiprobable signaling using C_s. Thus

$$CER(C_s) = \frac{T}{2^{R(C_s)}} . \tag{17}$$

Let $CER(f_0, ..., f_{T-1})$ denote the minimum constellation expansion ratio that can result from non-equiprobable signaling with probabilities $f_0, ..., f_{T-1}$. Then

$$CER(f_0, ..., f_{T-1}) = \frac{T}{2^{H(f_0, ..., f_{T-1})}} , \tag{18}$$

since the minimum CER occurs with the maximum rate, namely $H(f_0, ..., f_{T-1})$.

The constellation expansion ratio $CER(C_s)$ as defined above is independent of the dimension N. However if the 2-dimensional projections of the constellation Ω are identical, then the magnification factor $CER_2(C_s)$ for the size of the constituent 2-dimensional constellation is given by

$$CER_2(C_s) = CER(C_s)^{2/N} . \tag{19}$$

Consider now the factor $PAR_N(C_s)$ by which the ratio of peak to average power is magnified through non-equiprobable signaling using the shaping code C_s. Let P_{max} be the peak power of the subconstellation Ω_0 bounded by the region $\mathcal{R}$. The constellation Ω is bounded by $T^{1/N}\mathcal{R}$ and so the peak power is $T^{2/N}P_{max}$. On the other hand, Lemma 1 gives the average signal power for non-equiprobable transmission to be $\sum_{i=0}^{T-1}((i+1)^{2/N+1}-i^{2/N+1})f_iP_0$, where P_0 is the average power of the points in Ω_0. This means that the magnification factor $PAR_N(C_s)$ is given by

$$PAR_N(C_s) = \frac{T^{2/N}}{\sum_{i=0}^{T-1}((i+1)^{2/N+1}-i^{2/N+1})f_i}$$

$$= \left(\frac{T}{2^{R(C_s)}}\right)^{2/N} \times \left(\frac{2^{R(C_s)}}{\sum_{i=0}^{T-1}((i+1)^{2/N+1}-i^{2/N+1})f_i}\right)$$

$$= CER_2(C_s) \times \gamma_N(C_s) . \tag{20}$$

The magnification factor $PAR_N(C_s)$ depends only on the probabilities f_i and not on the rate $R(C_s)$. We shall write $PAR_N(f_0, ..., f_{T-1})$ for the magnification factor of any shaping code with probabilities f_i.

4. Multidimensional Cross Constellations

Multidimensional cross constellations were introduced by Wei [9] and then generalized by Forney and Wei [7]. First we consider the original construction of Wei. These 2^{m+1}-dimensional constellations allow transmission of $\beta = n + 1/2^m$ bits per 2-dimensional symbol. We divide a 2-dimensional signal constellation Ω into an inner constellation Ω_0 and an outer constellation Ω_1, where $|\Omega_0| = 2^n = 2^m|\Omega_1|$. We shall suppose this division is accomplished by scaling a basic region. The multidimensional cross constellation consists of all sequences $(z_1, ..., z_{2^m})$ of 2-dimensional signal points, such that at most one signal point is in Ω_1.

Since

$$\left|\{(z_1, ..., z_{2^m}) \mid z_i \in \Omega_0\}\right| = 2^{2^m n} , \quad \text{and}$$

$$\left|\{(z_1, ..., z_{2^m}) \mid z_j \in \Omega_1 \text{ and } z_i \in \Omega_0 \text{ for } i \neq j\}\right| = 2^m \times (2^n/2^m) \times (2^n)^{2^m-1} = 2^{2^m n+1} ,$$

the total number of signal points in $2^{2^m n+1}$ as required. The average transmitted signal power P is given by

$$P = f_0 P_0 + f_1 P_1 ,$$

where f_i is the frequency of subconstellation Ω_i and P_i is the average power of the signal points in Ω_i.

Direct calculation gives

$$f_1 = \frac{1}{2} \times \frac{1}{2^m} = \frac{1}{2^{m+1}} , \text{ and } f_0 = \frac{2^{m+1}-1}{2^{m+1}} ,$$

since the probability that some element of a block of 2^m symbols belongs to Ω_1 is 1/2. The continuous approximation predicts

$$\frac{1}{(2^m+1)|\Omega_0|/2^m} \left(|\Omega_0| P_0 + (|\Omega_0|/2^m) P_1 \right) = (2^m+1) P_0/2^m ,$$

so that $P_1 = (2^{m+1}+1) P_0/2^m$. Hence

$$P = \left(\frac{2^{m+1}-1}{2^{m+1}} \right) P_0 + \left(\frac{1}{2^{m+1}} \right) \times \left(\frac{2^{m+1}+1}{2^m} \right) P_0$$

$$= \left(\frac{2^{2m+1}+2^m+1}{2^{2m+1}} \right) P_0 .$$

Hence the biasing gain γ is given by

$$\gamma = 2^{1/2^m} \times \left(\frac{2^{2m+1}}{2^{2m+1}+2^m+1} \right) = \frac{2^{2m+1+1/2^m}}{2^{2m+1}+2^m+1} .$$

If we assume that the basic region is circular then the ratio PAR of peak to average power is given by

$$PAR = PAR(B_2) \times (1+1/2^m) \times \left(\frac{2^{2m+1}}{2^{2m+1}+2^m+1} \right)$$

$$= \frac{4(2^m+1)2^m}{2^{2m+1}+2^m+1} .$$

Evaluating these expressions at $m = 2$, 3, and 4 gives

	$m=2$	$m=3$	$m=4$
biasing gain	0.122	0.081	0.046
PAR	2.162	2.102	2.057

We conclude by making a few remarks about the generalization of multidimensional cross constellations by Forney and Wei. These constellations allow transmission of any number $\beta = n + d/2^m$ bits per 2-dimensional symbol, where n is an integer that is "large enough" and d is any integer in the range $0 \le d < 2^m$. The aim of the construction is an N-dimensional constellation with $2^{n(N/2)+d}$ points, where $N = 2^{m+1}$. The complexity depends on the number of non-zero entries in the binary expansion of d; it is easier to send 1/2 bit than it is to send 7/8 bit.

5. Binary Covering Codes

The non-equiprobable signaling scheme introduced by Calderbank and Ozarow partitions a low-dimensional signal constellation Ω into T annular subconstellations $\Omega_0, ..., \Omega_{T-1}$ with increasing average power. Signal points in the same subconstellation are used equiprobably, and a shaping code selects the subconstellation Ω_i with frequency f_i. In this direct approach to non-equiprobable signaling it is the frequencies f_i that are the fundamental quantities. If the shaping code is a block code then the result is an equiprobable signaling scheme in high dimensional space, constructed in a novel and useful way. But there is no reason to restrict shaping codes to be block codes.

We shall see that biasing gains produced by this method are superior to those produced by multidimensional cross constellations.

We shall also compare this method with equiprobable signaling schemes based on Voronoi regions of certain multidimensional lattices (presented in [6]). For comparable shape gain and constellation expansion ratio, the ratio of peak to average power of the non-equiprobable signaling scheme will turn out to be superior.

Our signal constellations may be viewed as the Cartesian product of a basic region with a shaping code. The complexity of addressing points is closer to that of addressing the N-cube (which is a particularly simple Cartesian product) than that of addressing the N-sphere or an N-dimensional Voronoi region.

Binary Shaping Codes. A natural way of constructing binary shaping codes is to use the greedy algorithm, which adds binary vectors in order of increasing Hamming weight. Table 4.1 lists the parameters of *greedy shaping codes* $\mathcal{G}_{v,k}$, where

$$\mathcal{G}_{v,k} = \{c \in \mathbb{F}_2^v \mid wt(c) \leq k\} .$$

We have included in Table 1 the total shape gain $\gamma = \gamma_2(\mathcal{G}_{v,k}) + \gamma(B_2)$, and the actual peak to average power ratio $PAR = PAR_2(\mathcal{G}_{v,k}) \times 2$, in the case when the basic 2-dimensional region is a circle. In Table 1 (taken from Calderbank and Ozarow [2]) we use $R_{v,k}$ to denote the rate of the code $\mathcal{G}_{v,k}$.

ν	k	f_0	$R_{\nu,k}$	$\gamma_2(\mathcal{G}_{\nu,k})$	$CER_2(\mathcal{G}_{\nu,k})$	$PAR_2(\mathcal{G}_{\nu,k})$	γ	PAR
6	2	.73	.74	.35	1.19	1.29	.55	2.58
7	2	.76	.69	.38	1.23	1.34	.58	2.68
8	2	.78	.65	.40	1.27	1.40	.60	2.80
9	2	.80	.61	.41	1.31	1.44	.61	2.88
10	2	.82	.58	.42	1.34	1.47	.62	2.94
11	3	.75	.71	.44	1.21	1.34	.64	2.68
12	3	.78	.69	.45	1.24	1.38	.65	2.76
13	3	.79	.66	.47	1.27	1.41	.67	2.82
14	3	.80	.63	.47	1.29	1.44	.67	2.88
15	3	.82	.61	.48	1.31	1.46	.68	2.92
16	4	.77	.71	.49	1.23	1.37	.69	2.74
17	4	.78	.69	.50	1.24	1.39	.70	2.78
18	4	.79	.67	.51	1.26	1.41	.71	2.82
19	4	.80	.65	.51	1.28	1.43	.71	2.86
20	5	.77	.72	.51	1.21	1.37	.71	2.74

Table 1. The parameters of greedy binary shaping codes.

Example 4. Consider the shaping code $\mathcal{G}_{12,3}$. There are 299 codewords so the rate $R_{12,3} = \log_2(299)/12 = 0.68$ (in practice we might create a rate 2/3 code by deleting codewords). The probability f_0 is given by

$$f_0 = \frac{\sum\limits_{j=0}^{3} \binom{12}{j}(12-j)}{12 \sum\limits_{j=0}^{3} \binom{12}{j}} = 0.78 \ .$$

The gain over equiprobable signaling with a spherical constellation is given by

$$\frac{2^{R_{12,3}}}{f_0 + 3(1-f_0)} = 0.45 \text{ dB} \ .$$

After adding the shape gain of the circle over the square, we end up with a total shape gain of 0.65 dB. This matches the shape gain afforded by equiprobable signaling with a constellation bounded by the Voronoi region of the Gosset lattice E_8 (see Conway and Sloane [4]).

Shaping through non-equiprobable signaling expands a 2-dimensional signal constellation by a factor $2/2^{0.08} = 2^{0.32} = 1.25$ with respect to equiprobable signaling. Recall that the Gosset lattice E_8 is given by

$$E_8 = \{(z_1, ..., z_8) \mid z_i \in \mathbb{Z}, \ i = 1, ..., 8 \ \text{ or } \ z_i \in \mathbb{Z} + 1/2, \ i = 1, ..., 8, \ \text{and} $$

$$z_1 + \cdots + z_8 \in 2\mathbb{Z}\} \ .$$

If $\mathcal{R}$ is the Voronoi region of E_8 then the fundamental volume $V(\mathcal{R}) = 1$. Since E_8 is fixed by coordinate permutations, all 2-dimensional projections of the Voronoi region $\mathcal{R}$ are identical. Let $\mathcal{R}_2$ denote this common (constituent) 2-dimensional projection. To see that $\mathcal{R}_2$ is the square shown in Fig. 4, simply use the fact that the lattice E_8 contains vectors $(\pm 2,0,0, ..., 0)$, $(0, \pm 2,0, ..., 0)$, and $(\pm 1,\pm 1,0, ..., 0)$. The volume $V(\mathcal{R}_2) = 2$. To construct a signal constellation Ω of size 2^{8n} drawn from the integer lattice $\mathbb{Z}^8$ we scale the Voronoi region $\mathcal{R}$ by a factor 2^n, and take

$$\Omega = \{x \in \mathbb{Z}^8 \mid x \in 2^n \mathcal{R}\} .$$

The 2-dimensional projection $\Omega_{[2]}$ of Ω is given by

$$\Omega_{[2]} = \{x \in \mathbb{Z}^2 \mid x \in 2^n \mathcal{R}_2\} .$$

To the accuracy of the continuous approximation, the number of signal points in $\Omega_{[2]}$ is the volume $V(2^n \mathcal{R}_2)$ of the region $2^n \mathcal{R}_2$ which is given by $V(2^n \mathcal{R}_2) = 2.2^{2n}$. Thus, shaping through equiprobable signaling using a constellation (drawn from the integer lattice $\mathbb{Z}^8$) bounded by the Voronoi region of the Gosset lattice E_8 requires a constellation expansion ratio of 2.

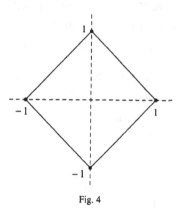

Fig. 4

Returning to shaping through non-equiprobable signaling, the ratio of peak to average power is given by

$$PAR(B_2) \times \frac{2}{f_0 + (1-f_0)3} = 2.76 ,$$

assuming that the basic region is a circle. For the Voronoi region of E_8, the ratio of peak to average power is given by

$$PAR_2(\mathcal{R}) = \frac{12960}{1858} = 6.97 .$$

By realizing E_8 as a ternary lattice (instead of a binary lattice) it is possible to reduce the margin of victory; the CER becomes 1.73 and the PAR becomes 4.65.

Shaping through Binary Covering Codes. For small v and k it is not too difficult to map data bits to codewords in the shaping code $\mathscr{G}_{v,k}$. For larger v and k, we might replace the set $\mathscr{G}_{v,k}$ by a set of coset representatives for a binary code C. For example, the set $\mathscr{G}_{7,3}$ is the set of coset leaders for the repetition code of length 7. In this case C is a perfect binary covering code; the covering radius is 3, the spheres of radius 3 about the codewords are disjoint, and together these spheres exhaust 7-dimensional binary space. We do not require that the code C be perfect, only that it be good; if

$$|C|\left(\sum_{i=0}^{r}\binom{v}{i}\right)\sim 2^{v}$$

then we need $D(x,C)\le r$ for almost all vectors $x\in \mathbb{F}_2^{v}$. Since the constellation expansion ratio should be as small as possible, the information rate of the covering code C should be low. The complexity of mapping data bits to codewords is then determined by the complexity of decoding the covering code. This is because the mapping is done by generating vectors that are distinct modulo the covering code C, and then decoding these vectors to coset leaders. This method is the same as that described by Conway and Sloane for addressing points in a Voronoi constellation (see [3]). When a coset contains more than one vector with low Hamming weight it is possible to introduce an opportunistic secondary channel.

Example 5. Here the binary covering code C is the [16,5] Reed-Muller code R_4. The rows of Table 2 (which is taken from Sloane and Dick [8]) give the weight distribution of the cosets of R_4. We choose one coset leader from each coset to obtain a shaping

Number of Cosets	Weights																
	0	1	2	3	4	5	6	7	8	9	10	11	12	13	14	15	16
1	1								30								1
16		1						15		15						1	
320			1				7		16		7				1		
560				1		3		12		12		3		1			
840					2		8		12		8		2				
35					4				24				4				
448						6		10		10		6					
28							16				16						

Table 2. Cosets of the Reed-Muller code R_4.

code C_s with rate 11/16. The probability f_0 of the symbol is 0 is

$$f_0 = \frac{1\times 16 + 16\times 15 + 320\times 14 + 560\times 13 + (840+35)\times 12 + 448\times 11 + 28\times 10}{2048\times 16}$$

$$= \frac{6231}{8192} = 0.76062\ldots,$$

and the biasing gain $\gamma(C_s)$ is given by

$$\gamma(C_s) = \frac{2^{R(C_s)}}{3-2f_0} = 0.3706\ldots \text{ dB}.$$

The constellation expansion ratio is approximately 1.24, and the ratio of peak to average power is approximately 2.70, assuming that the basic region is a circle.

Different coset leaders representing the same coset can be used to support an opportunistic secondary channel. When the encoder selects a coset $R_4 + z$ with multiple coset leaders, we have the opportunity to transmit an extra $t(z)$ bits by our choice of coset leader. Table 3 gives the weight distribution of this opportunistic shaping code C'_s.

Number of Cosets	Weight of Coset Leader	Number of Coset Leaders in C_s
1	0	1
16	1	16
320	2	320
560	3	560
840	4	840×2
35	4	35×4
448	5	448×4
28	6	28×16

Table 3. The weight distribution of an opportunistic shaping code.

The rate of this shaping code is

$$R(C'_s) = 11/16 + \frac{840}{2048} \times \frac{1}{16} + \frac{35}{2048} \times \frac{1}{8} + \frac{448}{2048} \times \frac{1}{8} + \frac{28}{2048} \times \frac{1}{4} = 0.7460...$$

of which approximately 5% is probabilistic. The frequency f_0 is unchanged, so the biasing gain $\gamma(C'_s)$ is given by

$$\gamma(C'_s) + \frac{2^{R(C'_s)}}{3 - 2(6231/8192)} = 0.5468... \text{ dB}.$$

Berlekamp and Welch [1] used the symmetry group of the Reed-Muller code R_5 to partition the 2^{26} cosets into 48 equivalence classes. They then used a simple computer program to enumerate the weights of the 64 vectors in each of the 48 cosets. The biasing gain of the corresponding shaping code is 0.439 dB, the constellation expansion ratio is 1.139, and the ratio of peak to average power is 2.520. This coset enumeration is a computational tour de force and we refer the reader to the original paper for details.

References

[1] E. R. Berlekamp and L. R. Welch, Weight distributions of the cosets of the [32,6] Reed-Muller code, IEEE Trans. Inform. Theory, vol. IT-18, 1972, 203-207.

[2] A. R. Calderbank and L. H. Ozarow, Non-equiprobable signaling on the Gaussian channel, IEEE Trans. Inform. Theory, vol. IT-36, 1990, 726-740.

[3] J. H. Conway and N. J. A. Sloane, "A fast encoding method for lattice codes and quantizers," *IEEE Trans. Inform. Theory*, vol. IT-29, 1983, 820-824.

[4] J. H. Conway and N. J. A. Sloane, *Sphere packings, Lattices, and Groups,* Springer-Verlag, New York, 1988.

[5] G. D. Forney, Jr., Coset codes – Part I: Introduction and geometrical classification, *IEEE Trans. Inform. Theory*, vol. IT-34, 1988, 1123-1151.

[6] G. D. Forney, Jr., "Multidimensional constellations – Part II: Voronoi constellations," *IEEE J. Select. Areas Commun.*, vol. SAC-7, 1989, 941-958.

[7] G. D. Forney, Jr. and L. F. Wei, Multidimensional signal constellations – Part I: Introduction, figures of merit, and generalized cross constellations, *IEEE J. Select Areas Commun.*, vol. SAC-7, 1989, 877-892.

[8] N. J. A. Sloane and R. J. Dick, On the enumeration of cosets of first order Reed-Muller codes, Proceedings IEEE International Conference on Communications, Montreal, June 1971, Vol. 7, p. 36-2 to 36-6.

[9] L. F. Wei, Trellis coded modulation with multidimensional constellations, *IEEE Trans. Inform. Theory*, vol. IT-33, 1987, 483-501.

CODING AND MODULATION FOR THE GAUSSIAN CHANNEL, IN THE ABSENCE OR IN THE PRESENCE OF FLUCTUATIONS

CODAGE ET MODULATION POUR LE CANAL GAUSSIEN, SANS OU AVEC FLUCTUATIONS

Gérard Battail, Hélio Magalhães de Oliveira and Zhang Weidong

Ecole Nationale Supérieure des Télécommunications,
Département Communications (also URA 820 of C.N.R.S.)
46 rue Barrault, 75634 PARIS CEDEX 13, France

Abstract

Looking for systems which combine coding and multilevel modulation whose Euclidean distance distribution is close to that which results in the average from random coding, we consider the combination of an MDS code over a large-size alphabet and a one-to-one mapping of the alphabet into a symmetric constellation e.g., phase modulation. Its performance in the presence of additive Gaussian noise can be predicted from that of random coding, provided the signal-to-noise ratio is small enough. The results exhibit the sphere hardening phenomenon whether or not amplitude fluctuations are present. Weighted demodulator output and soft decoding should be effected in order to achieve this performance. Such decoding can be done in principle according to previous works by Fang and Battail. A prohibitive complexity can be avoided only at the expense of strict optimality.

INTRODUCTION AND SUMMARY OF RESULTS

We were recently led to question the relevance of the minimum distance criterion when applied to long block codes i.e., where many errors are likely to occur in each codeword [1]. As a better criterion, we proposed a proximity measure of the code distance distribution with respect to that which results in the average from random coding. The criticism of the minimum distance criterion, as well as the proposed one, were formulated in the case of the Hamming distance, but they are also valid for the Euclidean metric.

Systems which combine coding and multilevel modulation e.g., trellis-coded modulations to be used over the narrow-band Gaussian channel [2], were designed for the Euclidean metric which is natural for this channel but they used the minimum distance criterion. We found it interesting to consider coded-modulation systems whose Euclidean distance distribution is good according to our own criterion.

Looking for such a system, we first notice that the distance distribution of the maximum-distance-separable codes (MDS), a class which especially includes the Reed-Solomon codes, is close to that which results in the average from random coding if the alphabet size and the code dimension are large [3]. Therefore, these codes are good according to our criterion, as far as the Hamming distance is concerned. In order to define a coded-modulation system using these codes whose Euclidean distance distribution meets our criterion, we must define the mapping of the code symbols into a constellation.

Let us consider a system using an MDS code over a large-size alphabet (whose cardinality is denoted by q), associated with a symmetric constellation of same size. We are specially interested in the case of phase modulation because of its robustness against non-linearities and amplitude fluctuations. Let first the mapping of the code symbols into this constellation be random and vary according to the position in the word.

The distance distribution of MDS codes and the random character of the mapping of the code symbols into the constellation result in a distribution of the Euclidean distances close to that obtained using random coding and nonrandom mapping, provided the alphabet size and the code dimension are large enough. Furthermore, the Euclidean distance distribution is not actually modified if the mapping from the code alphabet to the constellation is nonrandom and does not vary with the symbol position. Performance close to that of random coding may thus be expected from this fully deterministic scheme (we already arrived at a similar conclusion via a different approach [4], but the coded-modulation system here is completely explicit). Decoding is much less complex, however, than if random coding were used. Moreover, in the case of phase modulation, this scheme easily achieves rotational invariance in order to cope with an arbitrary phase reference, unlike trellis-coded modulation.

The performance thus obtained can be predicted from that of random coding, using an extension of Shannon's results [5], provided that the signal-to-noise ratio is small enough (in the range where coding is actually useful). The availability of tables is very helpful for literal computations and, of course, numerical results are easily obtained using digital computers.

These results, as expected, exhibit the "sphere hardening" phenomenon : for a large enough signal-to-noise ratio, the word error probability is extremely small ; if the SNR decreases, it increases suddenly and approaches 1. This transition is the steeper, the larger the number of dimensions. It occurs at an SNR close to $2^{2C} - 1$, where C is the channel capacity in binary units. This behaviour is observed whether or not amplitude fluctuations are present.

Achieving such performance actually demands that maximum likelihood decoding be employed. In other words, the point representing a codeword to be chosen is the closest to the one which represents the received signal (according to the Euclidean distance). Weighted demodulator output and soft decoding are thus necessary. The algorithm of Fang and Battail enables optimum soft decoding of nonbinary linear codes e.g., Reed-Solomon ones [6-8]. However, a prohibitive complexity can be avoided only by limiting the maximum number of candidate words tried for decoding each received word, at the expense of strict optimality. Another problem which needs further work is quantization in the received signal space.

OUTLINE OF A COMMUNICATION SYSTEM USING MDS CODES

In general, MDS codes need not be linear. However, the most important members of the family i.e., Reed-Solomon (RS) and generalized RS codes, are linear, so that we may restrict ourselves to linear codes. Then, the distance distribution reduces to the weight distribution.

The weight distribution of an MDS code over the field of q elements can be found in many textbooks and is recalled in [3]. Let n be the code length and k its dimension, then the number A_j of words of Hamming weight j is :

$$A_0 = 1,$$

$$A_i = 0, \quad 0 < i < n-k+1,$$

$$A_j = \binom{n}{j} \sum_{i=0}^{j-1-(n-k)} (-1)^i \binom{j}{i}(q^{j-i-(n-k)} - 1), \quad n-k+1 \leq j \leq n. \tag{1}$$

Interestingly, for given values of n and k, this weight distribution is uniquely determined by the fact the code is MDS.

Equality (1) can be rewritten as

$$A_j = q^{-(n-k)}\binom{n}{j}(q-1)^j \tag{2}$$

$$- \binom{n}{j}[\sum_{i=j-(n-k)}^{j} (-1)^i \binom{j}{i}q^{j-i-(n-k)} + \sum_{i=0}^{j-1-(n-k)} (-1)^i \binom{j}{i}] .$$

The first term of the right hand side is the average weight which results from the random choice of q^k q-ary n-tuples. We shall assume in the remainder of this paper that the second term is negligible with respect to the first one at least for the largest values of A_j, which occurs if q and k are large enough. The Hamming distance distribution is then close to that which results from random coding, so the MDS code is not only good according to the conventional criterion (since it achieves the largest possible minimum distance $d = n - k + 1$) but also according to the one which we proposed [1].

Notice that the choice of an MDS code limits the word length to a maximum equal to $q + 1$, except if q is even and k or $n - k = 3$, in which case the maximum length equals $q + 2$, if we accept the MDS conjecture [9]. In any case, the maximum code length is close to the alphabet size. Therefore, the use of a long code also means that of a large alphabet which, according to (2), results in a distance distribution close to that of random coding for reasonable values of k.

Let us assume that the q-point constellation to be used is symmetric in the sense that the Euclidean distance distribution between some of its points and the other ones does not depend on the point considered (e.g., phase modulation i.e., in 2 dimensions, the set of points on a circle at angles $2i\pi/q$ with respect to the x axis, i an integer such that $0 \leq i \leq q-1$). In order to define the mapping from the code alphabet into the points of this constellation, let us first assume that symbol 0 is mapped into the point labelled 0 in the constellation, while the other symbols are one-to-one mapped into the other points at random, and differently at each position in the word.

Let us define the Euclidean weight as the squared Euclidean norm, an additive quantity. Consider the Euclidean weight distribution of the combination of MDS coding with the mapping just defined. The closeness of the Hamming weight distribution (2) to the average random weight results in a number of zero Euclidean weights close to that which would have been obtained in the average by random coding and any deterministic mapping. The use of a random and varying mapping for the nonzero symbols further results in a Euclidean weight distribution close to that which would result in the average from random coding where each point of the constellation is chosen at random, independently from the previous choices, at each of the n symbol positions in the word and where q^k words are so chosen, independently from each other, as the codebook.

If now we assume that the one-to-one mapping of the code alphabet into the constellation is non-random and does not vary with the symbol position, the exact computation of the Euclidean distance distribution appears as a very difficult problem but statistical simulation can be used to estimate it. For instance, we considered Reed-Solomon codes over GF(16), with $n = 15$. A large number of information vectors were obtained by random choice of each of their k components with uniform probability over GF(16). Each of these vectors was encoded according to the code already defined. We assumed either a random and variable mapping of the code alphabet into the 16-point phase modulation constellation, either an arbitrary mapping common to all positions, or the "natural" mapping of the symbols 0, 1, α, α^2, $\cdots$ α^{14} into the angles 0, $\pi/8$, $2\pi/8$, $\cdots$ $15\pi/8$, respectively (α denotes a primitive element of GF(16)). Whatever the mapping chosen, it remained the same for all words. The Euclidean weight corresponding to each word was computed and a histogram of the weights was drawn. A sample of results thus obtained is given in Figure 1. Although q and k are only moderately large, the Euclidean weight distribution thus obtained appears very close to that of random coding.

Figure 1. Squared Euclidean distance distribution of the combination of the (15,8) Reed-Solomon code over GF(16) with 16-phase modulation.

The mapping of the alphabet symbols into the 16 phases was : (1) random and varying with the symbol location ; (2) random but the same for all symbols ; (3) deterministic. For comparison purpose, the squared distance distribution of randomly chosen 15-uples was also drawn (4). The curves show a histogram where 100,000 words were classified into 100 equally spaced intervals of the full range of possible Euclidean distances. The points are too close to each other to be separated on the graph.

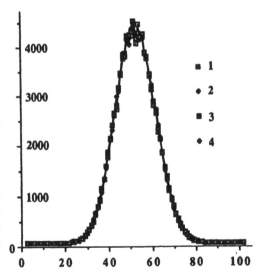

These results also show that almost no chànge in the Euclidean distance distribution results if the one-to-one mapping of the code alphabet into the constellation is nonrandom and does not vary with the symbol position. Besides the evidence provided by simulation, we believe this statement is true, although we are unable to exhibit a formal proof of it. In brief, this would result from the many symmetries of the MDS codes, which mimic the average regularity of randomness.

PERFORMANCE PREDICTION USING RANDOM CODING

We defined in the previous section a nonrandom coded-modulation system where the Euclidean distance distribution is close to that of random coding i.e., which constitutes some discrete approximation of the random coding without any restriction to a finite alphabet, as studied by Shannon [5]. If the average length of the noise vector is large with respect to the average distance between the points of the constellation i.e., if the signal-to-noise ratio is small enough, the discrete location of the signal points may be ignored and Shannon results provide an estimate of the performance the system can achieve.

Average error probability

Consider random coding where M points are chosen at random in an n-dimensional Euclidean signal space, with uniform density inside the hypersphere corresponding to a limited power, in the presence of additive Gaussian noise. Its exact average error probability was computed by Shannon [5], resulting in :

$$P_e = 1 + \int_0^\pi [1 - \omega_n(\theta)]^{M-1} Q_n'(\theta) \, d\theta , \tag{3}$$

where :

$Q_n(\theta)$ is defined as the probability that the point which represents the received signal is moved by the noise outside a cone with vertex at the origin, whose axis is colinear with the transmitted vector and whose half-angle is denoted by θ. We notice that the derivative of $Q_n(\theta)$, denoted by $Q_n'(\theta)$, is negative ;

$\omega_n(\theta) = \Omega_n(\theta)/\Omega_n(\pi)$, where $\Omega_n(\theta)$ is the solid angle of this cone i.e.,

$$\Omega_n(\theta) = C(n) \int_0^\theta \sin^{n-2}x \, dx, \tag{4}$$

where $C(n)$ does not depend on θ.

We rewrite (3) as

$$P_e = 1 - \int_0^\pi F_n(R;\theta) \, G_n(\rho;\theta) \, d\theta \tag{5}$$

where

$$F_n(R;\theta) \overset{\Delta}{=} [1 - \omega_n(\theta)]^{M-1} , \qquad (6)$$

where $R \overset{\Delta}{=} \dfrac{1}{n} \ln(M)$ is the information rate per dimension (in nats), and

$$G_n(\rho;\theta) \overset{\Delta}{=} -Q'_n(\theta) \qquad (7)$$

where $\rho = S/N$ is the channel signal-to-noise ratio. Both factors depend on the number of dimensions n. The first one does not depend on the noise but on the number M of codewords. According to the above definition, it can be expressed in terms of the rate R, which therefore was written in argument of F_n in (6). The second factor does not depend on the code but on the channel noise. We now examine each of these two factors.

First factor of the integrand in (5)

The integral in (4) can explicitly be found in [10], which results in :

$$\omega_n(\theta) = \frac{1}{\pi} \left[\theta + 2 (-1)^{n/2-1} \sum_{i=0}^{n/2-2} (-1)^i \frac{C_{n-2}^i}{C_{n-2}^{n/2-1}} \frac{\sin[(n-2i-2)\theta]}{n-2i-2} \right], \quad n \text{ even}, \qquad (8a)$$

$$\omega_n(\theta) = \frac{1}{2} \left[1 - \frac{\displaystyle\sum_{i=0}^{(n-3)/2} (-1)^i C_{n-2}^i \frac{\cos[(n-2i-2)\theta]}{n-2i-2}}{\displaystyle\sum_{i=0}^{(n-3)/2} (-1)^i C_{n-2}^i \frac{1}{n-2i-2}} \right], \quad n \text{ odd}. \qquad (8b)$$

These expressions are exact. If we look for an approximation of $F_n(R;\theta)$, we may write $\theta = \sin^{-1}(x)$ and $\omega_n(\sin^{-1}(x)) = a_{n-1} x^{n-1} + \varepsilon x^{n-1}$, where ε approaches 0 as θ vanishes. The first nonzero term $a_{n-1} x^{n-1}$ is actually the dominant term of a series which converges provided $x < 1$ i.e., for any $\theta < \pi/2$.

If we furthermore neglect 1 with respect to M, we may write :

$$F_n(R;\theta) \approx [1 - a_{n-1} \sin^{n-1}\theta]^{\exp(nR)}.$$

From (8a) and (8b) it results that

$$a_{n-1} = \frac{2}{\pi(n-1)!} \sum_{i=0}^{n/2-2} (-1)^i \frac{C_{n-2}^i}{C_{n-2}^{n/2-1}} (n-2i-2)^{n-2}, \quad n \text{ even},$$

$$a_{n-1} = \frac{1}{2(n-1)!} \frac{\displaystyle\sum_{i=0}^{(n-3)/2} (-1)^i C_{n-2}^i (n-2i-2)^{n-2}}{\displaystyle\sum_{i=0}^{(n-3)/2} (-1)^{i+1} C_{n-2}^i / (n-2i-2)}, \quad n \text{ odd}.$$

It can be shown, using Euler's gamma function, that :

$$a_{n-1} = \frac{2^{n-2} \; \Gamma^2(n/2)}{\pi \; \Gamma(n)} \; , \tag{9}$$

regardless of the parity of n.

For n large enough, Stirling formula results in :

$$a_{n-1} \underset{.}{\approx} 1/\sqrt{2\pi n}, \tag{10}$$

so we have

$$F_n(R;\theta) \approx \exp\left[- \frac{\exp[n\,(R + \ln(\sin\theta))]}{\sin\theta \; \sqrt{2\pi n}}\right]. \tag{11}$$

For large n this function varies fast as θ increases, from nearly 1 to nearly 0, near the angle such that the argument in (11) equals -1 :

$$\theta \approx \sin^{-1}[(2\pi n)^{1/2(n-1)} \; \exp(-Rn/(n-1))]. \tag{12}$$

If we let n approach infinity, $F_n(R;\theta)$ becomes a step function with its transition at

$$\theta_R \approx \sin^{-1}[\exp(-R)], \tag{13}$$

a result already obtained by Shannon [5].

Second factor of the integrand in (5)

Shannon [5] has shown that

$$Q_n(\theta) = P(n-1, \sqrt{nS/N}, \sqrt{n-1}\cot\theta), \tag{14}$$

where S and N denote the average power of signal and noise, respectively, and $P(\cdot,\cdot,\cdot)$ denotes the non-central t-distribution i.e.,

$$P(m,a,t) = \Pr\left[z + a \le t \; \sqrt{\frac{1}{m} \sum_1^m x_i^2}\right], \tag{15}$$

where z and x_i, $1 \le i \le m$, are mutually independent Gaussian random variables, with zero mean and variance 1 ; a and t are constant parameters.

Absence of amplitude fluctuations

We first assume the received signal amplitude is constant. The variable r defined as $r = m^{-1/2}\sqrt{\sum x_i^2}$ has a χ^2 distribution i.e., its density is

$$p_r(r) = \frac{m^{m/2}}{2^{m/2-1}\Gamma(m/2)} \; r^{m-1} \; \exp(-mr^2/2) \; . \tag{16}$$

Using this density to express the probability in (15) results in :

$$P(m,a,t) = \int_{t}^{+\infty} p_u(u) \, du, \tag{17}$$

where the following expression of $p_u(u)$ can be found with the help of [10] :

$$p_u(u) = \frac{(n-1)^{(n-1)/2} \exp(-a^2/2)}{(u^2+n-1)^{n/2}\sqrt{\pi}} \sum_{i=0}^{\infty} b_i \, z^i \,, \tag{18}$$

where $z = au / \sqrt{2(u^2+n-1)}$. The coefficients of the series are :

$$b_0 = \frac{\Gamma(\frac{n}{2})}{\Gamma(\frac{n-1}{2})}, \quad b_1 = n - 1 \,, \tag{19}$$

and, for m a positive integer :

$$b_{2m} = b_0 \frac{n(n+2)(n+4)\cdots(n+2m-2)}{1.3.5\cdots(2m-1)} \frac{1}{m!} = b_{2(m-1)} \frac{n+2m-2}{m(2m-1)} \,,$$

$$b_{2m+1} = b_1 \frac{(n+1)(n+3)(n+5)\cdots(n+2m-1)}{3.5.7\cdots(2m+1)} \frac{1}{m!} = b_{2m-1} \frac{n+2m-1}{m(2m+1)} \,.$$

We thus may write

$$Q_n(\theta) = \int_{\sqrt{n-1}\cot\theta}^{+\infty} p_u(u) \, du \tag{20}$$

and

$$G_n(\rho;\theta) = - Q_n'(\theta) = \sqrt{n-1} \, (1 + \cot^2\theta) \, p_u(\sqrt{n-1}\cot\theta). \tag{21}$$

For n large enough, the density (16) can be approximated by the Gaussian density which has same mean μ and variance v. These are shown to be :

$$\mu = \sqrt{\frac{2}{n-1}} \, b_0 \,, \tag{22}$$

where b_0 is given by (19), and

$$v = 1 - \mu^2. \tag{23}$$

Applying Stirling formula to (19) shows that b_0 varies approximately as $(1 - 3/4n) \sqrt{n/2}$, so the corresponding values of μ and v are $\mu = (1 - 1/4n)$ and $v = 1/2n$. Thus, as n approaches infinity, μ approaches 1 and v vanishes.

Replacing (16) by its Gaussian approximation transforms (18) into :

$$p_u(u) \approx \frac{1}{2\pi\sqrt{v}} \exp(-\frac{\mu^2}{2v} - \frac{a^2}{2}) \frac{1}{u^2 + 1/v} + \tag{24}$$

$$+ \frac{1}{\sqrt{\pi v}} \ \frac{(au + \mu/v)}{[2(u^2 + 1/v)]^{3/2}} \ \exp[-\frac{(a - \mu u)^2}{2(u^2 v + 1)}] \ \text{erfc}[-\frac{au + \mu/v}{\sqrt{2(u^2 + 1/v)}}].$$

An approximation of the factor $G_n(\rho;\theta)$ results by substituting (24) for $p_u(u)$ in (17) and (14). Its maximum occurs at approximately

$$\theta_g = \tan^{-1} \frac{\mu \sqrt{n-1}}{\sqrt{n\rho}} = \tan^{-1} (b_0 \sqrt{\frac{2}{n\rho}}) \tag{25}$$

and its variance is equal to :

$$v_g = \frac{(n-1)[n\rho(1-\mu^2) + \mu^2]}{[(n-1)\mu^2 + n\rho]^2} = \frac{n(n-1)\rho - 2b_0^2(n\rho-1)}{(2b_0^2 + n\rho)^2}, \tag{26}$$

As n approaches infinity, the factor $G_n(\rho;\theta)$ becomes the Dirac density $\delta(\theta - \theta_C)$, since its variance vanishes and

$$\theta_g \rightarrow \tan^{-1}(\frac{1}{\sqrt{S/N}}) = \sin^{-1}(\frac{1}{\sqrt{1 + S/N}}) = \theta_C.$$

The angle θ_C is thus such that

$$\sin(\theta_C) = \exp(-C) , \tag{27}$$

where $C = \frac{1}{2} \ln(1 + S/N)$ is the channel capacity, expressed in nats per dimension. If $\theta_R > \theta_C$, where θ_R is given by (13) (which obviously implies $R < C$), then according to (3) the error probability vanishes. On the other hand, it approaches 1 if $\theta_R < \theta_C$ (or $R > C$).

Presence of amplitude fluctuations

In the presence of amplitude fluctuations, the parameter a in (15) becomes a random variable. We shall only consider the case where we assume plane constellations, with the signal amplitude distributed according to the Rayleigh density. Moreover, we assume perfect interleaving in the sense that the amplitudes associated with the symbols of a codeword are mutually independent random variables. We then have $n = 2l$, where l is the code length, and the probability density of a is :

$$p_a(a) = \frac{1}{2^{l-1} \rho^l \Gamma(l)} \ a^{2l-1} \ \exp(-a^2/2\rho) , \tag{29}$$

where ρ denotes the average signal-to-noise ratio, so the error function becomes, instead of (14) :

$$Q_n(\theta) = \int_0^\infty P(n-1, \ a, \ \sqrt{n-1} \cot\theta) \ \frac{a^{2l-1}}{2^{l-1} \rho^l \Gamma(l)} \ \exp(-a^2/2\rho) \ da. \tag{30}$$

The mean and variance of the random variable a are :

$$\mu_a = \frac{\Gamma(l+1/2)}{\Gamma(l)} \ \sqrt{2\rho} = \frac{(2l - 1) \sqrt{\rho}}{\sqrt{2} \ b_0}, \tag{31}$$

346

and

$$v_a = 2\rho[l - \frac{\Gamma^2(l+1/2)}{\Gamma^2(l)}] = 2\rho[l - \frac{(2l-1)^2}{4\,b_0^2}], \qquad (32)$$

where b_0 is given by (19), with $n = 2l$.

As n grows indefinitely, v_a vanishes and the density (29) approaches a Dirac density at $a = \sqrt{2l\rho}$ i.e., the same as in the absence of fluctuation. The same asymptotic behaviour as already described thus occurs, whether or not fluctuations are present.

All these results have been summarized in Figure 2, which shows the first factor in (5) $F_n(R;\theta)$ as given by its approximation (11) and the second one $G_n(\rho;\theta)$ computed according to (21) in the absence of fluctuations or numerically computed in the presence of Rayleigh fluctuations.

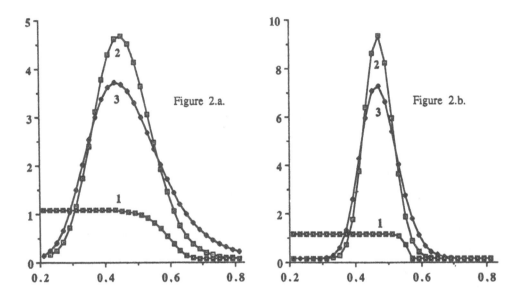

Figure 2. First and second factor in the integrand of equality (5).

Figure 2.a. The number of dimensions of the signal space is $n = 16$. The first factor (1) was plotted for $R = \ln2$; the second one for SNR = 6.0 dB with no fluctuations (2) and with Rayleigh fluctuations (3). Since the capacity is larger than the rate R, the integral of the product of the factors is close to 1 and thus the error probability is small. The angle θ, in radians, was plotted on the x-axis.

Figure 2.b. Same as Fig. 2.a except that the number of dimensions is now $n = 64$.

WEIGHTED DECODING OF MDS CODES

The conventional hard decoding of codes does not take into account the Euclidean distance of the signal representing the codeword with respect to the received signal. No loss of information would result only if weighted decoding is performed i.e., which takes into account the *a priori* probabilities of the symbols (which is equivalent to use the Euclidean metric).

We already described an algorithm for the optimum weighted decoding of linear block codes [6,7]. It can be thought of as a kind of sequential decoding but the finite context of blocks avoids the overflow problems encountered with convolutional codes. Although binary codes were mainly considered in [6,7], the algorithm can be extended to deal with nonbinary codes. This case was studied with more detail in [8], assuming some preliminary quantization of the received signal space in order to convert the actually continuous (e.g., Gaussian) channel into a discrete one.

Let us denote by A and B the channel input and output alphabets, respectively, with $\text{card}(A) = q < \text{card}(B) = Q$. We may assume that B results from quantization of the continuous channel output. The channel is then characterized by the qQ transition probabilities $\Pr(b_j|a_i)$ that $b_j \in B$ is received when $a_j \in A$ is transmitted.

We define the hard decision $\text{h}(j)$ on b_j as the symbol of A such that

$$\Pr(b_j|\text{h}(j)) = \max_{a \in A} \Pr(b_j|a) \qquad (33)$$

and introduce the non-negative quantities

$$v(a_i, b_j) = \log[\Pr(b_j|\text{h}(j))] - \log[\Pr(b_j|a_i)], \qquad (34)$$

where the logarithms are to an arbitrary base.

Now, if $c^i \in A^n = [c_1^i \ \cdots \ c_n^i]$ denotes the i-th codeword and $r \in B^n = [r_1 \ \cdots \ r_n]$ denotes a received vector, the generalized distance of r and c^i is defined as

$$Z(r, c^i) = \sum_{j=1}^{n} v(r_j, c_j^i). \qquad (35)$$

Let us assume that the code is in systematic form. Since the generalized distance is additive with respect to the symbols, we may write (35) as the sum of two terms namely

$$Z(r, c^i) = Z_u(r, c^i) + Z_s(r, c^i), \qquad (36)$$

where $Z_u(r, c^i)$ is the contribution of the information symbols and $Z_s(r, c^i)$ that of the check symbols. Optimum decoding consists of determining i^* such that $Z(r, c^{i^*}) < Z(r, c^i)$ for any $i \neq i^*$. $Z_u(r, c)$ is immediately determined once the information vector u corresponding to c is given. On the other hand, $Z_s(r, c)$ can be obtained only after the check symbols of c are computed i.e., after u has been encoded. The algorithm is designed in order to minimize the number of encodings necessary to determine i^*.

The received symbols are first reordered by decreasing reliability of the corresponding hard decisions. The first k symbols after reordering are taken as the information symbols, which is always possible for MDS codes. The vector $u^h = [h(r_1) \cdots h(r_k)]$ determines a first codeword c^h such that $Z_u(r, c^h) = 0$. Encoding u^h enables computing $Z(r, c^h)$ according to (36) and a threshold is set at this value. Now, other information vectors u, corresponding to code vectors c, are tried in the order of increasing $Z_u(r, c)$ and the corresponding $Z(r, c)$ is computed. If $Z(r, c)$ is found for some c smaller than the current threshold, then the new one is set to this value. The corresponding codeword c becomes the last provisional candidate to optimum decoding. The provisional candidate becomes the finally decoded word when the next $Z_u(r, c)$ exceeds the current threshold.

The complexity of this algorithm is small in the average, but exactly optimum decoding can result in a large amount of computation for certain (unfrequently met) received vectors. Therefore, suboptimum decoding should be preferred in any practical situation. It can result from stopping to try codewords before it becomes sure that the best one was found.

CONCLUSION

It is known that MDS codes are good. It is our opinion that they are still better than currently believed and that, when combined with a proper multilevel modulation, they can behave very similarly to random coding, thus closely approaching the capacity of the channel. The key to their efficient use is weighted decoding since it can approach maximum likelihood decoding which chooses the signal point closest to the received one according to the Euclidean metric. Complexity can be kept reasonable only at the expense of strict optimality.

Although the principle of such weighted decoding has already been elaborated, some work remains to be done before truly usable algorithms can be designed. For instance, proper quantization of the received signal space should provide an acceptable compromise between complexity and performance degradation. We hope that the prediction of achievable performance presented in this paper will prompt research in this direction.

References

[1]   G. BATTAIL, Construction explicite de bons codes longs, Annales Télécommunic., vol. **44**, n° 7-8, July-August 1989, pp. 392-404

[2]   G. UNGERBOECK, Channel coding with multilevel/phase signals, IEEE Trans. on Inf. Th., vol. **IT-28**, n° 1, Jan. 1982, pp. 55-67

[3]   K.-M. CHEUNG, More on the decoder error probability for Reed-Solomon codes, IEEE Trans. on Inf. Th., vol. **35**, n° 4, July 1989, pp. 895-900

[4]   H. MAGALHAES de OLIVEIRA and G. BATTAIL, A capacity theorem for lattice codes on Gaussian channels, SBT/IEEE International Telecommunications Conference, Rio de Janeiro, Brazil, 3-6 Sept. 1990

[5]   C.E. SHANNON, Probability of error for optimal codes in a Gaussian channel, BSTJ, vol 38, n° 3, May 1959, pp. 611-656

[6]   G. BATTAIL, Décodage pondéré optimal des codes linéaires en blocs I.- Emploi simplifié du diagramme du treillis, Annales Télécommunic., vol. 38, n° 11-12, Nov.-Dec. 1983, pp. 443-459

[7]   G. BATTAIL and J. FANG, Décodage pondéré optimal des codes linéaires en blocs II. - Analyse et résultats de simulation, Annales Télécommunic., vol. 41, n° 11-12, Nov.-Dec. 1986, pp. 580-604

[8]   J. FANG, Décodage pondéré des codes en blocs et quelques sujets sur la complexité du décodage, PhD Thesis, ENST, 18 March 1987

[9]   R.M. ROTH and A. LEMPEL, On MDS codes via Cauchy matrices, IEEE Trans. on Inf. Th., vol. 35, n° 6, Nov. 1989, pp. 1314-1319

[10]   I.S. GRADSHTEYN and I.M. RYZHIK, Tables of integrals, series, and products, Academic Press, 4-th edition, 1965

COMPARISON OF TWO MODULATION-CODING SCHEMES
FOR LOW-RATE DIGITAL LAND MOBILE RADIOCOMMUNICATION

R. Sfez (*), J.C. Belfiore (**), K. Leeuwin (*)(**), A. Fihel (*)

(*) TRT-PHILIPS, 5, rue Réaumur, 92352 Le Plessis Robinson CEDEX
(**) TELECOM-PARIS, 46, rue Barrault, 75634 Paris CEDEX 13

1. Introduction

The aim of this paper is the comparison of two modulation-coding schemes S_1 and S_2 for low-rate digital land mobile communications:

- S_1: 2/3-rate 16-state Ungerboeck's Trellis-Coded Modulation (TCM) 8-PSK, whose high bandwidth efficiency is a key argument *a priori* [1];

- S_2: QPSK modulation combined with a 3/4-rate 16-state convolutional punctured code [2] (this combination is very BER efficient over the additive white Gaussian noise (AWGN) channel, for a fairly high coding rate, hence the motivation for it).

Both schemes are possibly suitable for the future Mobile Digital Trunked Radiocommunication System (MDTRS). Actually, TCM and QPSK have been proposed as candidate schemes for this system [3,4]: comparing them under significant conditions for this application is the main motivation for this paper.

A priori, MDTRS is a low-traffic density, medium-to-low rate system (modulation gross rate should be lower than 40 kbit/s), intended for the private European market. As for the radio aspect, complexity of the transmitters, bandwidth efficiency, coverage, and channel reuse should be of prime interest. Therefore, schemes S_1 and S_2, which involve the same decoding complexity, should be compared with respect to efficiency criteria C_1 and C_2:

- C_1: (frequency reuse / bandwidth) performances (in order to maximize spectrum efficiency);

- C_2: (coverage / bandwidth) performances (for economical reasons, coverage should be as important as frequency reuse).

For this comparison, the same information bit rate, and the same reliability (in terms of the probability that a given BER after decoding is ensured) should be assumed for both schemes.

This paper is divided into three main parts.

Basic assumptions and modellings are defined in the first part.

In the second part, we assume a memoryless Rayleigh fading channel. Computer simulation results are provided, comparing the BER performances of schemes S_1 and S_2. We define the efficiency criteria C_1 and C_2: using both the BER results obtained, and the modellings assumed in the first part, S_1 and S_2 are compared as regards to these criteria.

In the last part, we account for the memory of the mobile radio channel. Assumptions are made on the frequency band, the modulation gross rate, the vehicle speed, and on finite interleaving delay: in these (more realistic) conditions, the comparisons of the second part are repeated.

2. Basic assumptions and modellings

We assume that in a real mobile radio environment, the received power can be written as follows [5]:

$$P = K \, P_0 \, d^{-\alpha} \, M^2 \, R^2 , \tag{1}$$

where P_0 denotes a reference level, K is a constant factor (depending on antenna heights, carrier frequency, ..),
d is the distance between the transmitter and the receiver, and α is a real number (typically of between 3 and 4,
when the receiver is far enough from the transmitter).

Following the Okumura-Hata modelling [6], the term $K \, P_0 \, d^{-\alpha}$, gives the average received power at the
distance d from the transmitter. M is a random variable, log-Normally distributed, modelling the slow fluctuations due to shadowing: $10 \log(M)$ is Gaussian, with standard deviation σ typically, of between 5 and 7 dB.
Lastly, R is a normalized Rayleigh random variable, which modelizes the fast instantaneous fluctuations due to
multipaths: Rayleigh modelling can be considered as a worst case (no direct path). R and M are assumed to be
independant.

A non frequency-selective, Rayleigh fading channel is assumed, as a perfect coherent detection. Then, the
received complex signal, after sampling, can be written as

$$y_j = | \rho_j | x_j + w_j , \tag{2}$$

where $\{\rho_j\}$ is a zero mean, Gaussian random process, $\{x_j\}$ is the transmitted signal sequence, and $\{w_j\}$ is an
AWGN process. In the following, according to the case considered, the term w_i will be interpretated as local
noise, or as the sum of co-channel interferences.

This paper is restricted to the base-to-mobile link. We assume that a cell area can be approximated by a
discus with radius r, whose center is the transmitting base station. In each cell, a mobile receiver N is assumed
to be randomly and uniformly located: in a polar coordonate system admitting a base station B as the origin, the
pdf of the coordonates d and θ of receiver N are then:

$$p(\theta) = \frac{1}{2\pi} , \quad 0 \le \theta < 2\pi ,$$

$$\tag{3}$$

$$p(d) = \frac{2 \, d}{r^2} , \quad 0 \le d < r$$

These parameters are detailed on Figure 1 (on the example of an hexagonal cellular network).

3. Comparison over the memoryless Rayleigh fading channel

In this section, we assume ideal interleaving, or ideal frequency hopping, so that two any samples of random process $\{\rho_j\}$, in (2), are statistically independant: this first approach allows the evaluation of ideal performances over the Rayleigh fading channel.

3.1. BER performances

BER curves for schemes S_1, S_2, and for single QPSK, are drawn on Figure 2. In the computer simulations involved, channel state information (CSI) was assumed in any case: the maximum-likelihood (ML) sequence $\{\hat{x}_j\}$ (such that the Euclidean distance between sequences $\{\rho_j \hat{x}_j\}$ and $\{y_j\}$ is the smallest) was detected (using the Viterbi algorithm in the case of S_1 or S_2).

On Figure 2, BER is drawn as a function of Eb/N_0 (in décibels), where Eb is the energy of one information bit. Denoting by E the energy of one transmitted symbol, signal-to-noise ratio (SNR) E/N_0 is

$$E/N_0 = \rho\, q\, Eb/N_0\,,$$

where 2^q is the number of modulation levels, and ρ is the coding rate of the combined error-correcting code.

In the following, we denote by $E/N_0^{(i)}(\tau)$ the SNR value corresponding to a given BER τ for the scheme S_i, $i=1,2,$. For example, we identified by computer simulatons that, for $\tau = 10^{-2}$, $E/N_0^{(1)}(\tau) = 11.1$ dB, and $E/N_0^{(2)}(\tau) = 9.8$ dB;

The way these results translate into channel reuse or into coverage (according to the case considered) is of prime interest.

3.2. Comparison as regards to (frequency reuse / bandwidth) performances

A criterion accounting for bandwidth efficiency and frequency reuse performances can be derived from spectrum efficiency.

Spectrum efficiency is defined as "the carried traffic for the unit sizes of the radio bandwidth and service area" [7]. Assuming a regular cellular system, spectrum efficiency can be written as, [8],

$$\eta = \frac{a_c}{S\, 2W\, n_g\, n_z}\,, \tag{4}$$

where a_c is the carried traffic per cell, S is the cell area, W is the channel spacing, n_z is the number of channels per cell, and n_g is the cluster size (i.e. number of cells per cluster).

We now make two assumptions:

- comparing modulation-coding schemes only, we assume the same carried traffic per cell and the same number of channels per cell for the two schemes;

- we assume that co-channel interferences are the only source of "noise": then, as we shall see later, reliability does not depend on the cell area.

With the assumptions above, we derive from (4) that the criterion to be maximized is the ratio $1/(W\, n_g)$.

Let us assume hexagonal cells: without loss of generality, the minimum reuse distance (between a base station and its closest co-channel interferers), can be set to 1. Then, the cluster size takes the value:

$$n_g = \frac{1}{3\, r^2}\,,$$

so that the efficiency criterion to be maximized should be the following:

$$C_1 = \frac{r^2}{W} \tag{5}$$

C_1 is the contribution, from a communication scheme, to spectrum efficiency, from which it is derived. It is noticeable that this criterion was proposed by French [9] in order to compare modulations for cellular mobile radiocommunication.

We now consider the base-to-mobile link, and account for n co-channel interferers $I_1, ..., I_n$, (which are the n closest co-channel base stations). Following (1), the total instantaneous interferent power is then written as

$$K P_0 \sum_{i=1}^{n} d_i^{-\alpha} M_i^2 R_i^2 ,$$

where d_i is the distance between receiver N and interferer I_i, M_i and R_i modelize the slow and fast fadings, respectively, on the (undesirable) link between N and I_i. Part of these parameters is detailed on Figure 1, assuming an hexagonal cellular network using 7-cell clusters.

As a reasonable approximation, we identify the total co-channel interference as the AWGN w_j of modelling (2). Then, locally, the average carrier-to-interference ratio (C/I), at any mobile receiver,

$$C/I = \frac{d^{-\alpha} M^2}{\sum_{i=1}^{n} d_i^{-\alpha} M_i^2} \tag{6}$$

(without power control) can be identified as local SNR, E/N_0.

For any scheme S_i, $i=1,2$, we define reliability $R_i(\tau)$ as the probability that a cutoff BER τ is ensured:

$$R_i(\tau) = \Pr \left\{ E/N_0 > E/N_0^{(i)}(\tau) \right\} , \tag{7}$$

where, in this case, E/N_0 is defined by (6).

Basically, reliability depends on the ratio $1/r^2$, the specified cutoff BER τ, the communication scheme, the additional techniques involved (possibly power control, sectorial antennas, ..) and a few parameters: channel load l, standard deviation of slow shadowings σ, and factor α.

The calculation of reliability is described in appendix 2.

Reducing the $1/r^2$-factor implies a reduction of cluster size, hence an increase of efficiency (5), but at the price of a reliability impairment. Therefore, a trade-off has to be carried out between, say, spectrum efficiency, and transmission quality.

Provided both the cutoff BER τ and the desired reliability, schemes S_1 and S_2 can be compared with respect to
- bandwidth efficiency,
- the $1/r^2$-factor involved to ensure BER τ with the desired reliability,
all other parameters being set equal for both schemes.

Firstly, Monte-Carlo calculations of reliability are carried out as described in appendix 2. All the results provided in the following were obtained by assuming an attenuation factor $\alpha = 3.5$, a standard deviation of slow shadowings $\sigma = 6$ dB, a correlation factor of 0.7 for slow shadowings, a channel load of $l=1$. Furthermore, we have assumed the use of 120° sectorial antennas in each base station, as described in Figure 3. In this technique, any hexagonal "cell" is divided into three sectors using three different frequencies f_1, f_2, f_3: the resulting effect is that only part of the n interferers considered in (6) can actually be interfering, hence an improved channel reuse.

Lastly, for Monte-Carlo simulations, the locations of $2 \cdot 10^3$ mobile receivers were randomly picked, and the ten nearest co-channel rings were taken into account.

The conditions above are close to those assumed in [10-12].

Assuming a cutoff BER of $\tau=10^{-2}$, Figure 4 allows to compare the reliabilities met by S_1 and S_2, as functions of the $1/r^2$-factor. On the one hand, in order to meet a reliability of 0.9, a $1/r^2$ factor of 6.75 is involved by S_1, whereas a factor of only 5.8 is involved by S_2; on the other hand, S_1 is more bandwidth efficient than S_2 by a factor 4/3. From these considerations, it can be drawn that scheme S_1 is better efficient than S_2 (following (5)) by a factor 1.15.

3.3. Comparison as regards to (coverage / bandwidth) performances

For economical reasons, reduction of the base station density and of infrastructure cost are of fundamental importance, so that coverage may be of prime interest, instead of channel reuse. In that case, the criterion to be maximized should be:

$$C_2 = \frac{S}{W} , \qquad (8)$$

where S is the coverage area, and W is the channel spacing.

In this section, we consider a single cell. A Rayleigh fading channel is assumed following (2), where w_j is a local noise. Using both modelling (1) and definition (7), and assuming the same reliability and information bit rate for both schemes, we obtain that the gain in coverage area allowed by scheme S_1, with respect to S_2, is:

$$\left[\frac{r_1}{r_2}\right]^2 = (4/3)^{2/\alpha} \left[\frac{E/N_0^{(2)}(\tau)}{E/N_0^{(1)}(\tau)}\right]^{2/\alpha} , \qquad (9)$$

where r_1 and r_2 are the coverage radii allowed by schemes S_1 and S_2, respectively.

Provided a cutoff BER τ, the SNR values $E/N_0^{(1)}(\tau)$ and $E/N_0^{(2)}(\tau)$ can be identified by computer simulations, and the comparison of coverage areas can be carried out, following (9). Furthermore, taking into account that scheme S_1 is more bandwidth efficient than S_2 by a factor 4/3, a comparison with respect to complete criterion C_2 is made: assuming that $\alpha=3.5$ and $\tau=10^{-2}$, it appears that S_1 is better efficient than S_2, following (8), by a factor 1.33.

4. Comparison over the Rayleigh fading channel with memory

The comparisons of section 3 are now repeated, but with the further (more realistic) assumption of a finite interleaving, so that the channel memory, related to the Doppler frequency, has to be taken into account.

In that case, the transmission modelling (2) is still valid, but the autocorrelation function of the envelope of the received signal is given by:

$$R_{|\rho|}(kT) = J_0(2\pi f_d kT) , \qquad (10)$$

where J_0 is the zero order Bessel function of the first kind, f_d denotes the Doppler frequency, and T denotes one symbol duration [13].

In this section, this was simulated as follows. The two (correlated) Gaussian components of process $\{\rho_j\}$ were obtained as the output of a recursive filter (approximating the Doppler "filtering"), applied on two uncorrelated Gaussian random processes [14]: this allows a good approximation of the theoretical envelope autocorrelation.

From now on, we shall assume a carrier frequency of 900 MHz, a vehicle speed uniformly distributed between 10 km/h and 130 km/h, a gross rate of 16 kbit/s, and a channel (interleaving / de-interleaving) delay of

100 ms (for computer simulations, a (16,25) convolutional interleaving [15] is assumed).

The transmitter scheme is detailed on Figure 5.

In Figure 6, BER curves related to schemes S_1, S_2, and to single QPSK are drawn as an example, with the assumptions above, and for a vehicle speed of 10 km/h: from this, we can conclude that, due to finite interleaving, the channel memory severely degrades the performance of any of the encoded schemes; the channel cutoff rate is too low, so that schemes S_1 and S_2 are less BER efficient than single QPSK, and cannot be compared under these unacceptable conditions.

In order to reduce the effects of the channel memory, multi-channel transmission [16] of the order M can be assumed, as described on Figure 7: after interleaving, M any consecutive encoded symbols (with rate 16 kbit/s) are transmitted on M sub-carriers (on each of them, the modulation rate is of only 16/M kbit/s). The M subcarriers are spaced by at least once the channel coherent bandwidth, so that the fadings on M consecutive transmitted symbols are independant.

From now on, the transmission scheme of Figure 7 is assumed, with multi-channels of limited order (M=2): as we shall see, it enables the encoded schemes S_1 and S_2 to provide significant coding gains, with respect to single QPSK.

4.1. BER performances

Figures 8 to 12 provide the BER curves related to schemes S_1, S_2, and to single QPSK, assuming vehicle speeds of 10, 40, 70, 100, and 130 km/h, respectively.

In any case, it is noticeable that, for a BER of around 10^{-2}, the encoded schemes are better BER efficient than single QPSK (contrary to Figure 6): hence the interest for multi-channel transmission.

As a second comment, assuming a cutoff BER of 10^{-2}, it would appear that, except for Figure 11 (where schemes S_1 and S_2 provide the same performance), S_1 is significantly more BER efficient than S_2: this was not the case when a memoryless Rayleigh fading channel was assumed (see Figure 2).

From the assumption that vehicle speed is uniformly distributed, the comparisons of section 3 can be repeated for each speed, and then a global comparison (averaged on speed), can be achieved. For a cutoff BER of 10^{-2}, we have identified, for each of the vehicle speeds assumed in the computer simulations, the corresponding SNR $E/N_0^{(1)}(\tau)$ and $E/N_0^{(2)}(\tau)$: then, global comparisons is possible, assuming that vehicle speed can take the value 10, 40, 70, 100 or 130 km/h with a probability of 1/5.

4.2. Comparison as regards to (frequency reuse / bandwidth) performances

We now compare schemes S_1 and S_2 under the same conditions as in section 3.2: assuming a cutoff BER of 10^{-2}, Figure 13 provides reliability as a function of the $1/r^2$-factor, for both schemes (Figure 13 can be compared with Figure 4, which was related to the memoryless Rayleigh fading channel).

For a reliability of 0.9, we have obtained that a factor $1/r^2 = 8.5$ is involved by scheme S_2, whereas a factor of 9 is involved by scheme S_1. Taking into account that S_1 is more bandwidth efficient that S_2 by a factor 4/3, it appears that, scheme S_1 is better efficient than S_2, following (5), by a factor 1.26.

4.3. Comparison as regards to (coverage / bandwidth) performances

In this section, we compare schemes S_1 and S_2 under the same conditions as in section 3.3: from (9), we derive that the average gain in coverage area allowed by scheme S_1, with respect to S_2, is given by:

$$\frac{r_1^2}{r_2^2} = (4/3)^{2/\alpha} < \left[\frac{E/N_0^{(2)}(\tau)}{E/N_0^{(1)}(\tau)} \right]^{\frac{2}{\alpha}} > , \qquad (11)$$

where the brackets $< . >$ implies an average on vehicle speed. Taking into account both the result of calculation (11) and the bandwidth efficiencies of schemes S_1 and S_2, it appears that scheme S_1 is better efficient than S_2, following (8), by a factor 1.46.

5. Conclusion

This paper was intended to compare Trellis-Coded Modulation (TCM) 8-PSK (S_1) and conventional QPSK combined with a convolutional code (S_2) for digital land mobile radiocommunication.

The two schemes, which involve the same decoding complexity, were compared as regards to (frequency reuse / bandwidth) efficiency (criterion C_1) and (coverage / bandwidth) efficiency (criterion C_2).

With the assumption that user quality depends on the BER after decoding (not on the statistics of residual errors), we have considered a cutoff BER of 10^{-2}, and a desired reliability of 0.9. The same user bit rate was assumed for both schemes.

Over the memoryless Rayleigh fading channel, BER curves were obtained by computer simulations (assuming ideal coherent detection), and then translated into channel reuse performances, and coverage performances: from this, we have concluded that scheme S_1 is better efficient than S_2 by a factor 1.15 (following criterion C_1), and by a factor 1.33 (following C_2).

Over the Rayleigh fading channel with memory (due to both finite interleaving constraints and the Doppler effect), we have assumed a multi-channel transmission of the order 2, in order to limit the impairment of the coding performances (due the channel memory). Two conclusions can be drawn:

- from the BER curves obtained, it would appear that in the presence of residual channel memory, the trellis code assumed gets better BER efficient than the convolutional code assumed (whereas this was not the case over the memoryless Rayleigh fading channel): a theoretical analysis of this phenomenon may not be easy *a priori*, since the design of good codes for Rayleigh fading channel with residual memory remains as an open problem;

- assuming that vehicle speed is uniformly distributed between 10 km/h and 130 km/h (with a carrier frequency of 900 MHz), it appears that scheme S_1 is better efficient than S_2 by a factor 1.26 (following C_1) and 1.46 (following C_2).

As a conclusion, it seems, at least, that under ideal conditions (perfect coherent detection), the bandwidth expansion by a factor 4/3 paid in case of scheme S_2 (with respect to S_1), is not recovered by the BER performances: hence the interest for TCM. More generally, may be we can raise the issue that increasing the bandwidth efficiency (by combining multi-level modulations with well-suited trellis codes) results in an improvement of efficiencies C_1 and C_2.

References

[1] G. UNGERBOECK, Trellis-Coded Modulation with Redundant Signal Sets, IEEE Com. Mag., vol. 25, No. 2, Feb. 1987

[2] J. BIBB CAIN, G.C. CLARK Jr., J.M. GEIST, Punctured Convolutional Codes of Rate (n-1)/n and Simplified Maximum-Likelihood Decoding, IEEE Inf. Th., vol. 25, No. 1, Jan. 1979

[3] Proposed Specification for the Pan-European Digital Trunked System, ETSI/STC-RES6.2(90)02, Source: France, London, 27-28 Feb. 1990

[4] Proposed Radio Interface for the New Digital Trunked Standard, Source: Sweden, 1990

[5] G. MAZZIOTTO, A. SEDRATI, Statistique de la Propagation Hertzienne à 900 MHz en Zone Urbaine, GRETSI, Nice, 20-24 May 1985

[6] Y. OKUMURA, E. OHMORI, T. KAWANO, K. FUKUDA, Field Strength and its Variability in VHF and UHF Land-Mobile Radio Service, Rev. of the Elec. Com. Lab. (Japan), vol. 16, No. 9-10, 1968, pp. 825-873

[7] D.N. HATFIELD, Measure of Spectrum Efficiency in Land-Mobile Radio", IEEE Trans. Electromagn. Comput., vol. 19, No. 3, pt. 2, pp. 266-268, Aug. 1977

[8] Y. NAGATA, Y AKAIWA, Analysis for Spectrum Efficiency in Single-Cell Trunked and Cellular Mobile Radio, IEEE Trans. Vehic. Tech., vol. 35, No. 3, Aug. 1987

[9] R. FRENCH, The Effect of Fading and Shadowing on Channel Reuse in Mobile Radio, IEEE Trans. Vehic. Tech. vol. 28, No 3, Aug. 1979

[10] J.E. STJERNVALL, Calculation of Capacity and Co-Channel Interference in a Cellular System, Nordic Seminar on Digital Land Mobile Radiocommunication, 5-7 Feb. 1985, Espoo, Finland

[11] J.L. DORNSTETTER, D. VERHULST, Cellular Efficiency with Slow Frequency Hopping: Analysis of the Digital SFH 900 Mobile System, IEEE on SAC, vol. 5, n° 5, June 1987

[12] D. BERTHOUMIEUX, M. MOULY, Spectrum Efficiency Evaluation Methods, International Conference on Digital Land Mobile Radiocommunication, June-July 1987

[13] W.C. JAKES, Microwave Mobile Communications, J. WILEY, New-York, 1974

[14] J.R. BALL, A Real-Time Fading Simulator for Mobile Radio, The Radio and Electronic Engineer, vol. 52, n° 10, pp. 475-478, Oct. 1982

[15] J.L. RAMSEY, Realization of Optimum Interleavers, IEEE Trans. Inf. Th., 16 n° 3, May 1970, pp. 338-345

[16] Y. KAMIO, Performance of Trellis-Coded Modulation Using Multi-Frequency Channels in Land Mobile Communications, Proc. 40th IEEE VT Conf., May 1990

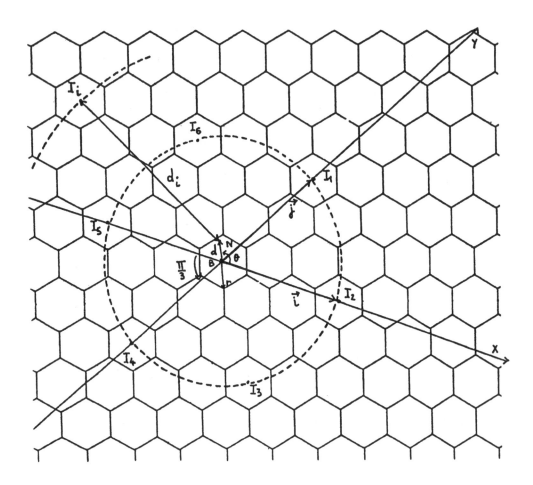

Figure 1 Hexagonal cellular network using 7-cell clusters

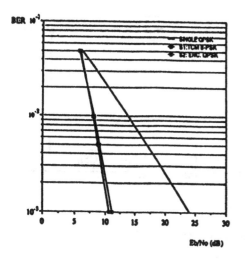

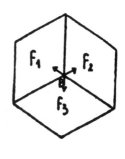

Figure 2 BER performance over the memoryless
Rayleigh fading channel

Figure 3 Sectorization of an hexagonal'
cell by 120° sectorial antennas

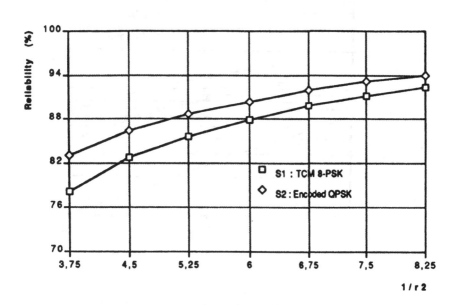

Figure 4 Reliability versus the $1/r^2$-factor

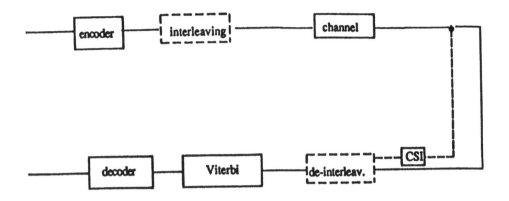

Figure 5 Transmission over one frequency channel

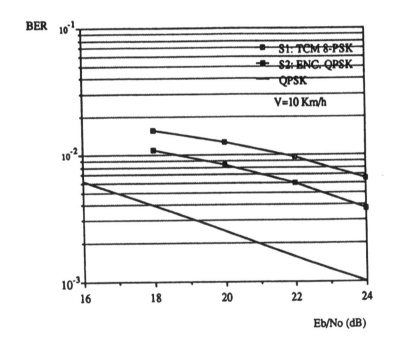

Figure 6 BER performance (vehicle speed = 10 km/h)

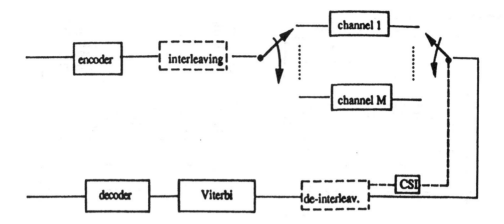

Figure 7 Multi-channel transmission

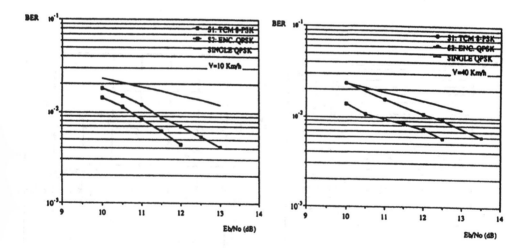

Figure 8 BER performance (2 frequency
channels, vehicle speed = 10 km/h)

Figure 9 BER performance (2 frequency
channels, vehicle speed = 40 km/h)

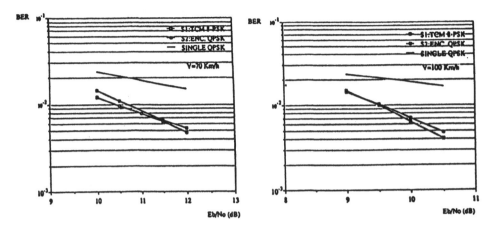

Figure 10 BER performance (2 frequency

channels, vehicle speed = 70 km/h)

Figure 11 BER performance (2 frequency,

channels, vehicle speed = 100 km/h)

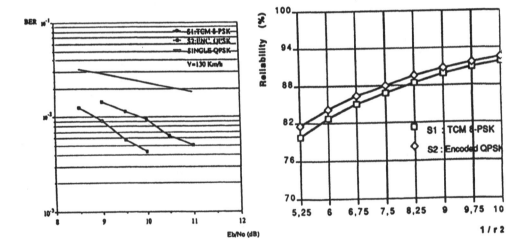

Figure 12 BER performance (2 frequency

channels, vehicle speed = 130 km/h)

Figure 13 Reliability versus the $1/r^2$-factor

vehicle speed = 10, 40, 70, 100

or 130 km/h with a probability of 1/5

APPENDIX 2
CALCULATION OF THE RELIABILITY

Let us first assume an hexagonal cellular network, with omni-directional antennas, and refere to Figure 1: using the coordonate system $(B, \vec{1}, \vec{j})$ depicted on that figure, any couple $(x_j, y_j) \in Z^2 - \{(0,0)\}$ defines the coordonates of a co-channel interferer I_j. The distance between mobile receiver N (with polar coordinates d and θ) and interferer I_j is

$$d_j = \left[x_j^2 + y_j^2 + x_j y_j + d^2 - d \left\{ (2x_j + y_j) \cos\theta + \sqrt{3} y_j \sin\theta \right\} \right]^{1/2} \tag{A.2.1}$$

Then, with the assumption that slow shadowings $M, M_1, \ldots, M_n$ are statistically independant, reliability can be written as

$$R_i(\tau) = \int_0^r p(d) \int_0^{2\pi} p(\theta) \int_{R^{*n}} \delta_j \, p(M_j) \, R_c \, \vec{d}M \, d\theta \, dd , \tag{A.2.2}$$

where and δ_j is a random variable: for $j = 1, \ldots, n$,

$$\Pr \{\delta_j = 1\} = l ,$$

$$\Pr \{\delta_j = 0\} = 1 - l \tag{A.2.3}$$

This means that co-channel interferers are active in $l\%$ of the time only (l is the channel load, accounting for traffic load and eventually, an activity ratio lower than 1). As for R_c, it denotes the conditional reliability:

$$R_c = \Pr \left\{ E/N_0 > E/N_0^{(i)}(\tau) \mid d, \theta, M_1, \ldots, M_n \right\}$$

$$= Q \left[\frac{5}{\sigma} \log \left[d^\alpha \sum_{j=1}^n d_j^{-\alpha} \, \delta_j \, M_j^2 \, E/N_0^{(i)}(\tau) \right] \right] , \tag{A.2.4}$$

where $Q(.)$ is the Q function:

$$Q(x) = \frac{1}{\sqrt{2\pi}} \int_x^\infty e^{-u^2/2} \, du \tag{A.2.5}$$

With the assumptions of an hexagonal cellular network with omnidirectional antennas, and of uncorrelated shadowings, calculation of the reliability is allowed by relations (A.2.1) to (A.2.5).

In practice, such a calculation is carried out by Monte-Carlo simulations:

- polar coordinates of receiver N are randomly picked (following (3), in case of a uniformly distributed location);

- for each interferer I_j, distance d_j can be calculated (according to (A.2.1), in case of an hexagonal cellular network);

- for each link between interferers I_j and mobile receiver N, (correlated or uncorrelated) shadowings M, $\{M_j\}$ and factors δ_j are randomly picked, so that the equivalent SNR E/N_0 can be calculated and compared with the cutoff SNR.

Repeating this process for numerous randomly located receivers N allows an evaluation of the reliability.

CHERNOFF BOUND OF TRELLIS CODED MODULATION OVER CORRELATED RAYLEIGH CHANNEL

K.LEEUWIN(*,**) - J.C.BELFIORE(*) - G.KAWAS KALEH(*)

(*) TELECOM PARIS - 46, rue Barrault - 75634 Paris cedex 13- FRANCE
(**) TRT, 5 avenue Réaumur, 92352 Le Plessis Robinson- FRANCE

ABSTRACT

We derive here a Chernoff upper bound for the pairwise error probability in the presence of an additive white Gaussian noise and a Rayleigh or Rice correlated fading. The bound is first used to obtain an upper bound for the bit error probability of a trellis-coded modulation system. The result is then employed to evaluate the performance of a multichannel system which represents a good strategy for reducing the time delay due to interleaving.

1. INTRODUCTION

On radio and satellite channels, system performances are degraded by Ricean or Rayleigh fadings resulting from multipath propagation. Performance can be improved by the appending of an encoder followed by an interleaver which decorrelates the fadings. In some cases, this is not sufficient and diversity has to be added. Generally, possible solutions have to be compared, and for this purpose, an error probability bound is helpful because it allows to avoid long and expensive simulations.

Some bounds have been derived in the past for Rayleigh or Rician channels. They are based on a discrete channel modeling [6] and, hence, are only applicable to hard decision systems. For soft decoding, available bounds [3] assume that the received samples are affected by uncorrelated fades. This implies that perfect interleaving should be made. In some cases, like mobile radio channel whose fading autocorrelation function is a slowly decreasing Bessel function, such interleaving cannot be achieved since it requires unallowed large time delay.

We derive in this paper a Chernoff bound on the error probability in the case of Rayleigh correlated fades. At the receiver, maximum-likelihood detection is assumed. The bound is valid for any type of convolutional encoding and linear modulation. We use trellis coded modulation as an application. In addition, and since this bound depends on the fading autocorrelation function, we apply it to evaluate the performances of sytems with partial interleaving and multi-channel transmission. Obviously, uncorrelated fading represents a special case of our bound.

Section 2 of this paper describes the general model under study. The calculation of the pairwise error probability bound is presented in section 3. In section 4, we give the bit error probability. Comparisons with computer simulations are presented in section 5.

2. TRANSMISSION MODEL

The general transmission model is depicted in figure 1.

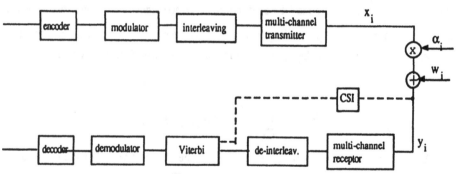

Figure 1. Block diagram of the transmission model

The input data bits are encoded by a convolutional encoder and mapped into modulation symbols. Several technics, like interleaving and multi-channel transmission, can then be used. In the channel, the ith transmitted symbol, denoted x_i, is multiplied by a complex Gaussian variable α_i; and a white Gaussian noise is added. Assuming coherent detection, the effect of the fading is limited to the multiplication of the symbol by the random variable $|\alpha_i|$. Therefore, the ith received baseband signal has the form:

$$y_i = |\alpha_i| \sqrt{2E_s}\, x_i + w_i \qquad i=1,2,...N \qquad (1)$$

This work was supported by T.R.T- LE PLESSIS ROBINSON- FRANCE, and ANRT where E_s is the symbol energy and w_i is an additive white Gaussian noise with zero mean and variance $2N_0$. The complex Gaussian variates α_i are correlated, with variance γ^2 (the envelope $|\alpha_i|$ is Rayleigh distributed). At the receiver, we assume perfect channel state information (CSI) i.e., the fading variables $|\alpha_i|$ are estimated perfectly. We also assume that a soft Viterbi decoder is used.

3. CHERNOFF BOUND OF THE PAIRWISE ERROR PROBABILITY IN A CORRELATED RAYLEIGH CHANNEL

An upper bound of the bit error probability, based on the union bound, is given by:

$$P_b \leq \sum_{x,t \,\in\, C} \sum a(x,t)\, p(x)\, P(\vec{x} \to \vec{t}) \tag{2}$$

where C is the set of all coded sequences, $p(x)$ is the probability of transmitting the sequence $\vec{x}$, $P(\vec{x} \to \vec{t})$ is the error probability of decoding the sequence $\vec{t}$ in place of the emitted sequence $\vec{x}$ ("Pairwise error probability"), and $a(x,t)$ is the number of resulting erroneous bits.

Assuming ideal CSI, the decision rule for the Viterbi detector is to minimize the following metric:

$$\underset{\vec{x} \,\in\, C}{\text{Min}} \sum_{i=1}^{N} |\, y_i - |\alpha_i| \sqrt{2E_s x_i}\,|^2 \tag{3}$$

Using this decoding rule, it can be shown that the Chernoff bound of the pairwise error probability, conditionned on the fading amplitude vector $\vec{\alpha} = (\alpha_1, \ldots, \alpha_N)$, is given by:

$$P\left(\vec{x} \to \vec{t}/\vec{\alpha}\right) \leq \prod_{i=1}^{N} e^{-\frac{E_s}{4N_0} |\alpha_i(x_i - t_i)|^2} \tag{4}$$

The pairwise error probability is therefore bounded as follows:

$$P\left(\vec{x} \to \vec{t}\right) \leq \int_{C^N} P\left(\vec{x} \to \vec{t}/\vec{\alpha}\right) p(\vec{\alpha})\, d\vec{\alpha} \tag{5}$$

where the probability density function of $\vec{\alpha}$ is given by:

$$p(\vec{\alpha}) = \frac{1}{(\pi \gamma^2)^N} \frac{1}{\det K} \exp\left(- \frac{{}^t\vec{\alpha} \; K^{-1} \; \vec{\alpha}^*}{\gamma^2}\right) \tag{6}$$

with K the NxN covariance channel matrix defined by: $K_{ij} = \dfrac{E(\alpha_i^* \; \alpha_j)}{\gamma^2}$

Substituting (4) and (6) in (5) gives:

$$P(\vec{x} \to \vec{t}) \leq \int_{C^N} e^{- \frac{E_s}{4N_0} \sum\limits_{i=1}^{N} |\alpha_i(x_i - t_i)|^2} \frac{1}{(\pi \gamma^2)^N} \frac{1}{\det K} \exp\left(- \frac{{}^t\vec{\alpha} \; K^{-1} \; \vec{\alpha}^*}{\gamma^2}\right) d\vec{\alpha} \tag{7}$$

The following step is to diagonalize simultaneously the two exponentional terms in (7) using three matrix transformations (see appendix). This leads to the following simplified integral:

$$P(\vec{x} \to \vec{t}) \leq \int_{C^N} e^{- \frac{E_s}{4N_0} \sum\limits_{i=1}^{N} |\alpha_i|^2 . \phi_i} \frac{1}{(\pi \gamma^2)^N} \exp\left(- \frac{\sum\limits_{i=1}^{N} |\alpha_i|^2}{\gamma^2}\right) d\vec{\alpha} \tag{8}$$

where ϕ_i (i=1,N) are the eigenvalues of the product matrix DK, with K the covariance channel matrix defined above, and D a diagonal matrix formed by the squared Euclidean distances between the correct codeword $\vec{x}$ and the incorrect codeword $\vec{t}$:

$$D = \begin{pmatrix} |x_1 - t_1|^2 & & \\ & \ddots & \\ & & |x_N - t_N|^2 \end{pmatrix}$$

The calculation leads to the expression:

$$P(x \to t) \leq \prod_{i=1}^{N} \frac{1}{1 + \dfrac{\Gamma \phi_i}{4}} \tag{9}$$

where $\Gamma = \dfrac{\gamma^2 E_s}{N_0}$ is the signal to noise ratio.

Notice that the case of uncorrelated fading (or perfect interleaving), treated by Divsalar and Simon [3], represents a particular case of our study. The corresponding Chernoff bound is obtained by replacing the covariance channel matrix K in (6) by the identity matrix. This leads to the same expression as in [3], i.e:

$$P(x \to t) \leq \prod_{i=1}^{N} \cfrac{1}{1 + \cfrac{\Gamma \left| x_i - t_i \right|^2}{4}} \tag{10}$$

4. EVALUATION OF THE BIT ERROR PROBABILITY BOUND

The evaluation of the bit error probability is derived using a transfert function approach. We use the method proposed by Wolf and Zehavi[2]. This method do not applied to all the trellis coded modulations, but is less complex than the general one.which needs a product state diagram.

The bit error probability of a $\frac{n-1}{n}$ rate code is given by:

$$P_b \leq \frac{1}{n-1} \frac{\partial T(W,L,I)}{\partial I} \Big|_{\substack{L=2^{1-n}, I=1 \\ W=W_0}} \tag{11}$$

with T(W,L,I) the modified transfert function, obtained through a state diagram where each branch is labeled with a *weight profile* (W_0).

The weight profiles of Ungerboeck 's eigth states trellis coded 8PSK modulation, adapted to the pairwise error bound derived above, are given by:

$$\cfrac{2}{1 + \cfrac{\Gamma \phi_i^1}{4}} + \cfrac{2}{1 + \cfrac{\Gamma \phi_i^2}{4}} \tag{12}$$

where ϕ_i^1 and ϕ_i^2 (i=1,N) are respectively the eigenvalues of the matrix $D^1 K$ and $D^2 K$. The elements of matrices D^1 and D^2 are evaluated following the method of Wolf and Zehavi; they are listed in table 1.

subset	error E^P	diagonal elements of D^1	diagonal elements of D^2
$A_c=\{0,2,4,6\}$	000	-	-
	001	δ_0	δ_0
	010	δ_1	δ_1
	011	δ_0	δ_2
	100	δ_3	δ_3
	101	δ_2	δ_2
	110	δ_1	δ_1
	111	δ_0	δ_2
$\hat{A}_c=\{1,3,5,7\}$	000	-	-
	001	δ_0	δ_0
	010	δ_1	δ_1
	011	δ_0	δ_2
	100	δ_3	δ_3
	101	δ_2	δ_2
	110	δ_1	δ_1
	111	δ_0	δ_2

table 1. Elements of the distance matrices D^1 and D^2

5. NUMERICAL RESULTS

We present in this section a comparison between analytical results obtained by using with the Chernoff bound and simulation results. We specify now the code, fading statistics and the communication systems used in this comparison.

We use Ungerboeck eight's states 8-PSK modulation code. Trellis-coded modulation has the advantage of achieving good coding gain without bandwidth expansion.

The baseband energy spectral density of the multiplicative fading variate $|\alpha_i|$, corresponding to the mobile radio channel, is given by [5]:

$$S_\alpha(f) = \begin{cases} \dfrac{1}{2\pi f_d} \dfrac{1}{\sqrt{1 - \dfrac{f - f0}{fd}}} & |f| \le f_d \\[6mm] 0 & |f| > f_d \end{cases} \qquad (13)$$

where f_d is the maximum Doppler frequency, and f_0 is the carrier frequency.

Taking the inverse Fourier transform of $S_\alpha(f)$, we get a Bessel autocorrelation function. The covariance channel matrix K is then equal to:

$$K_{ij} = \begin{cases} 1 & \text{if } i=j \\ J_0(2\pi f_d(i-j)T) & \text{else} \end{cases} \qquad (14)$$

where $J_0(.)$ is the zero order Bessel function of the first kind, and T is the symbol duration.

In the simulations, f_d is equal to 41.7 Hz, which corresponds to a vehicle speed of V=50 km/h at 900 MHz, and the bit rate is equal to 8 kbit/s. The complex Gaussian variables α_i are obtained at the output of a five-pole filter [4] feeded by a complex white Gaussian process.

Three communication systems are examined. The first system uses perfect interleaving which results in uncorrelated fadings. The second system is multichannel transmission without interleaving and the third one is multichannel transmission with partial interleaving.

It is useful to recall that there is an inherent difficulty with the union bound since it does not account for the intersection of the pairwise decision regions. The bound is asymptotically tight, but can diverge for low signal to noise ratios (SNR), because the contribution of the intersections is not negligible. In the correlated fading channel, the tightness of the bound is obtained for values of SNR much larger than that obtained for the Gaussian channel. This also implies that for channels with high-correlated fadings, tightness test requires very expensive simulation. Nevertheless, the particular cases tested here represents low-correlated fading.

The results are shown in Figures 3-5. The maximum lenght of the paths considered in the expression of the bit error probability upper bound (23) is indicated. The tightness of the bound is similar to what is usually obtained with the union bound in Rayleigh channel.

a) *Perfect interleaving*

The matrix K is the identity matrix and the pairwise error probability is given by (20). Analytical and simulation results are shown on Figure 3.

b) *Multi-channel transmission*

A multichannel system is presented here as a mean to overcome the difficulty of acheiving perfect interleaving in the presence of time delay constraints. The equivalent baseband transmission system is shown in Figure 2.

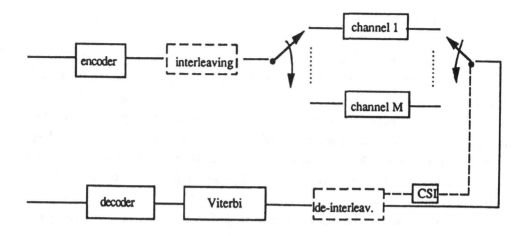

Figure 2: Multi-channel transmission scheme

In this system, M subchannels with uncorrelated fadings are allocated to one user. This can be done by having subchannels separated by more than the channel coherence bandwidth. The user's channel symbols are organized in blocks of size M symbols and the ith symbol of each block is transmitted sequentially on the ith subchannel. If the total transmission rate is $D=\frac{1}{T}$, the rate of each channel is D/M. Partial interleaving which satisfies time delay constraints can then be used. Another advantage of this method is to prevent intersymbol interference, resulting from selective fading, when the bandwidth of each subchannel is much less than the coherence bandwidth of the channel.

- Multi-channel transmission witout interleaving

The corresponding covariance matrix channel for M=4 is equal to:

$$K = \begin{pmatrix} 1 & 0 & 0 & 0 & K_{15} & 0 & \cdots \\ 0 & 1 & 0 & 0 & 0 & K_{26} & \cdots \\ 0 & 0 & 1 & 0 & 0 & 0 & \cdots \\ 0 & 0 & 0 & 1 & 0 & 0 & \cdots \\ K_{51} & 0 & 0 & 0 & 1 & 0 & \cdots \\ 0 & K_{62} & \cdots & & & & \end{pmatrix} \quad \text{and} \quad K_{ij}=J_0(2\pi f_d(i-j)T).$$

The results are shown in Figure 4.

- Multi-channel transmission with partial interleaving

We insert an interleaver between the encoder and the multichannel system as shown in Figure 2. We choose a convolutional interleaver [7], with parameters (n1,n2)=(8,25). This induces 100 ms delay, for 8 kbit/s data transmission. M is equal to 2 and 4. The results are shown respectively in figure 5 and 6.

6. CONCLUSION

We have derived a Chernoff upper bound for the pairwise error probability in the presence of an additive white Gaussian noise and a Rayleigh or Rice correlalted fading. The bound was first used to obtain an upper bound for the bit error probability of coded system using the transfert function of its state diagram. The bound derived can take into account different correlation statistics. It was employed to calculate the performance in the Rayleigh channel of an 8-state 8-PSK TCM using perfect interleaving and, also, using multichannel strategy. Multichannel system gives good performance when large time delay due to interleaving is not allowed.

Also, this bound can be used for the searching of codes adapted to correlated fadings channels.

AKNOWLEDGEMENTS

The authors wish to thank R.SFEZ for his help and his suggestions throughout this work.

APPENDIX

We need to simplify:

$$P\left(\vec{x}\to\vec{t}\right) \leq \int_{C^N} e^{-\frac{E_s}{4N_0}\sum_{i=1}^{N}|\rho_i(x_i-t_i)|^2}\frac{1}{(\pi\gamma^2)^N}\frac{1}{\det K}\exp\left(-\frac{{}^t\vec{\rho}\,K^{-1}\vec{\rho}^*}{\gamma^2}\right)d\vec{\rho}$$

This is accomplished by meaning of three matrix transformation.

The first variate transformation is given by: $\vec{\xi}={}^tU.\vec{\rho}$.

where U is a unitary (${}^tU.U^*=I$) transformation matrix derived from the eigenvectors of K.

Noting by Λ the diagonal matrix of the N eigenvalues of K, it follows

that: $\Lambda_{ij} = \begin{cases} \lambda_i & \text{if } i=j \\ 0 & \text{else} \end{cases}$, we have $K= U.\Lambda.{}^tU^* \Leftrightarrow \Lambda={}^tU^*.K.U$.

In other words, the transformed variate $\vec{\xi}$ is a gaussian complex vector whom covariance matrix is Λ.

Also, let D be the diagonal matrix formed with the distances $|x_i\text{-}t_i|^2$ (i=1,N):

$$D_{ij} = \begin{cases} |x_i\text{-}t_i|^2 & \text{if } i=j \\ 0 & \text{else} \end{cases}$$, we can rewrite the exponentional term in the Chernoff bound

under the following quadratic form: $\sum_{i=1}^{N} |\rho_i(x_i\text{-}t_i)|^2 = {}^t\vec{\rho} \, D \, \vec{\rho}^*$

Thus: ${}^t\vec{\rho}.D.\vec{\rho}^* = {}^t\vec{\xi}.{}^tU^*.D.U.\vec{\xi}^*$.

Secondly, given Ψ the square root matrix of Λ, $\Psi=\sqrt{\Lambda}$, ie $\Psi_{ij} = \begin{cases} \sqrt{\lambda_i} & \text{si } i=j \\ 0 & \text{sinon} \end{cases}$,

we make the next variate transformation: $\vec{\omega}=\Psi^{-1}.\vec{\xi}$.

$\vec{\omega}$ is a complex gaussian vector and its covariance matrix is the identity matrix.

Moreover, ${}^t\vec{\rho}.D.\vec{\rho}^* = {}^t\vec{\omega}.\Psi.{}^tU^*.D.U.\Psi.\vec{\omega}^*$; and noting $T=\Psi.{}^tU^*.D.U.\Psi$., we obtain: ${}^t\vec{\rho}.D.\vec{\rho}^* = {}^t\vec{\omega}.T.\vec{\omega}^*$.

T is hermitian, and thus is diagonalisable. Naming Φ the diagonal matrix of its eigenvalues, and naming S the unitary corresponding transformation matrix: $T= S.\Phi.{}^tS^* \Leftrightarrow \Phi= {}^tS^*.T.S$, the third variate transformation is defined as: $\vec{\alpha} = {}^tS.\vec{\omega}$. The covariance matrix of $\vec{\alpha}$ is still the identity matrix. The quadratic form simplifies to:

$${}^t\vec{\rho}.D.\vec{\rho}^* = {}^t\vec{\alpha}.{}^tS^*.T.S.\vec{\alpha}^* = {}^t\vec{\alpha}.\Phi.\vec{\alpha}^* = \sum_{i=1}^{N} |\alpha_i|^2.\phi_i$$

where ϕ_i (i=1,N) are the eigenvalues of $T=\Psi.{}^tU^*.D.U.\Psi$; ie $\Psi^{-1}.T.\Psi= {}^tU^*.D.U.\Lambda$; and of $U.(\Psi^{-1}.T.\Psi).{}^tU^*=DK$.

Hence, we obtain the simplified integral:

$$P(\vec{x}\rightarrow\vec{t}) \leq \int_{\mathbb{C}^N} e^{-\frac{E_s}{4N_0}\sum_{i=1}^{N} |\alpha_i|^2.\phi_i} \frac{1}{(\pi\,\gamma^2)^N} \exp\left(-\frac{\sum_{i=1}^{N} |\alpha_i|^2}{\gamma^2}\right) d\vec{\alpha}$$

REFERENCES

[1] G.UNGERBOECK, "Channel coding with multilevel phase signals", IEEE TRANS.INF.THEORY, vol. IT-28, Jan. 1982

[2] ZEHAVI and WOLF, "On the performance evaluation of trellis codes", IEEE TRANS. ON INF. THEORY , Vol-IT 33, Mars 87.

[3] D.DIVSALAR and M.K.SIMON, "The design of trellis coded MPSK for fading channels: Performance Criteria", "_____: Set Partitioning for optimum code design", IEEE TRANS. ON COMMUN., vol 36, Sept. 1988

[4] BALL, "A real time fading simulator for mobile Radio", THE RADIO AND ELECTRONIC ENGINEER Vol.52, n°10, Oct.88

[5] JAKES "Microwave Mobile Communication" [WILEY-INTERSCIENCE]

[6] KANAL and SASTRY
"Models for Channels with Memory and Their Applications to Error Control"
PROCEEDINGS OF THE IEEE, Vol. 66, n°7, July 1978

[7] RAMSEY
"Realization of Optimum Interleavers"
IEEE TRANS. ON INF. THEORY , Vol-IT 16, May 70.

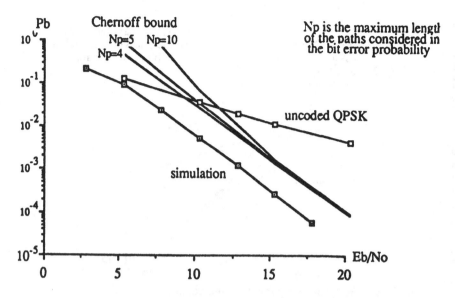

Figure 3. Bit error probability for Rayleigh channel with perfect interleaving

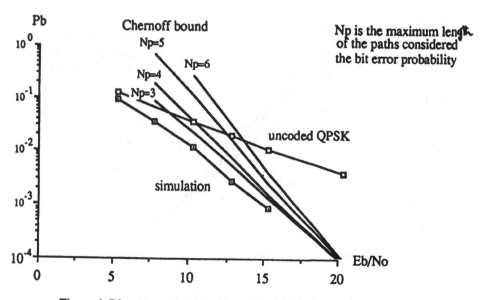

Figure 4. Bit error probability for Rayleigh channel
using multichannel system (M=4) without interleaving

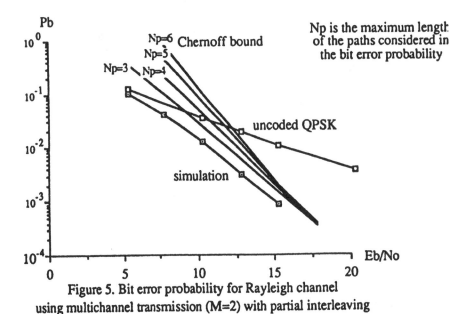

Figure 5. Bit error probability for Rayleigh channel
using multichannel transmission (M=2) with partial interleaving

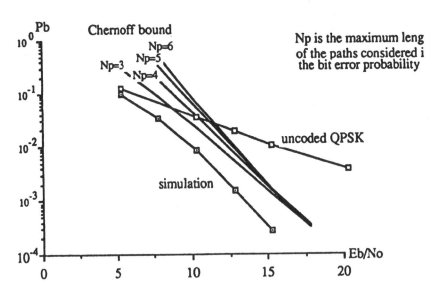

Figure 6. Bit error probability for Rayleigh channel
using multichannel system (M=4) and partial interleaving

SECTION 8

APPLICATIONS OF CODING

A HYBRID FEC-ARQ COMMUNICATION SYSTEM USING SEQUENTIAL DECODING

Marc M. DARMON, Philippe R. SADOT

ALCATEL TRANSMISSION FAISCEAUX HERTZIENS – ATFH
Military Transmissions Department
92301 Levallois-Perret FRANCE

ABSTRACT - This paper presents the results of a study which aimes at defining an error correcting scheme matched to the troposcatter (Rayleigh) channel, which would permit to insure very high reliability links for digital data transmissions.

Forward Error Correction (FEC) and interleaving methods are not compliant to the high variability of the fades durations, and insuring an excellent quality by these means is quite impossible, even with very efficient, complex and expensive codes associated with deep interleavers.

Usual Automatic Repeat reQuest (ARQ) systems are not satisfactory either, because of three reasons at least. First, an error detecting code would be overwhelmed and the retransmissions would be too numerous ; moreover, an error burst may be undetected and the residual BER can be relatively high. Secondly, each retransmission request increases the delay which may become incompatible with telephony applications or even diverge. In the third place, ARQ systems usualy assume that the feedback channel is noiseless ; in our applications the backward channel is also a Rayleigh channel, and the acknowledgement may sometimes be erased by fades.

We propose a hybrid ARQ+FEC system using a long constraint length convolutional code and the sequential decoding algorithm (Stack or MultiStack Algorithm), associated with a modified Go-Back-N ARQ protocol, able to cope with a two-way noisy channel. The data are sent in a continuous stream of blocks, according to an adaptive frame structure permitting to catch up the transmission delay due to the repeat requests. The protocol control is achieved by a modified HDLC structure.

The features of this system will be given for a particular implementation example : for a data rate of 1 Mbit/s and a rate 1/2 convolutional code of constraint length 35, the effective throughput rate is about 1/4, and a zero-error transmission is assured with a bounded transmission delay (< 250 ms) as soon as the average SNR is better than 1 dB in threefold diversity.

INTRODUCTION

We present the results of a study which aimes at defining an error correcting scheme matched to the troposcatter (Rayleigh) channel, which would permit to insure very high reliability links for digital data transmissions despite the fades. In fact this scheme will be also adapted to a jammed channel.

When the transmission channel is unidirectionnal, transmission errors correction is made by mean of a Forward Error Correction (FEC). But this error correction strategy is weakly adaptive and that is a drawback for the troposcatter channel and the jammed channel because the instantaneous SNR can be very variable. When a FEC strategy is applied, interleaving has to be used to decorrelate errors and this creates a transmission delay which is in direct proportion to the interleaving depth.

It is attractive to use a feedback channel, if possible, to reduce the complexity of the coding scheme and the interleaving depth. If there is a feedback channel, it can be more useful to applie an Automatic Repeat reQuest (ARQ) strategy. Data blocks are sent coded by an error detecting code ; if an error is detected the block is asked again, if not the block is acknowledged. The optimum is reached with a hybrid FEC-ARQ strategy, using an error correcting code and an ARQ protocol.

We had to find out a correcting code wich permits to reach very low Binary Error Rates ($< 10^{-9}$ for instance) and to detect the moments when the channel is so noisy that the code is not able to restitute the aimed quality. Convolutional codes, with a long constraint length, decoded by sequential algorithm (Stack Algorithm or MultiStack Algorithm) seem to be adapted. We have studied an ARQ protocol adapted to the troposcatter channel (the backward channel is also a Rayleigh channel)

I. The Hybrid Strategy

Forward Error Correction (FEC) and interleaving methods are not compliant to the high variability of the fades durations, and insuring an excellent quality by these means is quite impossible, even with very efficient, complex and expensive codes associated with deep interleavers.

Usual Automatic Repeat reQuest (ARQ) systems are not satisfactory either, because of three reasons at least. First, an error detecting code would be overwhelmed and the retransmissions would be too numerous ; moreover, an error burst may be undetected and the residual BER can be relatively high. Secondly, each retransmission request increases the delay which may become incompatible with telephony applications or even diverge. In the third place, ARQ systems usualy assume that the feedback channel is noiseless ; in our applications the backward channel is also a Rayleigh channel, and the acknowledgement may sometimes be erased by fades.

We propose a hybrid ARQ+FEC system using a long constraint length convolutional code and the sequential decoding algorithm (MultiStack Algorithm), associated with a modified Go-Back-N ARQ protocol, able to cope with a two-way noisy channel and to catch up the delays due to retransmissions.(cf Part II.).

The data are sent in a continuous stream of blocks, according to an adaptive frame structure permitting to catch up the transmission delay due to the repeat requests.

II. The ARQ protocol and the Frame structure

1. The ARQ protocol

Three basic types of ARQ schemes can be used : the stop-and-wait ARQ, the go-back-N ARQ and the selective-repeat ARQ. The complexity of the go-back-N ARQ is halfway between the others. In a go-back-N ARQ system, blocks are transmitted continuously. Between the transmission of one block and the reception of the acknowledgement of this block, the transmitter sends N-1 other blocks. If a negative acknowledgement is received, the transmitter backs up to the block that is negatively acknowledged and resends that coded block and all the blocks after that one.

We have modified the Go-Back-N ARQ to cope with a two-way noisy troposcatter channel and to control the transmission delay. When the transmission conditions are bad the delay due to retransmissions increases and has to be caught up when the channel quality becomes good. Here are the main principles of this new protocol :

– If the code detects that it will not be able to restitute the aimed quality (stack overflow if we use a stack sequential decoder), the block is asked again.

– When a retransmission request is received, we don't take it into account if the corresponding block has been sent less than N blocks before : This avoids to send again blocks which might be correctly received.

– If a retransmission request is received after many positive acknowledgements, we don't take into account the N possible following retransmission requests. If a block is badly received, it is asked again till it is correctly received. So, the transmitter will receive N successive retransmission requests but only the first one has to be answered.

– When the transmitter is late* it sends two blocks together each time instead of only one. Of course some room is allowed in the frame (cf II.2) for the additionnal block.

– If a block sequence is badly received by A, A doesn't know whether the acknowledgements from B are positive or negative. A has to consider negative one acknowledgement out of N and positive the N-1 others. This rule avoids the transmitter to send non expected blocks.

2. The Frame structure

The frame consists of three blocks.

FSW	CONTROL	INFO

```
        48 bits      512 bits
```

– The FSW block is a Frame Synchronization Word to synchronise the blocks

– The Control (CONTROL) block contains all informations necessary for the ARQ protocol : positive or negative acknowledgment, sent out block number, expected block number ...

Here is the constitution of CONTROL block. This a HDLC compatible structure ; in particular, using a HDLC flag permits to use the usual HDLC integrated circuits.

FLAG	ACK	C	N(S)	N(R)

```
8 bits   1 bit 1 bit  11 bits   11 bits
```

* First bit (ACK) is the acknowledgement of the last received block. It is 1 if the block has been well received and 0 if the block is badly received.

* Next bit (C) indicates how many bits from INFO block are real data bits : 256 or 512.

* Next 11 bits (N(S)) number the sent INFO block (modulus 2048).

---

* i.e that the number of the block he sends out is quite lower than the number of the block he would send if there had not been any retransmission request.

* Next 11 bits (N(R)) number the INFO block (modulus 2048) which is expected from the correspondent

* The FLAG is a synchronisation sequence. It is in fact the Frame Synchronization Word of CONTROL blocks .

Let us precise the signification of ACK,C,N(S) et N(R).

* If ACK is equal to 1, that means that the last received block has been correctly received** , that it was the block the receiver was waiting for and that the block the receiver is now expecting is block number N(R). This means that all blocks before N(R) have been correctly received. If ACK is equal to 0, this means a bad reception or a bad decoding, and it will be considered as a retransmission request of blocks starting from N(R).

* If C is equal to 0, this means that the transmitter is not late. There is no transmission delay because of retransmissions. The INFO block contains 256 data bits and 256 filling bits. If C is equal to 1, the transmitter is late and sends 512 data bits in each INFO block. In this case, the receiver has to take into account the whole INFO block, and not only the first 256 bits. The transmitter is considered late when the number N(S) of the block he sends is lower than $OPT - 4$, where OPT is the number of the block he would send if there had not been any retransmission.

– The Information (INFO) block contains the information bits. Note that the number of data bits in this block depends on the transmitter delay. If the transmitter is late (cf. II.1), the INFO block contains twice 256 data bits. If the transmitter is not late, it sends 256 data bits and 256 filling bits.

The CONTROL block contains 48 bits and the INFO block 512. Then both blocks are coded by an error correcting code. For instance, we can use the convolutional code of rate 1/2 and constaint length 35. This systematic code (polynomial 71446626555 in octal notation) has got a free distance of 19. CONTROL and INFO blocks become a set of 1190 bits (1190 = 2x[48+512+35]). The last 2x35 bits consist of the coded queue of the code.

FSW	CONTROL + INFO Coded

1190   bits

At the receiver end, bits between Frame Synchronization Words are sent to the decoder. If the decoding is considered as reliable, the 48 first bits are the CONTROL block. The protocol makes use of the values of ACK, C, N(S) and N(R).

III. The coding scheme

1. Convolutional codes

To get a very good reliability, we have to code the set CONTROL+INFO with a very efficient correcting code. If we use convolutional codes decoded by sequential algorithm, we can get an arbitrarily low error rate by increasing the code constraint

---

** A block is considered well received if the decoding of the CONTROL+INFO set is estimated reliable (the stack did not overflow).

length. If we use the systematic 1/2 code, the polynomial of which is 71446162655 (octal notation), the error rate is lower than 10^{-9} (the simulations give zero error) when the decoding is successful. If the decoding is not successful (that means the stack has overflown), the block is asked again. We will see that the sequential decoding using Stack Algorithm gives a signal when the decoder is not able to insure the aimed quality ; that is usefull for the hybrid FEC-ARQ protocol.

2. Sequential decoding

We will not give much details of the Stack Algorithm to execute a sequential decoding of convolutional codes. The convolutional code may be represented through a code tree, which is an expanded version of the trellis diagram, in which every path is totally distinct from every other path. The Stack Algorithm is a tree-searching algorithm, where each observed node is recorded in an ordered stack according to its likelihood function metric. If the stack overflows, that means that the block is quite noisy and the decoding has to be stopped. The block will be asked again. The stack overflow is the indication that the decoder will not be able to restitute the aimed quality.

It is usual to use a stack depth equal to 150% of the block length. If the stack depth is chosen larger, a lower number of blocks will be asked again, but the error rate will increase because we will try to decode very noisy blocks. So we can adjust the stack depth to achieve a trade-off between reliability and transmission delay. The table below shows the influence of the stack depth on the number of Automatic Retransmission reQuests :

Eb/No	stack depth	BER before decoding	after decoding	Nb d'ARQ (/400)
0 dB	150%	0,082	$< 10^{-6}$ (0 errors)	29
-0,6 dB	150%	0,097	$< 10^{-6}$ (0 errors)	191
- 1 dB	150%	0,108	$< 10^{-6}$ (0 errors)	342
- 1 dB	200%	0,108	$< 10^{-6}$ (0 errors)	228
- 1 dB	300%	0,108	$< 10^{-6}$ (0 errors)	121
- 1 dB	400%	0,108	$< 10^{-6}$ (0 errors)	81
- 1 dB	500%	0,108	$< 10^{-6}$ (0 errors)	68
- 1 dB	600%	0,108	$< 10^{-6}$ (0 errors)	56
- 2 dB	600%	0,136	$< 10^{-6}$ (0 errors)	374
- 2 dB	1000%	0,136	$5. \, 10^{-4}$ (32 errors)	337

IV. Performances of the system.

This system has been tested on a simulated Rayleigh channel. The simulations have shown no transmission errors, which means a Binary Error Rate lower than 10^{-9}. The performances of the system are in fact represented by the evolution of the transmission delay. In a usual ARQ system the delay, due to retransmissions, increases. Here this delay can be bounded, and so this system may be adapted to the transmission of TDMA signals containing telephony and digital datas.

The table below shows the evolution of the delay according to the average SNR and the diversity.

Average Signal/Noise ratio	diversity	maximum delay
0 dB	1	divergence
2 dB	1	divergence
3 dB	1	divergence
4 dB	1	3,3 secondes
0 dB	2	divergence
0,7 dB	2	divergence
1 dB	2	11 secondes
2 dB	2	1,5 secondes
3 dB	2	0,9 secondes
4 dB	2	0,5 secondes
5 dB	2	0,41 secondes
6 dB	2	0,22 secondes
1 dB	3	0,3 secondes

We see that if we consider that the maximum transmission delay for a telephone transmission is 0,3 seconds, this system is adapted for a zero-error transmission as soon as the average Signal to Noise Ratio is better than 1 dB in threefold diversity, for instance.

Two patents have been registered about this system and the modification of the ARQ protocol.

384

V. References

[1] Shu Lin, D.J. Costello, "Error Control Coding" Prentice Hall 1983

[2] Shu Lin, P.S. Yiu, "An Efficient Error Control Scheme for Satellite Communications", IEEE Trans on Comm. March 1980

[3] H. Yamamoto, K. Itoh, "Viterbi Decoding Algorithm for Convolutional Codes with Repeat Request", IEEE Trans on Inf. Theory, Sept. 1980

[4] A. Drukarev, D.J. Costello, "Hybrid ARQ Error Control Using Sequentiel Decoding", IEEE Trans on Inf. Theory, jul. 1983

[5] P.Y. Pau, D. Haccoun, "Sequential Decoding with ARQ", IEEE ISIT, Saint-Jovite 1983

[6] J.M. Wozencraft, I.M. Jacobs, "Principles of Communication Engineering", Wiley, New York, 1965

[7] G.D. Forney, E.K. Bower, "A High-Speed Sequential Decoder. Prototype Design and Test", IEEE Trans on Comm. March 1980

[8] F. Jelinek, "A Fast Sequential Decoding Algorithym Using a Stack", IBM J. Res. and Dev. Nov. 1969

[9] M. Cerdervall, R. Johanesson, K. Zigangirov, "Creeper. A New Algorithm for Sequential Decoding", IEEE ISIT Brighton June 1985

[10] D. Haccoun,"A branching Process Analysis of the Average Number of Computations of the Stack Algorithm", IEEE Trans on I.T. May 1984

[11] R. Johanesson, IEEE Trans on I.T., Jul. 1975, "Robustly Optimal Rate One-Half Binary Convolutionnal Codes"

[12] J.L. Massey, "Variable-Length Codes and the Fano Metric", IEEE Trans on I.T., Jan 1972

[13] G.D. Forney, "Convolutiuonal Codes III : Sequential Decoding", Information and Control, Jul. 1974

[14] S. Kallel, D. Haccoun, "Sequentiel Decoding with ARQ and Code Combining", IEEE Trans on Comm., n7, jul. 1988

[15] S. Lsaku, K. Hatori, "Performance Analysis of Some ARQ Protocols", Electronics and Communications in Japan, Vol. 65-8 n4, 1982

[16] G.C. Clark, J.B. Cain, "Error-Correction Coding for Digital Communications", Ed. Plenum Press 1981

[17] J.J. Metzner, "Improvements in Block-Retransmission Schemes", IEEE Trans on Comm., n2, Feb. 1979

[18] B. Arazi, "Improving the Throughput of an ARQ Stop-and-Wait Sceme for Burst Noise Channels", IEEE Trans on Comm., June 1976

A 30 Mbits/s (255,223) REED-SOLOMON DECODER

Jean-Luc POLITANO
Dominique DEPREY
ALCATEL ATFH - 55 rue Greffülhe - 92301 LEVALLOIS PERRET - FRANCE

ABSTRACT - The paper describes the VLSI implementation of a Reed-Solomon Coder-Decoder. This standard cell ASIC has been designed with VLSI Technology CAD tools on a SUN/4 workstation.

The chip includes two 256-byte ROM - table look-up for the inverse of field elements in $GF(2^8)$ - , one 512-byte RAM (buffer register) and the equivalent of 55 Kgates. The die size is 440 mils x 440 mils and the process used for manufacturing is CMOS 1.0 μ.

The Reed-Solomon code is a (N,N-32) block code of 8-bit symbols capable of correcting up to v symbol errors and ρ symbol erasures provided that $2v + \rho \leq 32$. N assumes values less than or equal to 255 (when N<255 a shortened code is generated).

The data rate achieved by the decoder depends on the dimension K of the code selected and equals K+256x34 Mbits/s (34 MHz is the maximum frequency corresponding to the industrial worst case). For example the (255,223)RS operates at 30 Mbits/s compared to the (64,32) RS operating at 4,2 Mbits/s.

RESUME - Cet article décrit l'implémentation d'un circuit VLSI codeur-décodeur de Reed-Solomon. Ce circuit ASIC précaractérisé a été développé en utilisant les outils CAO de VLSI Technology sur une station de travail SUN/4.

Le circuit comporte deux mémoires ROM de 256 octets - tabulation des inverses des éléments du corps $GF(2^8)$ - , une mémoire RAM de 512 octets (mémoire tampon) ainsi que l'équivalent de 55 Kportes. La taille de la puce est de 11,2 mm x 11,2 mm et sa fabrication fait appel à une technologie CMOS 1.0 μ.

Le code de Reed-Solomon est un code en bloc (N,N-32) dont les symboles sont des octets et capable de corriger jusqu'à v symboles faux et ρ symboles effacés dans la mesure où $2v + \rho \leq 32$. N peut prendre toutes les valeurs comprises entre 33 et 255 (N<255 correspond à un code raccourci).

Le débit utile du décodeur dépend de la dimension K du code sélectionné et est égal à K+256x34 Mbits/s (34 MHz est la fréquence maximum de fonctionnement en pire cas industriel). A titre d'exemple, le RS(255,223) autorise un débit utile de 30 Mbits/s alors que le RS(64,32) ne permet d'atteindre que 4,2 Mbits/s.

I INTRODUCTION

Within the context of an Alcatel N.V. project, ATFH had to study and implement a single chip Reed-Solomon coder-decoder; One can find many papers on this subject but only a few circuits have really been designed. Moreover only a few integrated circuits manufacturers are proposing at the moment Reed-Solomon CODECs and this leads several companies to develop their own chip.

When we started the study of this ASIC last quarter 1989, the maximum useful rate estimated for the (255,223)RS was bounded by 10 and 20 Mbits/s. In fact it turned out to be quite easy to reach 30 Mbits/s.

Code description [1],[2]

The Reed-Solomon code is a (N,N-32) with $33 \leq N \leq 255$.

These codes are defined on the Galois field GF(256). Each codeword contains K information symbols and 32 check symbols. The decoder presented in this paper is capable of correcting up to ν symbols errors and ρ symbols erasures provided that $2\nu + \rho \leq 32$.

The Galois field under consideration is defined by the primitive polynomial
$p(x) = x^8 + x^4 + x^3 + x^2 + 1$.

In order to support the architecture description, let us proceed to the algorithmic description of the application. Only the main equations are provided, without development.

The generator polynomial for the code is defined as :
$$G(x) = \prod_{i=1}^{32}(x-\alpha^i) = \sum_{j=0}^{32} G_j x^j \text{ where } \alpha \text{ is a primitive root of } p(x)$$
The RS code block is defined to be $C(x) = x^{32}.I(x) + [x^{32} I(x)] \bmod G(x)$ where $C(x)$ is the codeword represented as an N-1 order polynomial and $I(x)$ is the K information symbols represented as an order K-1 polynomial.

The main steps in the decoding procedure are :

1) Syndrome computation

If $V(x)$ is the received polynomial, the syndromes are defined by :
$S_k = V(\alpha^k)$ $1 \leq K \leq 32$
$$S(x) = \sum_{i=1}^{32} S_i x^{i-1} \text{ be the syndrome polynomial.}$$

2) Erasure positions computation

3) Berlekamp-Massey algorithm

The inputs of this block are the syndromes and the erasures positions and the outputs are the errata (errors as well as erasures) locator polynomial $\Lambda(x)$ and the errata magnitude polynomial $\Omega(x)$.

The key equation is $S(x).\Lambda(x) = \Omega(x) \bmod x^{32}$

4) Once the two polynomials $\Lambda(x)$ and $\Omega(x)$ are known, the magnitude of a given error is found as follows :

Let α^i be a zero of $\Lambda(x)$ (i.e. $\Lambda(\alpha^i) = 0$). The error magnitude at location -i is :

$$Y_{-i} = \frac{\Omega(\alpha^i)}{\Lambda'(\alpha^i)}$$ where $\Lambda'(x)$ is the first derivative of $\Lambda(x)$ with respect to x.

The synoptic of the circuit can then be represented as follows :

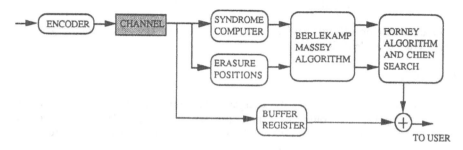

I I ARCHITECTURE

ENCODER

Encoding the (N,K) Reed-Solomon (cyclic code) in systematic form consists of three steps. [1],[2]
  - multiply the information polynomial $I(x)$ by x^{32}
  - divide $x^{32}.I(x)$ by $G(x)$ to obtain the remainder $R(x)$ and form the code word $x^{32}.I(x) + R(x)$
These three steps can be accomplished with a division circuit which is a linear 32-stage shift register with feedback connections based on the generator polynomial : $G(x) = 1 + G_1x + G_2 x_2 + ... + G_{32} x^{32}$

The control block receives the clock signal, the reset signal and the selected value of K. It generates all control signals for the encoder, a "shout" signal to get information symbols from an external FIFO and a "shin" signal to write codeword symbols into another external FIFO.
Let CLKE be the clock signal applied on the control block. The information data rate at the input of the encoder is $K+256$.CLKE and the codeword data rate at the output is $(K+32)+256$.CLKE.

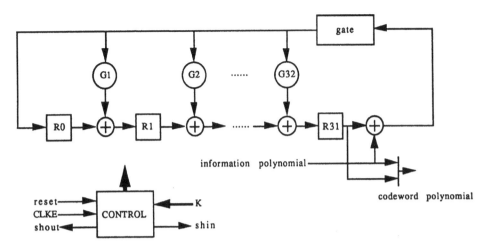

The clock rate for all the registers is CLKE÷8 and this part of design requires almost five equivalent Kgates.

DECODER

The control block receives the clock signal, the reset signal and the selected value of K. It generates all control's signals for the decoder, a "shout" signal to get received symbols from an external FIFO and a "shin" signal to write decoded symbols into another external FIFO.

Let CLKD be the clock signal applied on the control block. The received data rate at the input of the decoder is $(K+32)+256.CLKD$ and the data rate at the output is $K+256.CLKD$.

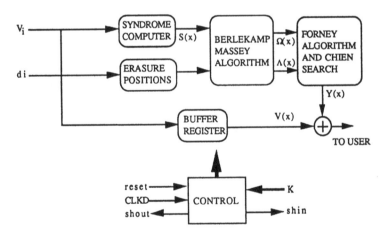

The RS decoder architecture uses a three stages pipeline decoding technique [3],[4],[5], and the timing chart for this high speed RS decoder can be represented as follows :

syndrome generation	codeword NO.1	codeword NO.2	codeword NO.3
Berlekamp-Massey algorithm		codeword NO.1	codeword NO.2
Forney algorithm			codeword NO.1
error correction			codeword NO.1

The delay between inputs and outputs of the decoder equals lapse of time allowed to decode two words (i.e. : 256x8.CLKD)

a) Syndromes computer

A systolic architecture which is well suited for a VLSI implementation has been selected. The computation of S_k is simply the evaluation of the received polynomial on the value α^k. An efficient way to evaluate a polynomial is to use the Horner rule : $S_k=((V_{n-1}\alpha^k + V_{n-2})\alpha^k + ...)$

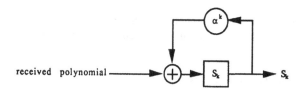

All 32 syndromes are calculated simultaneously with 32 circuits operating in parallel. The clock rate for all the 32x8 D-type flipflop registers is CLKD+8 and this part of design requires almost five equivalent Kgates.

b) Erasure positions

Each received symbol V_i entering the decoder is associated with a binary flag di wich indicates whether the corresponding symbol is an erasure or not. The erasure block must translate these bits into a set of positions which are elements of GF(256).
The hardware ressources used to compute these 32 possible positions can be represented as follows :

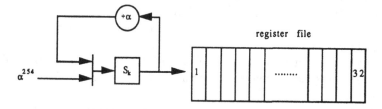

When di is set, the corresponding position is stored into the register file of depth 32. The clock rate for all the registers is CLKD+8 and this part of design requires almost three equivalent Kgates.

c) Berlekamp-Massey algorithm

When syndromes and erasure positions are ready all these bytes are loaded into registers.

The error and erasure locator polynomial is initialized with $\Lambda(x)=1$ and the errata evaluator polynomial $\Omega(x)$ is initialized with $S(x)$.[5]

Let ρ be the number of erasures. The first ρ iterations generate the erasure locator polynomial and modify the $\Omega(x)$ polynomial according to the equation :

$$\Omega(x)=S(x).\Lambda(x) \bmod x^{32}$$

An index r is used to count out the first ρ iterations while the erasure polynomial is being computed and then continues to count out the iterations of the Berlekamp-Massey algorithm, stopping when $r=32$.

In order to respect the data rate, the discrepancy relative to the iteration r has to be computed during the iteration $(r-1)$.

This block includes a 256-byte ROM which is a table look-up for the inverse of the discrepancy and the equivalent of 18 Kgates. All the registers are shifted with the clock signal CLKD+2.

The numbers of erasures and errors for each processed codeword are available as outputs of the chip. They can be used to optimize the external erasure decision criterion.

d) Forney algorithm and Chien search

The next step is to compute the estimated error polynomial $Y(x)$.[5]

This block evaluates three polynomials simultaneously : The errata location polynomial $\Lambda(x)$, the errata magnitude polynomial $\Omega(x)$, and the first derivative of the errata location polynomial $\Lambda'(x)$. The architecture is based on three independant blocks.

One block, the $\Lambda(x)$ block, searches for the zeros of the errata location polynomial. The $\Lambda'(x)$ block which only loads the odd coefficients of $\Lambda(x)$, evaluates the first derivative of $\Lambda(x)$ and $\Omega(x)$, which has a data path totally separate from the other two parts, evaluates $\Omega(x)$ for all α^i.

In order to respect the data rate, $\Lambda(x)$, $\Lambda'(x)$, and $\Omega(x)$ are evaluated in parallel for eight consecutive powers of α.

All the registers are shifted with a CLKD+2 MHz clock, and every 32 clock cycles, 16 new field elements - $\Omega(\alpha^i),\Omega(\alpha^{i+1}),...,\Omega(\alpha^{i+7}),\Lambda'(\alpha^i),\Lambda'(\alpha^{i+1}),...,\Lambda'(\alpha^{i+7})$ - and 8 binary flags - $\Lambda(\alpha^i)$, $\Lambda(\alpha^{i+1})$, ... , $\Lambda(\alpha^{i+7})$ - are computed and stored into $(16 \times 8 + 8)$ D-type flipflops.

These values are multiplexed, $\Lambda'(\alpha^i)$ going through the second ROM of the design to get $\Lambda'(\alpha^i)^{-1}$, a general multiplier computes $\Omega(\alpha^i).\Lambda'(\alpha^i)^{-1}$. This value is reset when $\Lambda(\alpha^{-i})\neq 0$.

Finally, the buffer memory is read out symbol-by-symbol and "exclusive-or'ed" with the outputs Y_{-i}.

This block includes a 256 bytes ROM which is a table look-up for the inverse of $\Lambda'(\alpha^i)$ and the equivalent of 24 Kgates.

III CONCLUSION

Finally, the chip was fast prototyped in 12 weeks by VLSI Technology, leading the completion of this high performance ASIC chip in less then 10 months. The chip has been succesfully tested and exhibits performances corresponding to the specifications.

The main features of this chip are :

34 MHz clock rate (industrial worst case)
120 pins in a CPGA package
Die size : 125 mm^2
55 Kgates
2 ROM of 256 bytes
1 RAM of 512 bytes
5 volts supply
power : 1.5 W (industrial worst case)

With the same technology, CMOS 1.0 µ, one can develop a single chip (255,223) RS coder-decoder processing data rates up to 100 Mbits/s. The increase of the gate count would be less than 25 % with a more parallel structure (more multipliers and basic gates but almost the same number of D-type registers).

Chip Layout :

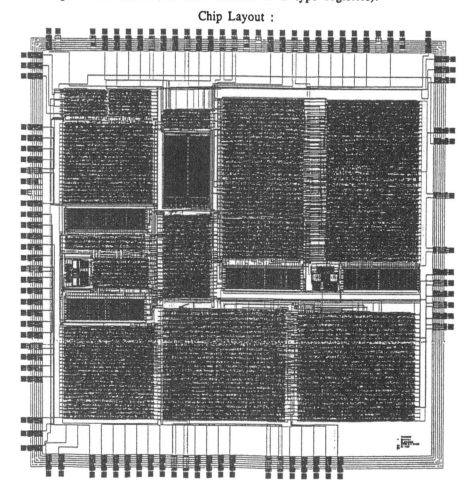

References :

1. Berlekamp, E., R., "The Technology of Error-Correcting Codes", Proceedings of IEEE, Vol. 68, No. 5, May 1980.

2. Berlekamp, E., R., Algebraic Coding Theory, McGraw Hill, New-York : 1968.

3. Peterson, W. and Welson, N., Error Correcting Codes, 2nd Edition, The MIT Press, Cambridge, MA : 1972.

4. K. Y. Liu, "Architecture for VLSI Design of Reed-Solomon Decoders", IEEETC, Vol. c-33, pp.178-189, Feb. 1984.

5. Georges C. Clark and J. Bibb Cain, Error Correcting Coding For Digital Communication, New-York, NY, Plenum Press, 1981.

Lecture Notes in Computer Science

For information about Vols. 1–429
please contact your bookseller or Springer-Verlag

Vol. 473: I.B. Damgård (Ed.), Advances in Cryptology – EUROCRYPT '90. Proceedings, 1990. VIII, 500 pages. 1991.

Vol. 474: D. Karagiannis (Ed.), Information Syetems and Artificial Intelligence: Integration Aspects. Proceedings, 1990. X, 293 pages. 1991. (Subseries LNAI).

Vol. 475: P. Schroeder-Heister (Ed.), Extensions of Logic Programming. Proceedings, 1989. VIII, 364 pages. 1991. (Subseries LNAI).

Vol. 476: M. Filgueiras, L. Damas, N. Moreira, A.P. Tomás (Eds.), Natural Language Processing. Proceedings, 1990. VII, 253 pages. 1991. (Subseries LNAI).

Vol. 477: D. Hammer (Ed.), Compiler Compilers. Proceedings, 1990. VI, 227 pages. 1991.

Vol. 478: J. van Eijck (Ed.), Logics in AI. Proceedings, 1990. IX, 562 pages. 1991. (Subseries in LNAI).

Vol. 480: C. Choffrut, M. Jantzen (Eds.), STACS 91. Proceedings, 1991. X, 549 pages. 1991.

Vol. 481: E. Lang, K.-U. Carstensen, G. Simmons, Modelling Spatial Knowledge on a Linguistic Basis. IX, 138 pages. 1991. (Subseries LNAI).

Vol. 482: Y. Kodratoff (Ed.), Machine Learning – EWSL-91. Proceedings, 1991. XI, 537 pages. 1991. (Subseries LNAI).

Vol. 483: G. Rozenberg (Ed.), Advances In Petri Nets 1990. VI, 515 pages. 1991.

Vol. 484: R. H. Möhring (Ed.), Graph-Theoretic Concepts In Computer Science. Proceedings, 1990. IX, 360 pages. 1991.

Vol. 485: K. Furukawa, H. Tanaka, T. Fullsaki (Eds.), Logic Programming '89. Proceedings, 1989. IX, 183 pages. 1991. (Subseries LNAI).

Vol. 486: J. van Leeuwen, N. Santoro (Eds.), Distributed Algorithms. Proceedings, 1990. VI, 433 pages. 1991.

Vol. 487: A. Bode (Ed.), Distributed Memory Computing. Proceedings, 1991. XI, 506 pages. 1991.

Vol. 488: R. V. Book (Ed.), Rewriting Techniques and Applications. Proceedings, 1991. VII, 458 pages. 1991.

Vol. 489: J. W. de Bakker, W. P. de Roever, G. Rozenberg (Eds.), Foundations of Object-Oriented Languages. Proceedings, 1990. VIII, 442 pages. 1991.

Vol. 490: J. A. Bergstra, L. M. G. Feljs (Eds.), Algebraic Methods 11: Theory, Tools and Applicatlons. VI, 434 pages. 1991.

Vol. 491: A. Yonezawa, T. Ito (Eds.), Concurrency: Theory, Language, and Architecture. Proceedings, 1989. VIII, 339 pages. 1991.

Vol. 492: D. Sriram, R. Logcher, S. Fukuda (Eds.), Computer-Aided Cooperative Product Development. Proceedings, 1989 VII, 630 pages. 1991.

Vol. 493: S. Abramsky, T. S. E. Maibaum (Eds.), TAPSOFT '91. Volume 1. Proceedings, 1991. VIII, 455 pages. 1991.

Vol. 494: S. Abramsky, T. S. E. Maibaum (Eds.), TAPSOFT '91. Volume 2. Proceedings, 1991. VIII, 482 pages. 1991.

Vol. 495: 9. Thalheim, J. Demetrovics, H.-D. Gerhardt (Eds.), MFDBS '91. Proceedings, 1991. VI, 395 pages. 1991.

Vol. 496: H.-P. Schwefel, R. Männer (Eds.), Parallel Problem Solving from Nature. Proceedings, 1991. XI, 485 pages. 1991.

Vol. 497: F. Dehne, F. Fiala. W.W. Koczkodaj (Eds.), Advances in Computing and Intormation - ICCI '91 Proceedings, 1991. VIII, 745 pages. 1991.

Vol. 498: R. Andersen, J. A. Bubenko jr., A. Sølvberg (Eds.), Advanced Information Systems Engineering. Proceedings, 1991. VI, 579 pages. 1991.

Vol. 499: D. Christodoulakis (Ed.), Ada: The Choice for '92. Proceedings, 1991. VI, 411 pages. 1991.

Vol. 500: M. Held, On the Computational Geometry of Pocket Machining. XII, 179 pages. 1991.

Vol. 501: M. Bidoit, H.-J. Kreowski, P. Lescanne, F. Orejas, D. Sannella (Eds.), Algebraic System Specification and Development. VIII, 98 pages. 1991.

Vol. 502: J. Bārzdiņš, D. Bjørner (Eds.), Baltic Computer Science. X, 619 pages. 1991.

Vol. 503: P. America (Ed.), Parallel Database Systems. Proceedings, 1990. VIII, 433 pages. 1991.

Vol. 504: J. W. Schmidt, A. A. Stogny (Eds.), Next Generation Information System Technology. Proceedings, 1990. IX, 450 pages. 1991.

Vol. 505: E. H. L. Aarts, J. van Leeuwen, M. Rem (Eds.), PARLE '91. Parallel Architectures and Languages Europe, Volume I. Proceedings, 1991. XV, 423 pages. 1991.

Vol. 506: E. H. L. Aarts, J. van Leeuwen, M. Rem (Eds.), PARLE '91. Parallel Architectures and Languages Europe, Volume II. Proceedings, 1991. XV, 489 pages. 1991.

Vol. 507: N. A. Sherwani, E. de Doncker, J. A. Kapenga (Eds.), Computing in the 90's. Proceedings, 1989. XIII, 441 pages. 1991.

Vol. 508: S. Sakata (Ed.), Applied Algebra, Algebraic Algorithms and Error-Correcting Codes. Proceedings, 1990. IX, 390 pages. 1991.

Vol. 509: A. Endres, H. Weber (Eds.), Software Development Environments and CASE Technology. Proceedings, 1991. VIII, 286 pages. 1991.

Vol. 510: J. Leach Albert, B. Monien, M. Rodríguez (Eds.), Automata, Languages and Programming. Proceedings, 1991. XII, 763 pages. 1991.

Vol. 511: A. C. F. Colchester, D.J. Hawkes (Eds.), Information Processing in Medical Imaging. Proceedings, 1991. XI, 512 pages. 1991.

Vol. 512: P. America (Ed.), ECOOP '91. European Conference on Object-Oriented Programming. Proceedings, 1991. X, 396 pages. 1991.

Vol. 513: N. M. Mattos, An Approach to Knowledge Base Management. IX, 247 pages. 1991. (Subseries LNAI).

Vol. 514: G. Cohen, P. Charpin (Eds.), EUROCODE '90. Proceedings, 1990. XI, 392 pages. 1991.

Vol. 516: S. Kaplan, M. Okada (Eds.), Conditional and Typed Rewriting Systems. Proceedings, 1990. IX, 461 pages. 1991.